REVIEWS in MINERALOGY   Volume 20

MODERN POWDER DIFFRACTION

D.L. BISH AND J.E. POST, editors

Authors:

DAVID L. BISH
Earth and Environmental Sciences Division
Los Alamos National Laboratory, MS D469
Los Alamos, New Mexico 87545

LARRY W. FINGER
Geophysical Laboratory
2801 Upton Street, N.W.
Washington, D.C. 20008

SCOTT A. HOWARD
Department of Ceramic Engineering
University of Missouri-Rolla
Rolla, Missouri 65401

RON JENKINS
JCPDS
International Centre for Diffraction Data
1601 Park Lane
Swarthmore, Pennsylvania 19081

JEFFREY E. POST
Department of Mineral Sciences
Smithsonian Institution, NHB 119
Washington, D.C. 20560

KIMBERLY D. PRESTON
Department of Ceramic Engineering
University of Missouri-Rolla
Rolla, Missouri 65401

ROBERT C. REYNOLDS, JR.
Department of Earth Sciences
Dartmouth College
Hanover, New Hampshire 03755

DEANE K. SMITH
Department of Geosciences
The Pennsylvania State University
239 Deike Building
University Park, Pennsylvania 16801

ROBERT SNYDER
New York State College of Ceramics
Alfred University
Alfred, New York 14802

ROBERT VON DREELE
Los Alamos Neutron Scattering Center
Los Alamos National Laboratory, MS H805
Los Alamos, New Mexico 87545

Series Editor:

PAUL H. RIBBE
Department of Geological Sciences
Virginia Polytechnic Institute and State University
Blacksburg, Virginia 24061

Published by
THE MINERALOGICAL SOCIETY OF AMERICA
Washington, D.C.

COPYRIGHT 1989
MINERALOGICAL SOCIETY OF AMERICA
Printed by BookCrafters, Inc., Chelsea, Michigan.

REVIEWS IN MINERALOGY

(Formerly: SHORT COURSE NOTES)

ISSN 0275-0279

Volume 20: *MODERN POWDER DIFFRACTION*

ISBN 0-939950-24-3

ADDITIONAL COPIES of this volume as well as those listed below may be obtained at moderate cost from the MINERALOGICAL SOCIETY OF AMERICA, 1625 I Street, N.W., Suite 414, Washington, D.C. 20006 U.S.A.

Volume 1: Sulfide Mineralogy, 1974; P. H. Ribbe, Ed. 284 pp. Six chapters on the structures of sulfides and sulfosalts; the crystal chemistry and chemical bonding of sulfides, synthesis, phase equilibria, and petrology. ISBN# 0-939950-01-4.

Volume 2: Feldspar Mineralogy, 2nd Edition, 1983; P. H. Ribbe, Ed. 362 pp. Thirteen chapters on feldspar chemistry, structure and nomenclature; Al,Si order/disorder in relation to domain textures, diffraction patterns, lattice parameters and optical properties; determinative methods; subsolidus phase relations, microstructures, kinetics and mechanisms of exsolution, and diffusion; color and interference colors; chemical properties; deformation. ISBN# 0-939950-14-6.

Volume 4: Mineralogy and Geology of Natural Zeolites, 1977; F. A. Mumpton, Ed. 232 pp. Ten chapters on the crystal chemistry and structure of natural zeolites, their occurrence in sedimentary and low-grade metamorphic rocks and closed hydrologic systems, their commercial properties and utilization. ISBN# 0-939950-04-9.

Volume 5: Orthosilicates, 2nd Edition, 1982; P. H. Ribbe, Ed. 450 pp. Liebau's "Classification of Silicates" plus 12 chapters on silicate garnets, olivines, spinels and humites; zircon and the actinide orthosilicates; titanite (sphene), chloritoid, staurolite, the aluminum silicates, topaz, and scores of miscellaneous orthosilicates. Indexed. ISBN# 0-939950-13-8.

Volume 6: Marine Minerals, 1979; R. G. Burns, Ed. 380 pp. Ten chapters on manganese and iron oxides, the silica polymorphs, zeolites, clay minerals, marine phosphorites, barites and placer minerals; evaporite mineralogy and chemistry. ISBN# 0-939950-06-5.

Volume 7: Pyroxenes, 1980; C. T. Prewitt, Ed. 525 pp. Nine chapters on pyroxene crystal chemistry, spectroscopy, phase equilibria, subsolidus phenomena and thermodynamics; composition and mineralogy of terrestrial, lunar, and meteoritic pyroxenes. ISBN# 0-939950-07-3.

Volume 8: Kinetics of Geochemical Processes, 1981; A. C. Lasaga and R. J. Kirkpatrick, Eds. 398 pp. Eight chapters on transition state theory and the rate laws of chemical reactions; kinetics of weathering, diagenesis, igneous crystallization and geochemical cycles; diffusion in electrolytes; irreversible thermodynamics. ISBN# 0-939950-08-1.

Volume 9A: Amphiboles and Other Hydrous Pyriboles—Mineralogy, 1981; D. R. Veblen, Ed. 372 pp. Seven chapters on biopyribole mineralogy and polysomatism; the crystal chemistry, structures and spectroscopy of amphiboles; subsolidus relations; amphibole and serpentine asbestos—mineralogy, occurrences, and health hazards. ISBN# 0-939950-10-3.

Volume 9B: Amphiboles: Petrology and Experimental Phase Relations, 1982; D. R. Veblen and P. H. Ribbe, Eds. 390 pp. Three chapters on phase relations of metamorphic amphiboles (occurrences and theory); igneous amphiboles; experimental studies. ISBN# 0-939950-11-1.

Volume 10: Characterization of Metamorphism through Mineral Equilibria, 1982; J. M. Ferry, Ed. 397 pp. Nine chapters on an algebraic approach to composition and reaction spaces and their manipulation; the Gibbs' formulation of phase equilibria; geologic thermobarometry; buffering, infiltration, isotope fractionation, compositional zoning and inclusions; characterization of metamorphic fluids. ISBN# 0-939950-12-X.

Volume 11: Carbonates: Mineralogy and Chemistry, 1983; R. J. Reeder, Ed. 394 pp. Nine chapters on crystal chemistry, polymorphism, microstructures and phase relations of the rhombohedral and orthorhombic carbonates; the kinetics of $CaCO_3$ dissolution and precipitation; trace elements and isotopes in sedimentary carbonates; the occurrence, solubility and solid solution behavior of Mg-calcites; geologic thermobarometry using metamorphic carbonates. ISBN# 0-939950-15-4.

Volume 12: Fluid Inclusions, 1984; by E. Roedder. 644 pp. Nineteen chapters providing an introduction to studies of all types of fluid inclusions, gas, liquid or melt, trapped in materials from the earth and space, and their application to the understanding of geological processes. ISBN# 0-939950-16-2.

Volume 13: Micas, 1984; S. W. Bailey, Ed. 584 pp. Thirteen chapters on structures, crystal chemistry, spectroscopic and optical properties, occurrences, paragenesis, geochemistry and petrology of micas. ISBN# 0-939950-17-0.

Volume 14: Microscopic to Macroscopic: Atomic Environments to Mineral Thermodynamics, 1985; S. W. Kieffer and A. Navrotsky, Eds. 428 pp. Eleven chapters attempt to answer the question, "What minerals exist under given constraints of pressure, temperature, and composition, and why?" Includes worked examples at the end of some chapters. ISBN# 0-939950-18-9.

Volume 15: Mathematical Crystallography, 1985; by M. B. Boisen, Jr. and G. V. Gibbs. 406 pp. A matrix and group theoretic treatment of the point groups, Bravais lattices, and space groups presented with numerous examples and problem sets, including solutions to common crystallographic problems involving the geometry and symmetry of crystal structures. ISBN# 0-939950-19-7.

Volume 16: Stable Isotopes in High Temperature Geological Processes, 1986; J. W. Valley, H. P. Taylor, Jr., and J. R. O'Neil, Eds. 570 pp. Starting with the theoretical, kinetic and experimental aspects of isotopic fractionation, 14 chapters deal with stable isotopes in the early solar system, in the mantle, and in the igneous and metamorphic rocks and ore deposits, as well as in magmatic volatiles, natural water, seawater, and in meteoric-hydrothermal systems. ISBN #0-939950-20-0.

Volume 17: Thermodynamic Modelling of Geological Materials: Minerals, Fluids, Melts, 1987; H. P. Eugster and I. S. E. Carmichael, Eds. 500 pp. Thermodynamic analysis of phase equilibria in simple and multi-component mineral systems, and thermodynamic models of crystalline solutions, igneous gases and fluid, ore fluid, metamorphic fluids, and silicate melts, are the subjects of this 14-chapter volume. ISBN # 0-939950-21-9.

Volume 18: Spectroscopic Methods in Mineralogy and Geology, 1988; F. C. Hawthorne, Ed. 698 pp. Detailed explanations and encyclopedic discussion of applications of spectroscopies of major importance to earth sciences. Included are IR, optical, Raman, Mossbauer, MAS NMR, EXAFS, XANES, EPR, Auger, XPS, luminescence, XRF, PIXE, RBS and EELS. ISBN # 0-939950-22-7.

Volume 19: Hydrous Phyllosilicates (exclusive of micas), 1988; S. W. Bailey, Ed. 698 pp. Seventeen chapters covering the crystal structures, crystal chemistry, serpentine, kaolin, talc, pyrophyllite, chlorite, vermiculite, smectite, mixed-layer, sepiolite, palygorskite, and modulated type hydrous phyllosilicate minerals.

Reviews in Mineralogy　　　　　　　　　　　　　　　　　Volume 20

MODERN POWDER DIFFRACTION

FOREWORD

David Bish and Jeffrey Post, the editors of this volume, convened a short course on advanced methods in powder diffractometry November 4 and 5, 1989, in conjunction with the annual meetings of the Mineralogical Society of America (MSA) and the Geological Society of America in St. Louis. This was the eighteenth such course organized by MSA, and this volume is the twenty-first book published in the *Reviews in Mineralogy* series, which began under the title *Short Course Notes* in 1974. The list of available titles appears on the opposite page; all but Volume 3 (out of print) are available from the Society at moderate cost.

Paul H. Ribbe
Series Editor
Blacksburg, VA

PREFACE AND ACKNOWLEDGMENTS

Beginning with the pioneering experiments conducted in 1913 and 1914, powder X-ray diffraction has been an important tool for studying minerals. Likewise, following the initial neutron and synchrotron X-ray diffraction experiments in the 1930's and 1960's, respectively, these methods also were quickly applied to the study of minerals. Today powder diffraction is a standard technique in many laboratories that is capable of providing a wealth of information in areas such as phase identification, crystal structure analyses, quantitative analyses, particle size and strain measurements, etc. In the last decade, several developments, including the increased availability of high-intensity, high-resolution neutron and synchrotron X-ray sources, have sparked a Renaissance in the field of powder diffraction. Perhaps the most significant changes, however, have occurred in the area of the standard laboratory X-ray powder diffractometer. Whereas until recently this instrument was a stand-alone device, typically providing data in the form of a strip-chart recording, new diffractometers are automated and interfaced to computers powerful enough to analyze data quickly and easily. Simultaneously, because of the availability of more powerful computers and of digital diffraction data, numerous advances were made and are continuing to be made in the analysis of data, including Rietveld refinement of crystal structures, profile fitting, and modeling of the diffraction patterns obtained from complex disordered structures such as the clay minerals.

In this "Reviews in Mineralogy" volume, we have attempted to: (1) provide examples illustrating the state-of-the-art in powder diffraction, with emphasis on applications to geological materials; (2) describe how to obtain high-quality powder diffraction data; and (3) show how to extract maximum information from available data. We have not included detailed discussions of certain specialized techniques, such as texture analysis or diffraction experiments under nonambient conditions, for example high-temperatures and high-pressures. In particular, the nonambient experiments are examples of some of the new and exciting areas of study using powder diffraction, and the interested reader is directed to the rapidly growing number of published papers on these subjects. Powder diffraction has evolved to a point where considerable information can be obtained from μg-sized samples, where detection limits are in the hundreds of ppm range, and where useful data can be obtained in milliseconds to microseconds. We hope that the information in this volume will increase the reader's access to the considerable amount of information contained in typical diffraction data.

We thank the authors of this volume for their willingness to participate in this endeavor and for the time and dedication required to produce this volume. We are also grateful to S. Guggenheim, E. Prince, and D. Cox for reviews. Paul Ribbe, the editor of this series, and his assistants, Marianne Stern and Margie Sentelle, ensured that the final preparation of this volume went smoothly and efficiently.

We are also very grateful to the International Centre for Diffraction Data and Siemens Analytical X-ray Instruments, Inc., for generous grants enabling us to reduce registration fees to encourage student participation.

David L. Bish	Jeffrey E. Post
Los Alamos, NM	Washington, D.C.

August 1, 1989

Table of Contents

Page	
ii	Copyright; Additional Copies
iii	Foreword
iii	Preface and Acknowledgments

Chapter 1 — R. C. Reynolds, Jr.
Principles of Powder Diffraction

1	INTRODUCTION
1	SCATTERING BY POINTS
3	SUMMATION OF AMPLITUDES
4	THE INTENSITY OF REFLECTION FROM AN INFINITE PLANE
6	THE INTENSITY OF REFLECTION FROM A FINITE PLANE
8	THE BRAGG LAW
9	DIFFRACTION FROM A SET OF REFLECTING PLANES
9	THE UNIT CELL AND THE INTERFERENCE FUNCTION
9	The unit cell
11	The interference function
13	CALCULATION OF THE THREE-DIMENSIONAL DIFFRACTION PATTERN
17	REFERENCES

Chapter 2 — R. Jenkins
Instrumentation

19	MEASUREMENT OF X-RAY POWDER DIFFRACTION PATTERNS—CAMERA METHODS
19	The Debye-Scherrer camera
21	The Gandolfi camera
21	The Guinier camera
23	MEASUREMENT OF X-RAY POWDER PATTERNS—POWDER DIFFRACTOMETERS
23	The Seemann-Bohlin powder diffractometer
24	The Bragg-Brentano parafocusing powder diffractometer
25	Systematic aberrations
25	*Axial divergence*
26	*Flat specimen*
26	*Absorption*
26	*Displacement errors*
27	EFFECT OF INSTRUMENT PARAMETERS
27	Focal spot dimensions of the tube
27	Effect of take-off angle
27	Effect of receiving slit width
29	Effect of axial divergence
29	Effect of divergence slit
31	*Choice of fixed divergence slit*
31	*Use of the variable divergence slit*
31	X-RAY DETECTORS
31	Quantum counting efficiency
32	Linearity
32	Proportionality
32	The gas proportional counter
33	The scintillation detector
34	Solid-state detectors
34	*The Si(Li) detector*
35	*The Ge(Li) detector*

35	*Cadmium telluride*
35	*Mercuric iodide*
35	Position-sensitive detectors
36	TECHNIQUES FOR THE SELECTION OF MONOCHROMATIC RADIATION
38	Use of the β-filter
39	Use of pulse-height selection
40	Use of monochromators
41	Use of solid-state detectors
43	RESOLUTION VERSUS INTENSITY
43	REFERENCES

Chapter 3   R. Jenkins
EXPERIMENTAL PROCEDURES

47	INTRODUCTION
49	FACTORS INFLUENCING d-SPACING ACCURACY
50	Influence of diffractometer misalignment
50	*The zero error*
50	*Errors in the 2:1 rotation of the goniometer*
50	Experimental 2θ errors
50	*Counting statistical limitations*
52	*Rate-meter errors*
54	*Step-scanning errors*
55	*Software-induced errors*
56	Problems in specimen placement
57	Problems in establishing the true peak position
57	*Problems with manual methods*
57	*Use of software peak-hunting programs*
60	*Treatment of the background*
60	*Treatment of the α_2*
60	Errors in conversion of 2θ to d
62	FACTORS INFLUENCING INTENSITY ACCURACY
63	Relative and absolute intensities
63	Structure-dependent factors
63	Instrument- and software-dependent factors
63	*Effect of the divergence slit*
65	*Influence of detector dead time*
65	*Influence of smoothing*
65	Specimen-dependent factors
66	Measurement-dependent factors
67	AUTOMATION OF QUALITATIVE PHASE ANALYSIS
67	Computer data-bases
67	Role of the personal computer
68	USE OF STANDARDS
68	Use of standards to establish quality of alignment
69	Use of external 2θ-alignment standards
69	Use of internal 2θ-alignment standards
69	Use of standards to model systematic relative intensity errors
69	REFERENCES

Chapter 4   D. L. Bish and R. C. Reynolds, Jr.
SAMPLE PREPARATION FOR X-RAY DIFFRACTION

73	INTRODUCTION
73	COMMON PROBLEMS IN SAMPLE PREPARATION

73	Grinding problems
74	Sample mounting
74	*Sample area*
75	*Sample thickness*
76	*Sample packing*
76	*Sample surface*
76	*Sample-height displacement*
77	METHODS OF GRINDING AND DISAGGREGATION
78	PARTICLE-SIZE REQUIREMENTS
78	Extinction
80	Particle statistics
82	Microabsorption
84	RANDOMLY ORIENTED POWDER PREPARATION
84	Side- and back-packed mounts
85	Solid filler materials
85	Use of a viscous mounting material or dispersion in a binder
87	Special methods for minimizing preferred orientation
87	*Tubular aerosol suspension*
87	*Liquid-phase spherical agglomeration*
88	*Spray drying*
89	PREPARING ORIENTED CLAY MINERAL AGGREGATES
91	Reconnaissance
91	Quantitative analysis
92	Crystal structure analysis
93	SPECIAL SAMPLE PREPARATION TECHNIQUES
93	Mounting small samples
97	REFERENCES

Chapter 5   **R. L. Snyder and D. L. Bish**

QUANTITATIVE ANALYSIS

101	INTRODUCTION
102	THE INTERNAL-STANDARD METHOD OF QUANTITATIVE ANALYSIS
102	Background
104	The generalized Reference Intensity Ratio
106	$I/I_{corundum}$
106	Measurement of I/I_c
107	Quantitative analysis with I/I_c - the *RIR* or Chung method
107	*Analysis when a known amount of corundum is added to an unknown*
107	*Analysis when a known amount of an internal standard is added to a sample*
108	*Analysis of a mixture of identified phases without the addition of an internal standard—the RIR or Chung method*
109	Constrained XRD phase analysis—generalized internal-standard method
111	THE INTERNAL-STANDARD METHOD—AN EXAMPLE
112	Internal standard selection
112	Diffraction line selection
115	Calibration of the Al_2O_3 standard
115	Calibration of the SiO_2 standard
115	Data collection and analysis
116	Running an unknown
120	EXAMPLE USE OF THE REFERENCE INTENSITY RATIO OR CHUNG METHOD
122	The Chung method using direct input of *RIR* and I^{rel} values
122	ABSORPTION-DIFFRACTION METHOD
122	Analysis of a mixture of known chemical composition

124	Analysis of a mixture of unknown chemical composition by measuring $(\mu/\rho)_j$
124	Analysis of a sample whose mass absorption coefficient is the same as that of the pure phase
124	Analysis of thin films or lightly loaded filters
124	Analysis of a binary mixture of two known phases
125	METHOD OF STANDARD ADDITIONS OR SPIKING METHOD
126	Amorphous phase analysis
126	ERROR PROPAGATION IN QUANTITATIVE ANALYSIS
128	FULL-PATTERN FITTING
129	Quantitative phase analysis using the Rietveld method
129	*Theory*
132	*Applications*
138	Quantitative phase analysis using observed patterns
142	REFERENCES

Chapter 6   R. C. Reynolds, Jr.
DIFFRACTION BY SMALL AND DISORDERED CRYSTALS

145	INTRODUCTION
145	DIFFRACTION BY SMALL CRYSTALS
145	Equi-dimensional parallelepiped crystals
147	Two-dimensional diffraction from turbostratic structures
150	Two-dimensional diffraction with some third-dimensional influence
155	ONE-DIMENSIONAL DIFFRACTION FROM SIMPLE CLAY MINERALS
155	Mathematical formulation
158	The $00l$ series and particle size effects
158	The $00l$ series and defect broadening
159	The $00l$ series and strain broadening
163	The effect of the continuous function $F(\theta)^2$ on peak shape and displacement
164	The pure diffraction line profile
165	The Lorentz factor for oriented clay mineral aggregates
170	MIXED-LAYERED MICAS AND CLAY MINERALS
170	Mathematical formulation
172	Interpretation of basal series diffraction patterns of mixed-layered clay minerals
175	Examples of mixed-layered clay minerals
178	DIFFRACTION FROM GLASS
178	Mathematical formulation
179	Examples of diffraction patterns from glass
181	REFERENCES

Chapter 7   D. K. Smith
COMPUTER ANALYSIS OF DIFFRACTION DATA

183	INTRODUCTION
184	TYPES OF NUMERICAL DIFFRACTION DATA
185	REFERENCE NUMERIC DATABASES FOR POWDER DIFFRACTION ANALYSIS
185	The Powder Diffraction File
185	*PDF-1*
186	*PDF-2*
186	*PDF-3*
187	NIST Crystal Data File
187	NIST/Sandia/ICDD Electron Diffraction Database
187	FIZ-4 Inorganic Crystal Structure Database
187	NRCC Metals Structure Database

188	Mineral Data databases
188	ANALYSIS OF d-I DATA
188	Processing of d-I data sets
189	Identification of mineral phases
190	*Mini- and main-frame routines*
190	*PC-based search routines*
190	Indexing of d-spacing data for an unknown
192	*Cell reduction and lattice evaluation*
193	Unit-cell refinements
194	Unit-cell identification
195	Quantitative analysis
195	Tests for preferred orientation
196	Structure analysis using integrated intensities
196	ANALYSIS OF DIGITIZED DIFFRACTION TRACES
196	Processing of raw digitized data
197	Location of peak positions and intensities
199	Crystal structure analysis
199	*Structure modeling*
200	*Structure refinement*
202	Quantification of crystallite parameters
202	*Quantitative phase analysis*
203	*Crystallinity*
203	*Crystallite size and strain*
204	File building and transfer
205	STRUCTURE DISPLAY PROGRAMS
206	ANALYTICAL PACKAGES
206	MISCELLANEOUS ROUTINES
207	GENERAL COMMENTS AND DISCLAIMER
207	SOURCES OF PROGRAMS
213	REFERENCES

Chapter 8   S. A. Howard and K. D. Preston
PROFILE FITTING OF POWDER DIFFRACTION PATTERNS

217	INTRODUCTION
220	Advantages of using refined line parameters
220	*Qualitative phase analysis*
221	*Indexing*
221	*Accurate lattice parameter determination*
221	*Quantitative phase analysis*
222	Development of the profile-shape-functions used in profile and pattern fitting
224	*The Rietveld method of pattern-fitting sructure-refinement*
226	*The PSF in Rietveld structure refinement algorithms*
228	*Profile-shape-functions in profile refinement of X-ray data*
231	The use of direct convolution functions
232	*Direct convolution products*
234	*The Voigt and pseudo-Voigt function*
240	Particle-size and strain analysis via direct- or approximated-convolution relations
241	Analysis using the Voigt function
243	Profile fitting of synchrotron diffraction data
243	Two-step structure refinement using a whole-pattern-fitting algorithm
244	CONSIDERATIONS IN PROFILE FITTING
244	General considerations in profile refinement
245	*Minimizing parameter correlation*
247	*The least-squares error criteria for refinement*

248	The choice of optimization algorithm
249	The convergence criterion for the optimization algorithms
250	Goodness of fit
250	Choice of profile-shape-function
251	Single-line PSFs
253	Accommodating satellite lines
253	Compound profile-shape-functions
255	Additional constraints on the profile shapes
257	Alpha-2 stripping
257	Digital filtering
259	Accommodating an amorphous profile in the pattern
261	Convolution-based profiles
261	W—the factor determining the number of lines in the instrument PSF
262	Development of the instrument profile calibration curves
262	The method of profile generation
263	Generation of the specimen profile
264	Numerical convolution of $(W*G)$ and S
264	Maintaining the definition of the convolute profile
266	Specimen broadening as a function of angle: particle size and strain constraints
268	Miscellaneous considerations
268	Background
269	Adding lines after the initial refinement
270	Discrepancies indicated by the difference patterns
271	SUMMARY
272	REFERENCES

Chapter 9 — J. E. Post and D. L. Bish
RIETVELD REFINEMENT OF CRYSTAL STRUCTURES USING POWDER X-RAY DIFFRACTION DATA

277	INTRODUCTION
279	RIETVELD METHOD IN PRACTICE
279	Data collection
281	Starting model
281	Computer programs
282	The diffraction profile
286	Unit-cell parameter refinement
288	A sample Rietveld refinement
290	Estimated standard deviations
293	EXAMPLES OF RIETVELD REFINEMENTS
294	Zeolites
294	Kaolinite
296	Hollandite
300	STRUCTURE SOLUTION USING RIETVELD METHODS
301	Todorokite
304	SUMMARY
305	REFERENCES

Chapter 10 — L. W. Finger
SYNCHROTRON POWDER DIFFRACTION

309	INTRODUCTION
309	SOURCE CHARACTERISTICS
312	Beam divergence/resolution

314	Photon output
316	Spectral distribution
317	Time structure of ring
318	HIGH-RESOLUTION EXPERIMENTS
318	Peak shape
320	Determination of lattice geometry
321	Extraction of intensities for structure solution
323	Non-isotropic broadening
324	LOW- TO MEDIUM-RESOLUTION EXPERIMENTS
324	Rapid measurement
324	Time-dependent studies
324	UTILIZATION OF SPECTRAL DISTRIBUTION
324	Experiments near absorption edges
326	"White" beam experiments
327	SAMPLE PREPARATION
329	PRESENT AND FUTURE OPPORTUNITIES
329	Operating beam-line locations and characteristics
329	New techniques
329	REFERENCES

Chapter 11   R. B. Von Dreele
NEUTRON POWDER DIFFRACTION

333	INTRODUCTION
333	THE NEUTRON—BASIC PROPERTIES
335	NEUTRON POWDER DIFFRACTION THEORY
338	REACTOR NEUTRON SOURCES
341	CONSTANT-WAVELENGTH POWDER DIFFRACTOMETER DESIGN
343	SPALLATION NEUTRON SOURCES
345	TIME-OF-FLIGHT POWDER DIFFRACTOMETER DESIGN
347	NEUTRON DETECTION
350	SPECIAL SAMPLE ENVIRONMENTS IN NEUTRON DIFFRACTION
351	RIETVELD REFINEMENT
352	Background intensity contribution
353	Bragg intensity contribution
354	Profile functions for CW and TOF
354	Reflection positions in powder patterns
354	TOF profile functions
355	Interpretation of TOF profile coefficients
357	CW profile functions
358	Interpretation of CW profile coefficients
359	Reflection peak widths
359	Systematic effects on the intensity
360	Extinction in powders
360	Powder absorption factor
361	Preferred orientation of powders
361	Other angle-dependent corrections
361	APPLICATIONS OF NEUTRON POWDER DIFFRACTION
362	Hydrogen atom location
363	Cation distribution
366	Combined X-ray/neutron Rietveld refinement
367	CONCLUSION
367	REFERENCES

1. PRINCIPLES OF POWDER DIFFRACTION
R. C. Reynolds, Jr.

INTRODUCTION

This chapter discusses the scattering of X-rays by points, the arrangement of those points into planes, the stacking of planes to form a unit cell, and finally the assembly of unit cells into three-dimensional crystals. The theoretical development is useful to the beginning student of X-ray powder crystallography, although the specialist will find little that is new.

The theory described is basic to X-ray diffraction models. Modeling is an approach that is used more and more routinely because of the easy availability of computing power that could not have been imagined 20 years ago. Indeed, procedures that once constituted major research efforts are applied daily for the identification and characterization of crystalline substances. Ideally, if one could measure the diffraction amplitudes (magnitudes and phases) scattered by the different planes of atoms in a crystal, one could use a mathematical "lens" to generate an image of the electron density and hence of the atomic structure. But except in rare and favorable situations, that cannot be accomplished because the *magnitude* of the scattering amplitudes can be measured but the *phases* of the scattered X-rays cannot. The mathematics of the diffraction problem thus cannot be readily inverted, and the nature of the atomic arrangement in crystals can only be deduced by trial-and-error procedures.

In this discussion I assume that an incident X-ray beam is parallel and strictly monochromatic with wavelength λ. The refractive index of matter is asumed to be unity and the mathematical treatment of the scattering functions of atoms involves real numbers, that is, the scattered amplitude has no imaginary component. The scattering by atoms is coherent. Compton, or incoherent scatter (scattering that involves an increase in wavelength) is easily measurable, but it is not involved in diffraction phenomena, and is ignored. The writer has borrowed heavily from James (1965) for the material given here.

SCATTERING BY POINTS

A single electron scatters as a point source, with the scattering amplitude being identical in all directions. The magnitude of scattering in a given direction (θ or 2θ) is described in units relative to the scattering from a single electron. Thus a magnitude of 12 for the diffraction angle 2θ means that a given configuration of electrons scatters like the constructive interference from 12 electrons. Intensity is the quantity measured by the diffraction device, and it is given by the magnitude of the amplitude squared. In the case of the complex scattering by a noncentrosymmetric arrangement of scattering points, the intensity is the product of the complex amplitude multiplied by the complex conjugate, a procedure that is described below.

Atoms are considered point sources of scattering even though their volume is finite and partially destructive interference takes place between the scattering of electrons within the atoms. At a diffraction angle of zero degrees, all scattering is in phase, so the scattering amplitude f_0 is equal to the number of electrons in the atom or the atomic number minus the positive charge or plus the negative charge for an ionic form of the atom. The scattering magnitude of an atom is, to a good approximation, proportional to the atomic number at small values of $\sin\theta/\lambda$. At higher diffraction angles, f_0 diminishes according to complicated functions of the electronic structure. Values for f_0 are best calculated by empirical functions that have been fit to the rigorously computed scattering factors. An excellent and easy to use polynomial approximation has been described by Wright (1973).

Atoms in a crystal structure are not fixed in space. Thermal effects cause vibration about their mean positions so that a given atom is represented by a three-dimensional scattering distribution whose shape may be complicated but is often referred to a triaxial ellipsoid whose three axial lengths are controlled by temperature and the nature of chemical bond strength in each of the three crystallographic directions. Atomic scattering factors are corrected by means of the Debye-Waller equation,

$$f = f_0 \, e^{(-B \sin^2\theta / \lambda^2)}, \qquad (1)$$

in which f is the scattering factor under experimental conditions, f_0 is the scattering factor at a temperature of absolute zero, and B is a parameter that is related to the mean amplitude of vibration normal to the diffraction direction for which f is applied. The temperature effect causes the amplitude of scattering to decrease monotonically with high diffraction angles, and the larger the value of B, the more rapid the diminution. Typical values of B for silicates at room temperature vary roughly between 0.5 and 2.5 $Å^2$, and these are not computed directly but are optimized by trial and error during a structure refinement.

Figure 1 represents the scattering factors (Wright, 1973) for Fe, Si, Al, and O that have been corrected for thermal vibration using B = 1.5 for the metals and B = 2.0 for oxygen. The scattering magnitudes are in expressed in electron scattering units and the diffraction angles refer to Cu$K\alpha$ X-rays. The values for Si, Al, and O are for "half-ionized" forms of the ions because of the significant covalent character of their bonds in silicates. The scattering factors decrease with increasing 2θ because of destructive interference within the atoms and because of thermal effects. The curves of Figure 1 show that Fe has a strong influence on peak intensities of silicate diffraction patterns because of its very different scattering factor compared to those of the elements that it substitutes for such as Mg, Al, and perhaps Si. Unfortunately, the very important Si-Al diadochy is difficult to characterize by X-ray diffraction methods because Si and Al have similar scattering factors. Quantification of Si-Al proportions and order-disorder in the pattern of

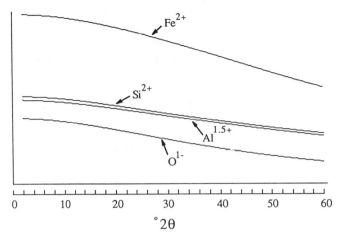

Figure 1. Atom scattering factors for Fe, Si, Al, and O (Cu$K\alpha$).

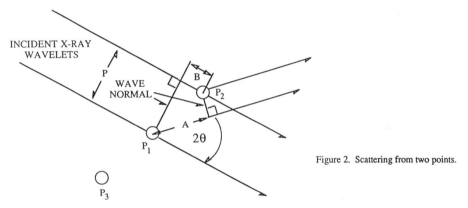

Figure 2. Scattering from two points.

substitution are based on the geometry of the cation coordination (bond lengths) because the ionic radii of Si and Al differ significantly, and a highly refined three-dimensional structure is required for such analyses.

SUMMATION OF AMPLITUDES

Figure 2 illustrates the procedure for adding amplitudes from scattering centers. In this case, atoms P_1 and P_2, whose scattering factors are f_1 and f_2, are separated from each other by P which is measured as their separation (in Å) in a direction normal to that of the incident X-ray wave. 2θ is the angle between the incident and the diffracted wave whose wavelength is λ (in Å). The wave normal from P_1 connects points of equal phase for the two incident wavelets, and the wave normal from P_2 connects points of equal phase for the scattered wavelets. A is the distance that the wavelet scattered from P_1 must travel to coincide with the wave normal of the diffracted wave which passes through P_2, and B is the distance that P_2 lies from the incident wave normal. Because B ≠ A there is a phase difference between the scattering centers which causes the magnitude of scattering to be less than $f_1 + f_2$. The path difference, for the diffraction angle 2θ, is equal to (A - B). If $(A - B)/\lambda$ is integral, then the scattered wavelets are perfectly in phase, and if $(A - B)/\lambda$ equals 0.5, 1.5, 2.5 etc., the scattering is out of phase and the resulting magnitude is zero. For the general case, we define the phase angle, ϕ, as the path difference times $2\pi/\lambda$, so for this example,

$$\phi = 2\pi (A - B) / \lambda \ . \tag{2}$$

The expression for adding amplitudes of scattered wavelets is given by

$$A = f_1 e^{i\phi_1} + f_2 e^{i\phi_2} \ , \tag{3}$$

where A is the resulting amplitude and $i = \sqrt{-1}$. We have selected P_1 as the origin for this calculation, so $\phi_1 = 0$ and the result is $A = f_1 + f_2 \exp(i\phi_2)$.

The exponential notations are cumbersome, and Euler's identity is used to convert them to their trigonometric forms,

$$f e^{i\phi} = f \cos \phi + f i \sin \phi \ . \tag{4}$$

The amplitude is $A = f_1 + f_2 \cos[2\pi(A - B) / \lambda] + i f_2 \sin[2\pi(A - B) / \lambda] \ .$

The intensity (I) is equal to A times its complex conjugate. The complex conjugate is the same expression with a minus sign replacing the plus sign that precedes i. Thus,

$$\left[f_1 + f_2 \cos(2\pi(A - B) / \lambda) + i f_2 \sin(2\pi(A - B) / \lambda) \right] \cdot$$

$$\left[f_1 + f_2 \cos(2\pi(A - B) / \lambda) - i f_2 \sin(2\pi(A - B) / \lambda) \right] =$$

$$\left[f_1 + f_2 \cos(2\pi(A - B) / \lambda) \right]^2 + \left[f_2 \sin(2\pi(A - B) / \lambda) \right]^2 \ . \tag{5}$$

Suppose now that the problem is changed a bit, and that another atom is present at position P_3 (Fig. 2) whose scattering factor is also equal to f_2. The atomic arrangement is now centrosymmetric, and the sum of the sines equals zero. The intensity under these conditions is

$$I = \{f_1 + 2f_2 \cos[2\pi(A - B) / \lambda]\}^2 \ . \tag{6}$$

Multiplication of f_2 by two is required to account for the scattering from the additional atom (P_3) whose position is equivalent to P_2.

THE INTENSITY OF REFLECTION FROM AN INFINITE PLANE

Figure 3 depicts the reflection condition from an infinitely long plane of atoms. [The horizontal plane normal to the figure is the width of the atomic plane and has identical geometry with respect to the incident and reflected wave and need not be considered.] The plane of atoms (points) indicated by the bold line consists of scattering centers so close together that they form a uniform plane. The "reflectivity" of this plane can be expressed in terms of the magnitude of scattering per unit of area, or per square Å. The phase lag between the two wavelets scattered by P_1 and P_2 is related to (A - B), and it is easy to see that if the angle of incidence equals the angle of reflection, (A - B) = 0, that is, there is no phase shift between these two or any other pair of points on this plane. Suppose now that the plane is rotated slightly within the plane of the figure, then (A - B) is finite, and if P_2 is moved away from P_1, some point will be reached at which (A - B) is equal to $\lambda/2$, resulting in the cancellation of the scattering from the two points. Consequently, for an infinitely long plane of scattering centers, oriented so that the angle of incidence is unequal to the angle of reflection, the amplitude of scattering from any point is cancelled by the scattering from another point, and the resulting magnitude or intensity of reflection from the entire plane is zero. An infinitely long plane thus diffracts only when the angles of incidence and "reflection" are identical. This requirement would seem to preclude measurable scattering from lattice planes in any crystalline powder because it is improbable that many or any atomic planes are *perfectly* aligned, but the stricture of this requirement is relaxed because of a number of unavoidable circumstances. The reflecting planes are finite in length and may be slightly bent according to continuous or periodic functions (see Cowley, 1961). For example, the small crystallite sizes of the clay minerals typically show this effect. In addition, the incident beam always has a degree of divergence and crossfire in the vertical plane as shown by Figure 3, though both of these effects are diminished by a large goniometer radius and a small X-ray tube focal spot. In any event, a first requirement for diffraction is that the planes of atoms or unit cells are oriented so that the angles of incidence and reflection are very nearly equal. In a powdered sample diffracting in the reflection mode, only those crystallites oriented so that a given *hkl* plane parallels the sample surface contribute to the intensity of a diffraction peak from plane *hkl*. Other crystals might as well not be there, except for the effects of their absorption of the incident and diffracted X-ray beams.

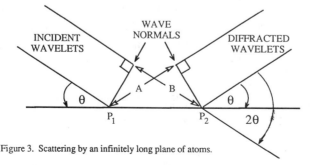

Figure 3. Scattering by an infinitely long plane of atoms.

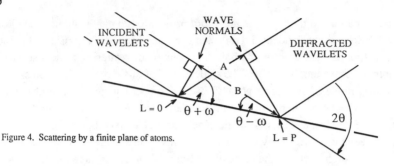

Figure 4. Scattering by a finite plane of atoms.

THE INTENSITY OF REFLECTION FROM A FINITE PLANE

Figure 4 shows the geometry for calculating the scattering from a plane of atoms of finite length. The plane has been tilted by the angle ω, causing unequal angles of incidence and diffraction. The origin for the calculation lies at $L = 0$, and in the following we neglect the imaginary component of the amplitude because we have chosen the point $L = 0$ so that it lies at the center of the plane of length $2P$, and consequently the integrated phase shift from $L = 0$ to $L = P$ has a mirror image from $L = 0$ to $L = -P$. The difference in phase between $L = 0$ and $L = P$ is given by $(2\pi/\lambda)(A - B)$, and from Figure 4, $A = P\cos(\theta + \omega)$ and $B = P\cos(\theta - \omega)$, so $(A - B)$ is equal to the following:

$$(A - B) = P[\cos(\theta + \omega) - \cos(\theta - \omega)] =$$

$$P[\cos\theta \cos\omega - \sin\theta \sin\omega - \cos\theta \cos\omega - \sin\theta \sin\omega] = -2P \sin\theta \sin\omega .$$

Multiplying by $(2\pi/\lambda)$ gives the phase angle, $\phi = -4\pi P\sin\theta \sin\omega/\lambda$. To obtain the amplitude scattered by a plane of length P, the cosine of the phase angle must integrated from $L = 0$ to $L = P$. The minus sign of ϕ is neglected for this integration because $\cos X = \cos -X$. The amplitude, A, is given by Equation 7.

$$A = \int_{L=0}^{L=P} \cos\left(\frac{4\pi \sin\theta \sin\omega}{\lambda} L\right) dL = \frac{\sin(4\pi P \sin\theta \sin\omega/\lambda)}{4\pi \sin\theta \sin\omega/\lambda} . \quad (7)$$

The intensity of scattering from a fixed area made up of discrete planes of length P requires that A^2 be normalized by dividing by P. Multiplying by $\sin\theta$ corrects for the incident X-ray energy received by a fixed area as a function of the mean angle of incidence θ. The overall result gives the intensity contributions of planes whose orientations range from $-\omega$ to $+\omega$, as a function of 2θ.

$$I(\omega) = \frac{\sin\theta}{P} \frac{\sin^2(4\pi P \sin\theta \sin\omega/\lambda)}{(4\pi \sin\theta \sin\omega/\lambda)^2} . \quad (8)$$

Crystallographers will recognize here the well-known function $\sin^2 NX / X^2$ which has many applications in diffraction theory.

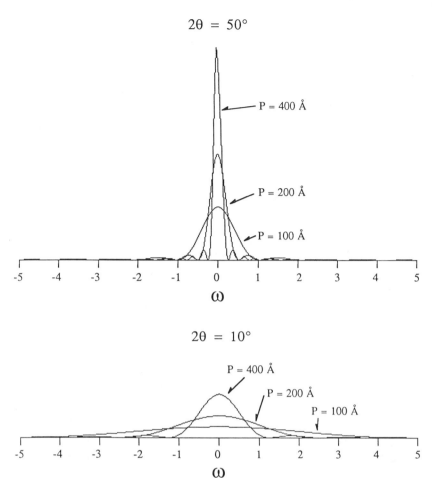

Figure 5. Relative intensities for crystallites that are rotated by the angle ω from the symmetrical reflection condition.

Figure 5 presents solutions of Equation 8. Trace A shows the effects of the plane length, P, on the intensity contributions from planes that are tilted by the angle ω. The intensity distributions become broader as P is diminished, and as the plane length approaches zero, the plane behaves as a point which scatters with equal magnitude in all directions. The integrated areas of these three "specular reflections" (as they would be named for the reflection of light from a mirror) are identical, and the heights of the reflections at ω = 0 (equal angles of incidence and reflection) are proportional to P. If we select 200 Å to represent a commonly used value for the lengths of clay mineral crystallites, we see that at 50°2θ, crystallites that are disoriented by only a few tenths of a degree or more make no significant contributions to a peak at that diffraction angle. I(ω) curves at 2θ = 10° for the same three plane lengths are shown on Figure 5B, which is plotted to the same scale as Figure 5A. The I(ω) distributions are broader because as the diffraction angle approaches zero, all scattering approaches the in-phase condition. Note that for P =

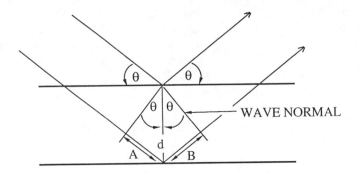

Figure 6. The Bragg Law for parallel planes.

200 Å, significant diffraction intensity arises from planes that are misaligned by 1.5°. The interesting conclusion to be derived from this discussion is that, as diffraction angle decreases, crystallites that are more and more misaligned make increased contribu-tions to the diffraction pattern. At present, there is no evaluation of the consequences of this effect on the intensities of low-angle superstructure reflections for oriented clay mineral aggregates.

THE BRAGG LAW

The diffraction principle developed by W. L. Bragg is the most general and powerful in diffraction theory, and uncharacteristically for a general theory, is one of the most intuitive and easy to understand. Figure 6 shows the diffraction from two scattering planes that are separated by the distance d (in Å) and intercept X-radiation of wavelength λ (in Å) at the incident angle θ. The experimental parameter 2θ is the angle between the diffracted and undeviated X-ray waves. The diagram shows only two scattering planes, but implicit here is the presence of many parallel, identical planes, each of which is separated from its adjacent neighbor by the spacing d. The locations of the planes are defined by a periodic function whose "wavelength" is related to the dimensions of the unit cell and the orientation of the plane hkl. The wave normals connect points of identical phase for the incident and diffracted waves. The distance (A + B) must equal a whole number of wavelengths for total constructive reinforcement to occur between the scattering from these planes, so we need to calculate the value of θ, for a given d, for which $(A + B) = n\lambda$, where n is integral. The direction of d is normal to the planes, and the wave normal is normal to the wavelets, so the angles opposite A and B are also θ. $\sin\theta = A/d = B/d$, so $(A + B) = 2d\sin\theta$. Constructive interference occurs when $(A + B) = n\lambda$, and the Bragg Law is

$$n\lambda = 2d\sin\theta \quad . \tag{9}$$

The integer n refers to to the order of the diffraction. For n = 1, (A + B) = 1, and for n = 2, (A + B) = 2 etc. Thus for planes such as, for example, the 112, separated from each other by 4 Å, d(112) = 4 Å, d(224) = 2 Å, d(336) = 1.333 Å etc. The 336 reflection can be thought of as the third order reflection of a 4 Å spacing or the first order reflection of a

1.333 Å spacing. The geometry of the unit cell is unknown for most powder diffraction work, or at best represents a conclusion rather than an a priori fact. The objective way to designate diffraction peaks is to set the integer n = 1 in the Bragg law and report the peak position as a spacing d, even though that value might not correspond to any real separation of equivalent atomic planes in the crystal.

DIFFRACTION FROM A SET OF REFLECTING PLANES

The derivation given above for the Bragg Law does not explain why crystals produce such sharp diffraction phenomena. Suppose that θ is increased by a small amount that leads to an increase in the quantity (A + B) by ε_0, and that the amplitude of the scattering for each of the planes is identical. Thus the phase (φ) between the two planes is equal to $(2\pi/\lambda)(A + B + \varepsilon_0)$. The cosine of φ is very close to the cosine of the phase angle for the Bragg condition. Hence, if ε_0 is small, only a slight reduction of magnitude results and significant intensity occurs at an angle different by ± ε_0 from the Bragg angle resulting in a broad diffraction peak. But there are additional, parallel planes, separated from the first by $2d$, $3d$, $4d$ etc. For the plane located at $2d$, $\varepsilon = 2\varepsilon_0$, and for $3d$, $\varepsilon = 3\varepsilon_0$ etc. Consequently, for a thick crystal and a finite value of ε_0, some plane will lie at the distance (nd) that is exactly out of phase with respect to the top plane of Figure 6, and the diffracted intensity will be zero at the angle slightly different from θ. The same argument can be made for any plane in a thick crystal; ideally, an infinitely thick *perfect* crystal would produce peaks with no width at all. In practice, however, perfect crystals are very rare, and even they produce reflections that are broadened by the finite width of the incident X-ray spectral distribution and other instrumental defocusing effects. If a crystal is thin, however, widely separated out-of-phase planes are precluded for a given value of ε_0, consequently, thinner diffracting domains produce broader peaks. The relation between line breadth and crystal thickness is given by the Scherrer equation, which follows from the Bragg Law (see Klug and Alexander, 1974).

$$\beta = K\lambda / T\cos\theta \quad , \tag{10}$$

where β is the width (in radians) of a reflection measured at half of its height, K is a crystal shape constant near unity and T is the thickness (in Å) of the coherent diffracting domain along a direction normal to the diffracting plane *hkl*. Instrumental broadening effects must be removed from the experimental diffraction profile to yield the pure diffraction breadth β, in order to apply this equation for practical measurements. Such procedures are described by Klug and Alexander (1974) and are illustrated in Chapter 6 of the present work.

THE UNIT CELL AND THE INTERFERENCE FUNCTION

The unit cell

Figure 7 is a modified version of Figure 6. Atoms have been substituted for the atomic

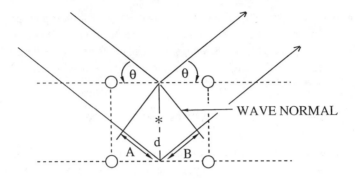

Figure 7. The Bragg Law for the unit cell.

planes and the atomic positions define a primitive, two-dimensional orthogonal unit cell with Z vertical with respect to the page. This arrangement is suitable for describing the production of the 00l or basal diffraction series. The term *structure factor*, represented by the symbol F, is used portray the amplitude of scattering from the unit cell in the direction θ, which lies in a plane normal to the atomic plane *hkl*. We can imagine the unit cell as a single point in space, but unlike an electron, a point that has a complicated directional scattering distribution. The location of the "point-source" unit cell in space is fixed by the origin that is selected for calculations of scattering amplitude, and the phases of scattering from all of the atoms are referred to that origin for which the phase angle is defined as zero. The reference point on Figure 7 for calculating the structure factor F(00l), is located at the asterisk, and that location is a center of symmetry. One quarter of an atom lies at each corner, and the reader may assume that for each atom, f = 1. The amplitudes are summed as follows: For the top atomic plane, the phase angle with respect to the center is $(2\pi/\lambda)2z\sin\theta$, where z is the atomic coordinate in the Z direction and is equal to $d/2$ where $d = c = d(001)$, and for the bottom atomic plane, ϕ is the same except that $z = -d/2$. F(001), then, is equal to

$$2(f/4)\cos[4\pi(c/2)\sin\theta/\lambda] + 2(f/4)\cos[4\pi(-c/2)\sin\theta/\lambda] =$$

$$4(f/4)\cos[4\pi(c/2)\sin\theta/\lambda] = f\cos(2\pi c\sin\theta/\lambda) \, ,$$

because $\cos X = \cos -X$. For centrosymmetric unit cells, F is conveniently calculated by omitting centrosymmetrically related atoms and by doubling their magnitude of scattering because of the $\cos X = \cos -X$ relationship. Because the origin is located at the center of symmetry, the sine terms sum to zero and, therefore, may be omitted. The scattering amplitude is expressed as a function of the continuous variable, θ, whereas F is defined in terms of the indices *hkl*. The conversion to the units *hkl* is easily made by use of the Bragg Law, $n\lambda = 2d\sin\theta$ where $n = l$ and $d = c$. $\sin\theta/\lambda = l/2c$, and the substitution of this into the cosine argument gives F(00l) = $f\cos(2\pi lz/c)$. Equation 11 applies to a centrosymmetric unit cell with any number of atoms.

$$F(00l) = \sum_j n_j f_j \cos(2\pi l\, z_j/c) \;, \tag{11}$$

where z_j is the distance in Å of each atomic layer from the center of symmetry, n_j is the number of atoms at z_j, and f_j is the scattering magnitude of each atom. Equation 11 is easily generalized into three dimensions by the expression,

$$F(hkl) = \sum_j n_j f_j \cos[2\pi(h\,x_j + k\,y_j + l\,z_j)] \;, \tag{12}$$

where x_j, y_j, and z_j refer to the three dimensional locations of the each atom expressed as *fractional coordinates of the unit cell* in the directions X, Y, and Z.

<u>The interference function</u>

Return now to the simple cell of Figure 7. A one-dimensional crystal of this "substance" can, for mathematical purposes, be visualized as a series of points, separated by the distance d (or c) along the Z direction. Each point scatters with amplitude $F(00l)$ in the direction l, and the summation of the amplitudes from these points to give the scattering amplitude for a crystal is calculated in exactly the same fashion as the calculation of F for an assemblage of atoms. Such a calculation leads to the interference function Φ, the derivation of which is given here according to the treatment by Brindley (1980), which is a particularly lucid derivation of this well-known expression.

Let ϕ be the phase difference between waves scattered from the first unit cell with respect to each of $N - 1$ others. The hypothetical crystal contains N layers. The resulting amplitude is

$$A = F + F\,e^{i\phi} + F\,e^{i2\phi} + F\,e^{i3\phi} + F\,e^{i4\phi} + \ldots F\,e^{i(N-1)\phi} \;. \tag{13}$$

F is factored out and the series can be expressed by the equation

$$A = F(1 - e^{iN\phi})/(1 - e^{i\phi}) \;. \tag{14}$$

The numerator and denominator are expressed as their trigonometric equivalents and are multiplied by their complex conjugates. For the numerator, this gives

$$[1 - \cos(N\phi) - i\sin(N\phi)]\,[(1 - \cos(N\phi) + i\sin(N\phi)] = 2[1 - \cos(N\phi)] \;,$$

and the denominator, calculated by the same method is equal to $2(1 - \cos\phi)$. We make use of the trigonometric identity $2\sin^2(N\phi/2) = [1 - \cos(N\phi)]$ to obtain

$$I = Lp\,|F|^2\,\frac{\sin^2(N\phi/2)}{N\sin^2(\phi/2)} = Lp\,|F|^2\,\frac{\sin^2(\pi l\,N)}{N\sin^2(\pi l)} \quad , \tag{15}$$

where I is the intensity and $|F|^2$ is the square of the magnitude of the scattering from a unit cell, and is equal to F^2 for a centrosymmetric unit cell. This one-dimensional example is arbitrarily derived for the 00l diffraction series. Division by N is required to produce the result for an infinitely thick aggregate of crystals, each of which has a thickness, N.

The quantity Lp has been introduced, which is is the Lorentz-polarization factor. For the *integrated* intensity from a randomly oriented powder,

$$Lp = \frac{1 + \cos^2 2\theta}{\sin\theta\,\sin 2\theta} \quad . \tag{16}$$

The Lorentz factor, which is the denominator of Equation 16, is usually written $1/\sin^2\theta\,\cos\theta$ for random powders. By trigonometric identity, $\sin 2\theta = 2\sin\theta\,\cos\theta$, and if the right side of this equation is substituted into the denominator of Equation 16 for the quantity $\sin 2\theta$, the two forms of the Lorentz factor are identical except for the "two", which can be neglected for studies involving relative intensities. There is an advantage in writing the Lorentz factor as James (1965) does (Eq. 16). This form allows the separation of the Lorentz factor into its two fundamental parts, $1/\sin 2\theta$, which is the single crystal factor for powders and single crystals, and Ψ, which is the powder ring distribution factor that varies from constant, for a single crystal, to $1/\sin\theta$ for a random powder. Ψ represents the fraction of crystallites in a powder that are oriented in such a way as to diffract, at each value of θ, into the rectangular aperture formed by the detector slit and the Soller slit parallel plates. It is discussed further in Chapter 6. The argument so far applies to the integrated intensity. The denominator of Equation 16 is $\sin 2\theta$ when it is used as a point-by-point correction for a 2θ-continuous diffraction profile (Cullity, 1978), because the single crystal term then is $1/\sin\theta$.

The important result of Equation 15 allows the diffraction intensity to be separated into three parts: the geometric factor, Lp, the unit cell quantity F^2, and the trigonometric quotient, called Φ, which is the interference function. Φ gives the diffraction pattern of a crystal structure for which all unit cells scatter as points with unit intensity. Equation 15 is easily generalized into three dimensions, viz.

$$I(hkl) = \frac{Lp(\theta)\,|F(hkl)|^2}{N_1 N_2 N_3}\,\frac{\sin^2(\pi h\,N_1)}{\sin^2(\pi h)}\,\frac{\sin^2(\pi k\,N_2)}{\sin^2(\pi k)}\,\frac{\sin^2(\pi l\,N_3)}{\sin^2(\pi l)} \quad , \tag{17}$$

in which N_1, N_2, and N_3 are, respectively, the number of unit cells in the X, Y, and Z directions, and Lp(θ) refers to the diffraction angle for the reflection hkl.

For very thick crystals containing no disorder, the interference function represented by the sine-squared quotients can be neglected, and the diffraction pattern can be described entirely by a set of intensities $I(hkl) = Lp |F(hkl)|^2$. For such thick crystals, the diffraction lines are almost entirely confined to diffraction angles at precisely the Bragg positions, and they are so sharp that each is overwhelmed by the small but finite breadth of the instrumental signature. The shapes of the different diffraction lines are almost identical because their character is fixed by the diffractometer and not by any intrinsic crystalline characteristic. But small and/or disordered crystals have broad lines, that is, diffraction is not restricted to the integral diffraction directions h, k, and l, and diffraction line breadth and shape are sample-specific parameters. The diffraction patterns for these must be computed treating h, k, and l as continuous variables, as is shown below and discussed in Chapter 7.

CALCULATION OF THE THREE DIMENSIONAL DIFFRACTION PATTERN

Equation 17 is easily solved for the intensity of diffraction in the direction hkl, but such a solution is significant only for a single crystal with unique orientation. We need the diffraction patterns of randomly oriented powders, and for these, the experimental data define an intensity distribution that is a continuous function of the diffraction angle 2θ. The intensity function must be summed over all values of hkl for which 2θ is identical. The calculation is repeated for each increment of 2θ to produce a model powder diffractogram. The procedure outlined is described by Brindley (1980).

The summation is done in reciprocal space. The complicated part is the calculation of the interference function, Φ, and a diagrammatic representation of the problem is shown by Figure 8, which represents a reciprocal lattice for an orthorhombic unit cell with $c > b$ and with a equal to approximately one-half of b. Only one-eighth of the reciprocal lattice is shown because the unit cell symmetry has been chosen to make the other octants identical. The orthogonal axes are plotted so that equal distances are represented by h/a, k/b and l/c. All distances are measured as $1/d$. The diffuse spheres of Figure 8 represent the interference function contributions to the X-ray intensity distributions about each of the reciprocal lattice points for which hkl is integral, and enough of these are labelled to demonstrate the idea. These look like electron distributions in space, but they are not. The reciprocal lattice spots are approximately triaxial ellipsoids, for the general case of a parallelepiped crystal whose dimensions are inversely proportional to the numbers of unit cells in the three directions X, Y, and Z, given respectively, by N_1, N_2, and N_3. Consider a single one-dimensional term from the interference function (Eq. 17).

$$\Phi = \frac{\sin^2(\pi h N_1)}{N_1 \sin^2(\pi h)} \qquad (18)$$

A plot of Φ vs h produces an essentially Gaussian peak shape whose height equals N_1 and whose breadth decreases as N_1 increases: the peak area is constant independent of N_1. At h(integral) $= \pm 1/N_1$, the numerator equals zero and the intensity is zero, defining the peak limits and the limits of the diffuse spots on Figure 8. Strictly speaking, the reciprocal lattice spots are spheres only for a crystal that is spherical, but we will ignore this shape factor here and assume that the spherical spots apply closely enough to an equidimensional parallelepiped. A comment is need on the behavior of Eqs. 15, 17, and 18 for values of h, k, or l that are integral. The quotient represented by, for example, Equation 18, is zero divided by zero when h is integral. The value of Φ at that point must be evaluated by application of limit theorem, but as a practical matter, a digital computer does essentially this if the calculation steps over the peak at small increments of h, for then h is a floating point number that never equals 0, 1, 2 etc.

The summation of the interference function maxima is conveniently accomplished by summing along the circumference of circles, all points on which are separated from the origin by $1/d$. The angle γ_1 is set to zero and Equation 17 is summed for $l = 0$ with h and k taking on the different values determined by the variation of γ_2 over the range of 0 to 90°. γ_1 is increased by an angular increment, defining a new circle with radius $R_1 = (1/d) \cos \gamma_1$, and a different value for l, and the summation is continued along the circumference of that circle. The process is continued until $\gamma_1 = 90°$, producing Φ for the point $1/d$. For the example of γ_1 selected in Figure 8, the summation along the circumference of the circle catches only the "outer" (large values of k) portion of the 011 diffraction spot, and that means that this intensity is all that contributes to Φ for the given γ_1 at the selected value of $1/d$. When the summations are completed, another d is selected and the process is repeated, building up the entire function $\Phi(d)$ for a randomly oriented powder aggregate. Each step of the summation is multiplied by the increments $\Delta\gamma_1$, $\Delta\gamma_2$, and the quantity $2\pi R_1$ which properly weights the area of reciprocal space represented by a given increment of γ_2 as a function of R_1.

The description given, and the schematic of Figure 8, apply only to the interference function, Φ. To get the actual diffraction intensity, we need to incorporate the effects of the Lorentz-polarization factor, and the scattering from unit cells in the directions h, k, and l. The method of summation shown by Figure 8 contains, implicitly, the point-by-point single crystal Lorentz factor, so this method requires multiplying by only $(1 + \cos^2 2\theta)/\sin\theta$. The values of $F(hkl)^2$ must be computed for each step in the summation and multiplied times the value of Φ for that step. Figure 9 depicts a completely hypothetical distribution of $F(hkl)^2$ values in reciprocal space. If you can visualize the three-dimensional point-by-point multiplication of the distributions of Figure 8 and Figure 9, you will perceive the distribution of diffraction intensity in reciprocal space, and it this intensity that is summed to provide the 2θ-continuous X-ray diffraction pattern.

The relations among γ_1 and γ_2, and h, k, and l are required. The derivations are not conceptually difficult, but their tedium precludes description here. The results are given in the equations below for those who may wish to develop a computer program for modeling three dimensional X-ray powder diffraction patterns. They apply to all crystal systems except for the triclinic.

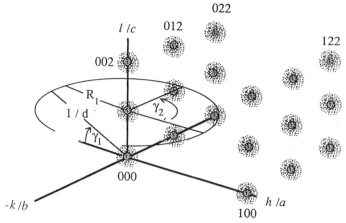

Figure 8. Reciprocal lattice representation of diffraction domains for a finite crystal.

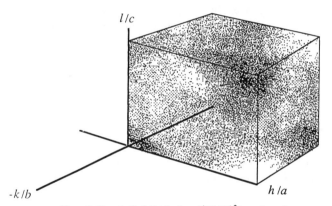

Figure 9. Hypothetical distribution of $|F(hkl)|^2$ in reciprocal space.

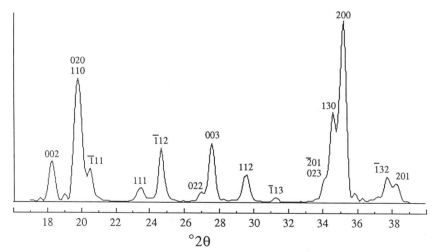

Figure 10. Calculated diffraction pattern for a randomly oriented, dehydrated $1M$ Na-montmorillonite powder (Cu$K\alpha$).

$$R_1 = (1/d) \cos \gamma_1$$
$$h = a\sin\beta \, (R_1 \cos \gamma_2 - \sin \gamma_1 / d\tan\beta^*)$$
$$k = R_1 b \sin \gamma_2$$
$$l = (c/d) \sin \gamma_1$$

The variables a, b, and c are the unit cell dimensions, β is the angle c makes with a, and $\beta^* = 180 - \beta$. For the orthogonal crystal systems, $\tan\beta^*$ is infinite. To avoid underflow errors, eliminate the last quotient in the expression for h.

$$h = aR_1 \cos \gamma_2 \ .$$

In the writer's experience, a likely source of error in these calculations lies in the assignment of atoms to their proper positions in the unit cell. If a minimum asymmetric unit of atoms is assigned, and equivalent sites generated using the appropriate symmetry operators, the calculated values for F^2 may not be valid when h, k, and l are nonintegral, even though F calculated for integral values passes all tests for consistency. Crystallographers who understand symmetry operations can doubtless solve this problem analytically, but for others, the safest course is to ignore equivalency of sites and to assign one atom or the proper fraction of one atom to all sites. In other words, assume that no atoms occupy special positions. Calculate a series of trial F values for nonintegral values of h, k, and l, and then, using the symmetry operations, eliminate as many sites as possible in order to minimize computation time.

Figure 10 shows a sample calculation. The diffraction pattern represents a dehydrated sodium-saturated $1M$ montmorillonite with $a = 5.21$ Å, $b = 9.00$ Å, $c = 9.90$ Å, and $\beta = 101.6°$. Idealized hexagonal symmetry was assumed for the atomic positions in each of the atomic planes, that is, there is no tetrahedral tilt or rotation. The crystals are roughly equant parallelepipeds for which $N_1 = 40$, $N_2 = 20$, and $N_3 = 20$ unit cells. False peaks in the diffraction pattern represent ripples in the interference function, and they are eliminated by summing the interference function over a distribution of values for N_1, N_2, and N_3.

Calculated three-dimensional diffraction patterns, such as the one shown by Figure 10, are largely useful for their instructional value. For research applications, the interesting calculations are those that apply to crystals with very different dimensions in the directions X, Y, and Z, for example thin plates, (micas and graphite), and fibers. In Chapter 6, the fundamental diffraction equation (Eq. 17) is modified so that various kinds of disorder can be introduced in one or more crystallographic directions. Such disorder can take the form of unit cell displacements that are systematically or randomly distributed in terms of both magnitude and direction. Large random displacements about nominal atomic positions lead to the characteristic diffraction patterns of glasses or, if the displacements are large enough, to the complete structural disorganization characterized by a low pressure gas. Another application lies in the study of coherent diffraction from aperiodic crystals, that is, crystals with two different kinds of unit cells that may be stacked randomly with respect to each other, or may be ordered according to some statistical scheme. All of these treatments are developed directly from Equation 17 by relatively simple mathematical operations, although available computer facilities finally limit what can be done.

REFERENCES

Brindley, G. W. (1980) Order-Disorder in clay mineral structures. Ch. 2 In: Crystal Structures of Clay Minerals and Their X-Ray Identification. Brindley, G. W. and G. Brown, eds. Mineralogical Society, London, 495 pp.

Cowley, J. M. (1961) Diffraction intensities from bent crystals. Acta Crystallogr. 14, 920-927.

Cullity, B. D. (1978) Elements of X-Ray Diffraction. Addison Wesley Publishing Co., Inc., New York, 555 pp.

James, R. W. (1965) The Optical Principles of the Diffraction of X-Rays. Cornell Univ. Press, Ithaca, New York, 664 pp.

Klug, H. P. and Alexander, L. E. (1974) X-Ray Diffraction Procedures. J. Wiley and Sons Inc., New York, 996 pp.

Wright, A. C. (1973) A compact representation for atomic scattering factors. Clays & Clay Minerals, 21, 489-490.

2. INSTRUMENTATION   Ron Jenkins

MEASUREMENT OF X-RAY POWDER PATTERNS
– CAMERA METHODS

Powder diffraction measurements are made by causing a beam of X-radiation to fall onto a suitably prepared specimen and measuring the angles at which a specific characteristic X-ray wavelength λ is diffracted. The diffraction angle θ, can be related to the interplanar spacing d, by the Bragg law:

$$n\lambda = 2d \sin\theta \, . \tag{1}$$

The simplest instrument used to measure powder patterns is a powder camera. Powder cameras were first developed in the early 1920s and while most powder diffraction work is carried out today with the powder diffractometer, there are several important film methods which are still popular because they offer valuable alternatives or complements to the diffractometer. Many factors can contribute to the decision to select a camera technique rather than the powder diffractometer for obtaining experimental data, including lack of adequate sample quantity, the desire for maximum resolution, and, possibly most important, the lower initial cost. Of the various camera types available, there are three commonly employed instruments. These are the Debye-Scherrer camera, the Gandolfi camera and the Guinier focusing camera. The Debye-Scherrer camera is the simplest and easiest to use of the powder cameras and probably in excess of 20,000 of these cameras have been supplied in the world. This number should be compared with about 15,000 powder diffractometers and about 2,000 Guinier cameras.

Figure 1 compares the essential features of the Debye-Scherrer camera and the powder diffractometer. The Debye-Scherrer camera uses a pin-hole collimator PC which allows radiation from the source S to fall onto the rod-shaped specimen SP. A secondary collimator, or beam-stop, SC is placed behind the sample to reduce secondary scatter of the beam by the body of the camera. The film FM is placed around the inside circumference of the camera body. A β filter FR is typically placed between the source S and the specimen to partially monochromatize the diffracted beam. The powder diffractometer uses a line source S of radiation which diverges onto the flat surface of the specimen SP via a divergence slit DS. A receiving slit RS is placed at the same distance from the center of the specimen as the source. This distance then becomes the radius of the goniometer circle G. A detector D (and often a monochromator) is placed behind the receiving slit to record the diffracted X-ray photons.

In the Debye-Scherrer powder camera the whole diffraction pattern is recorded simultaneously. In the powder diffractometer, the diffraction pattern is recorded sequentially by scanning the receiving slit through an angle 2θ while rotating the specimen through an angle θ. As indicated in Figure 1, the lines on the developed film from the camera correspond to the lines recorded on the powder diffractometer.

<u>The Debye-Scherrer camera</u>

The component parts of a typical Debye-Scherrer camera include a light-tight camera body and cover which hold the sample mount and film, a main beam stop, and an incident beam collimator (Fig. 1). A mount is provided which permits the camera body to rest on a track for stability during exposure. A filter is selected to remove most of the unwanted Kβ component from the incident beam while passing most of the Kα. During operation, the incident-beam collimator selects a narrow

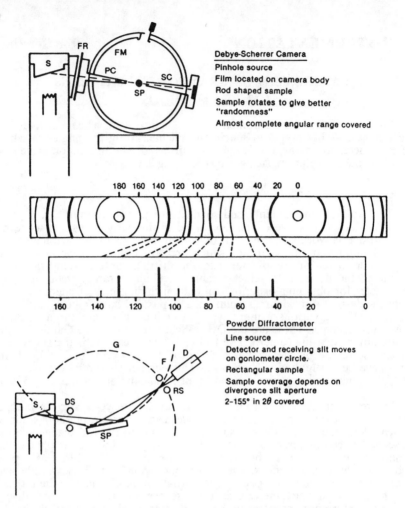

Figure 1. Measurement of powder patterns with the diffractometer and powder camera.

pencil of X-rays which passes through the sample in the camera center before striking the beam stop on the opposite end of the camera. The collimator normally selects a beam of about 1/2 mm diameter, and the camera is generally mounted on the point or spot focus of the X-ray tube to obtain the maximum beam intensity at the sample. The sample itself can be packed in a glass capillary or, if very small, mounted on the tip of a glass rod or other rigid material using rubber cement or other permanent adhesive. The sample is placed at the camera center and aligned to the main beam. The intensities of X-ray diffraction are bulk or mass-dependent phenomena, and the intensities of the diffracted beams are proportional to the crystalline sample volume irradiated by the beam. Thus exposure times have to be increased as the sample size decreases.

The line intensities from Debye-Scherrer films are normally estimated on a scale of one to ten by visual comparison with a series of lines of varying intensity registered on a standard exposed film. These standard films can be made by setting a flat film and slit in front of the direct or diffracted beam, inserting a series of attenuating metal foils for a series of increasing times, and moving the film a set amount after each exposure. Two factors which contribute to the low resolution of the Debye-Scherrer camera types are first, the incident-beam divergence due to a collimator which is typically designed to yield the narrowest beam with usable intensity, and second, the size of the capillary. Debye-Scherrer data are generally of significantly poorer quality than diffractometer data, mainly because the shapes of the lines in the film from Debye Scherrer camera are affected by both the absorption of the specimen and the physical size of the specimen. Other major sources of error in this type of camera arise from radius errors and difficulties in placing the specimen at the true center of the camera and the film at the true circumference. The problem of centering the specimen is further complicated due to the specimen eccentricity as it rotates about its own axis during exposure. Also, there are problems due to the divergence of the beam path through the camera. Finally, the recording film may suffer shrinkage during processing, leading to an apparent smaller camera radius. All of these factors limit the overall accuracy of the measured d-values to two to three times worse than similar data obtained with the diffractometer.

The Gandolfi camera

Although the Debye-Scherrer camera is able to handle very small specimens, this fact itself can give rise to problems when the specimen is so small that insufficient particles are present to give a random distribution. In some cases, one might even be asked to obtain a powder pattern from a small single crystal. The Gandolfi camera offers a good alternative to the Debye-Scherrer camera in these instances since this camera offers the additional feature of a second usable sample mount which is inclined at 45° to the normal sample rotation axis. During an exposure, the sample rotates not only about this 45° inclined axis, but the entire sample mount rotates about the normal sample axis. This provides additional randomization when a powder pattern is obtained from a very small number of particles or from a single crystal, and adequate patterns can be obtained from single crystals or powdered samples down to 30 μm in size.

The Guinier Camera

The basic drawback to both Debye-Scherrer and Gandolfi cameras is their lack of resolution. Where moderate to high resolution is required it is common to use a focusing camera, of which the Guinier camera is the most common. The sources of error in the Guinier camera (Edmonds, 1989) are not unlike those described for the Debye-Scherrer camera, with camera radius, eccentricity, X-ray beam divergence, and film shrinkage being the most important factors. Additional problems arising from specimen displacement and transparency are best compensated for by use of an internal standard. Where such techniques are employed, extremely accurate data can be obtained. The d-values obtained from diffractometers and Guinier cameras are usually of high quality, but in addition, one can also generally assume that even if an internal standard were not used, diffractometer and Guinier data are of higher quality than Debye-Scherrer data.

The Guinier geometry makes use of an incident-beam monochromator, which is simply a large single crystal cut and oriented to diffract the Kα1 component of the incident radiation. In addition to diffracting a very narrow wavelength band, the monochromator curvature is designed to use the whole surface of the crystal to diffract simultaneously, thus yielding a large diffracted intensity. As a consequence of this curvature and the orientation of the crystal itself, the monchromator not only separates Kα1 from Kα2 easily but converts the divergent incident beam into an intense convergent diffracted beam focused onto a sharp line. As monochromators are of the order of one cm wide, they are designed to use the line focus of a fine-focus X-ray tube in order to gain the most intensity. In order to gain maximum advantage from this high inherent resolution, it is usual to employ single-emulsion film. Under optimum conditions, lines as close as 0.01 mm can be resolved. For a

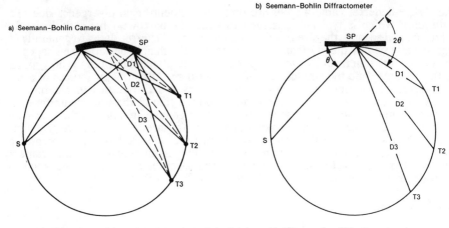

Figure 2. Geometric arrangement of the Seemann-Bohlin powder diffractometer.

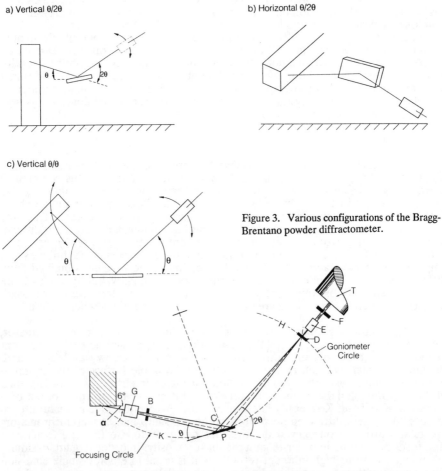

Figure 3. Various configurations of the Bragg-Brentano powder diffractometer.

Figure 4. Geometric arrangement of the Bragg-Brentano parafocusing geometry.

camera of 200 mm diameter, this corresponds to an ability to resolve lines from well-crystallized samples that are separated by as little as 0.005° 2θ.

Systematic errors in Guinier camera measurements arise from the Guinier geometry itself, and to camera machining and film shrinkage. In any case, the internal standard is necessary to get the most accurate data available from the Guinier camera. Another basic reason to use an internal standard concerns the range of diffraction data recorded on the film. The Guinier camera cannot record the entire range of possible reflections on one film, and occasionally several exposures may be necessary. Use of an internal standard greatly reduces many problems which may occur in the establishment of the complete diffraction pattern from the different data sets. For the front-reflection geometry the sample is used in transmission. Transmission geometry generally requires very thin samples so that absorption effects will not excessively attenuate the diffracted line intensities and sample thickness effects will not cause undue line broadening. The ideal sample thickness is camera dependent but is rarely larger than 0.1 mm Sample preparations normally require about 1 mg of well-ground sample, although special camera attachments permit smaller samples to be studied. Due to the sharpness of the diffracted lines, exposure times are relatively short and can be as little as 30 minutes or so for well crystallized materials. Longer exposure times can, of course, be expected for materials which are poorly crystalline.

Guinier line profiles can be digitized and computer processed to obtain accurate d-spacings. The measurement of line positions by least-squares fitting of a mathematical profile to digitized data across the profile is inherently more accurate than taking the position of maximum intensity because more data points are included. Using such techniques, lattice spacings with a precision of one part in 100,000 can be obtained (Edmonds et al., 1988). This precision is about an order of magnitude better than is being obtained in most diffraction laboratories today by any diffractometer technique.

MEASUREMENT OF X-RAY POWDER PATTERNS
- POWDER DIFFRACTOMETERS

The major problems involved with the use of powder cameras are relatively long exposure times, the need to work with dark room facilities, and difficulties in obtaining statistically precise count data. The idea of replacing the film by a photon detector dates back to the 1930s, such devices being called powder diffractometers. Early designs of the powder diffractometers by, among others, Le Galley (Le Galley, 1935) were non-focusing systems that utilized X-ray tube spot foci. These non-focusing systems required relatively long data collection times and gave rather poor resolution. The idea to use a focusing arrangement for the powder camera was proposed independently by both Seemann (Seemann, 1919) and Bohlin (Bohlin, 1920), even though the focusing arrangement proposed was not truly focusing because of the finite width of the source and specimen. Thus the word parafocusing was coined to describe the geometric arrangement, and Seemann-Bohlin parafocusing cameras have been available for many years. Parafocusing optics are also used in powder diffractometers. Of the various geometric arrangements developed, two have become popular, The Bragg-Brentano parafocusing system and the Seemann-Bohlin parafocusing system. The majority of commercially available powder diffractometers employ the Bragg-Brentano arrangement.

The Seemann-Bohlin powder diffractometer

Figure 2a shows the basic arrangement of the Seemann-Bohlin parafocusing geometry as typically used in cameras. Radiation from a point source S diverges onto the curved surface of the specimen SP. The diffracted radiation from each set of (hkl) planes converges onto the focusing circle at T1, T2, T3, etc. The distances from specimen to film are D1, D2 and D3 respectively. There are a number of different configurations used in the Seemann-Bohlin diffractometer geometry, including incident- or diffracted-beam monochromator, and stationary, reflection, or transmission specimen (Parrish, 1973); the movements employed in a typical system are as shown in Figure 2b. Here it is seen that the θ and 2θ are independently variable, and the distance from source to receiving slit is also variable. It is common practice to fix the angle of the specimen, giving a constant irradiation area. Use of a small θ value is especially useful for the measurement of thin-film specimens (Huang and Parrish, 1978).

The advantages offered by the Seemann-Bohlin geometry include higher absolute intensities without loss of resolution and the ability to work with a fixed specimen. The major disadvantage in the use of a powder diffractometer with the Seemann-Bohlin geometry is the difficulty in mounting the specimen such that it is co-concentric with the focusing circle. Additional mechanical problems may arise because of the large and variable distance from specimen to detector, especially where a diffracted-beam monochromator is employed. Use of an incident-beam monochromator makes it possible to move the specimen closer to the source slit, but this configuration is not effective in removing specimen fluorescence, and backgrounds may be high. The correction factors employed in Seemann-Bohlin diffractometers are rather more complex than in the simpler Bragg-Brentano arrangement (Parrish and Mack, 1967).

The Bragg-Brentano parafocusing powder diffractometer

Although there are many manufacturers of powder diffractometers, the majority of commercial systems employ essentially the Bragg-Brentano parafocusing geometry. A given instrument may provide a horizontal or vertical θ/2θ configuration, or a vertical θ/θ configuration (Fig. 3). The basic geometric arrangement is shown in Figure 4. A divergent beam of radiation coming from the line of focus L of the X-ray tube passes through the parallel plate collimators (Soller slits) G and a divergence slit assembly B and irradiates the flat surface of a specimen C. All of the rays diffracted by suitably oriented crystallites in the specimen at an angle 2θ converge to a line at the receiving slit D. Either behind or before the receiving slit may be placed a second set of parallel plate collimators E and a scatter slit F. A diffracted-beam monochromator may be placed behind the receiving slit at the position of the scatter slit. The axes of the line focus and of the receiving slit are at equal distances from the axis of the goniometer. The X-rays are detected by a radiation detector T, usually a scintillation counter or a sealed gas proportional counter. The receiving slit assembly and the detector are coupled together and move around the circle H about axis P in order to scan a range of 2θ (Bragg) angles. For θ/2θ scans, the goniometer rotates the specimen about the same axis as the detector but at half the rotational speed. The surface of the specimen thus remains tangential to the focusing circle K. In addition to being a device which accurately sets the angles θ and 2θ, the goniometer also acts as a support for all of the various slits and other components which make up the diffractometer. The purpose of the parallel-plate collimators is to limit the axial divergence of the beam and hence partially control the shape of the diffracted line profile. It follows that the center of the specimen surface must be on the axis of the goniometer and this axis must also be parallel to the axis of the line focus, divergence slit, and receiving slit.

Some of mechanical requirements of the parafocusing geometry are fulfilled in the design and the construction of the goniometer, whereas the others are met during the alignment procedure, for example, the take-off angle α of the beam, which is the angle between the anode surface and the center of the primary beam. Depending on the target of the X-ray tube and the voltage on the tube, the intensity increases and resolution decreases with increasing take-off angle. For normal operation the take-off angle is typically set at 3-6°. Two circles are generated by the parafocusing geometry: the focusing circle, indicated by the dashed line K in Figure 4, and the goniometer circle, indicated by the dashed line H. Note that the source L, the specimen P, and the receiving slit D all lie on the focusing circle, which has a variable radius. The specimen lies at the center of the goniometer circle, which has a fixed radius. In the alignment of the diffractometer, great care must be taken in the setting of the goniometer zero and the θ/2θ movements, since errors in these adjustments each may introduce errors in the observed 2θ values. Even with a properly aligned diffractometer all measurements will be subject to certain errors and the more important of these will be discussed below including the axial divergence error, the flat specimen error, the transparency error, and the specimen displacement error (Wilson, 1963; Parrish and Wilson, 1965). Other inherent errors may also be present but are generally insignificant for most routine work. These include refraction and effects of focal line and receiving slit width.

In order to establish the magnitude of potential errors in the diffractometer, it is good practice to prepare a calibration curve of observed against theoretical 2θ values, using a well-characterized material, after the alignment of the diffractometer. As an example, Figure 5 shows a typical calibration curve for a quartz specimen measured over the range of 20 to 110° 2θ. The zero-error line shows the theoretical 2θ values, that is, the values based on the actual "d" spacings of the material. The practical 2θ curve is the theoretical curve corrected for inherent aberrations, including axial divergence, flat-specimen error and transparency, but not including specimen displacement. The points shown are actual experimental points and indicate a good agreement with the predicted practical curve. When aligning the diffractometer it is useful to allow a range in the deviation of the fit of the data points to the practical curve defining an acceptable range of spread. The dotted lines on the figure show limits of +/-0.03°, but the value chosen in a particular laboratory will, of course, depend upon the accuracy being sought.

Systematic aberrations

Axial divergence. The axial-divergence error is due to divergence of the X-ray beam in the plane of the specimen. Referring again to Figure 4, radiation from the line source L diverges through the divergence slit B to the specimen C. This divergence slit does not, however, limit the divergence of the beam in the plane of the specimen, and so this is controlled by placing a collimator, consisting of thin molybdenum foils, between the source and the divergence slit (Soller slits). A second collimator is generally placed between the specimen and the receiving slit. In addition to producing some asymmetric broadening of the diffraction profile in the low 2θ direction, axial divergence introduces a decreasing negative error in 2θ up to 90°, then an increasingly positive error beyond 90°. The form of the axial-divergence error for a single set of Soller slits (Klug and Alexander, 1974a) is as follows:

$$\Delta 2\theta = -h^2(K_1 \cdot \cot 2\theta + K_2 \cdot \csc 2\theta / 3R^2), \qquad (2)$$

in which h is the axial extension of the specimen and R the radius of the goniometer circle. K_1 and K_2 are both constants determined by the specific collimator characteristics. The effect of the axial divergence on the profile shape can only be minimized by decreasing the axial divergence allowed by the collimator - either by decreasing the blade spacing in the collimator block or increasing the collimator length. Either process will also lead to a lowering of the absolute count rate from a given specimen.

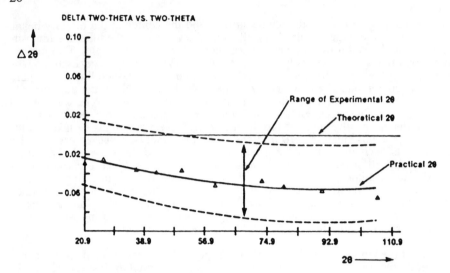

Figure 5. Calibration curves for α-SiO$_2$ showing theoretical, practical and experimental curves.

<u>Flat specimen</u>. The flat-specimen error occurs because the surface of a specimen is not co-concentric with the goniometer focusing circle but rather forms a tangent to it. The flat-specimen error causes asymmetric broadening of the diffracted line profile towards low 2θ angles. The form of the flat-specimen error is as follows:

$$\Delta 2\theta = -1/6\alpha^2 \cot\theta , \qquad (3)$$

in which α is the angular aperture of the divergence slit. Since the radius of the focusing circle decreases with increase in Bragg angle, the flat-specimen error also increases. Where fixed divergence slits are employed, the practical compromise is to use a slit which is wide enough to give reasonable intensities, while still keeping the flat specimen aberrations within acceptable limits. For most qualitative work where scans are started at around 8° 2θ, this generally means using a 1/2° divergence slit.

<u>Absorption</u>. The specimen-transparency error occurs because the incident X-ray photons penetrate to many atomic layers below the surface of the analyzed specimen. Accordingly, the average diffracting surface lies somewhat below the physical surface of the specimen. This error increases with decreasing absorption of the X-ray beam by the specimen, i.e., with decreasing linear absorption coefficient. For organic materials and other low absorbing specimens, where the linear absorption coefficients are very small, the transparency error can lead to angular errors of as much as a tenth of a degree. For this reason, low absorbing specimens are often deliberately prepared in the form of a thin film, using a low-background holder, to reduce the penetration effect.

<u>Displacement errors</u>. The specimen-displacement error arises due to practical difficulties in accurately placing the sample on the focusing circle of the goniometer. The form of the specimen-displacement error is given as follows:

$$\Delta 2\theta = -2s(\cos\theta/R) , \qquad (4)$$

where s is the displacement of the specimen from the focusing circle. Although the sample holder itself can be accurately referenced against a machined surface on the goniometer, it is difficult to pack the powdered sample into the holder such that its surface is level with the surface of the holder. The specimen-displacement error is by far the largest of the four aberrations and its major effects are to cause asymmetric broadening of the profile towards low 2θ values and to give an absolute shift in peak 2θ position, equal to about 0.01° 2θ for each 15 μm displacement.

EFFECT OF INSTRUMENT PARAMETERS

Focal spot dimensions of the tube

The maximum rating of the X-ray tube depends mainly upon the ability of the anode to dissipate heat, thus the specific loading, watts/mm^2, of the anode is an important parameter. Table 1 lists data from three different focal area copper anode tubes and shows clearly the relationship between focal area and specific loading. Fine-focus tubes are usually employed for powder diffractometry depending upon the resolution/intensity compromise required and these have typical loadings of about 350 watts/mm^2. It is not generally possible to choose just any combination of mA and kV for a given X-ray tube because the filament temperature must stay within certain limits to ensure optimum tube life, contamination rates, etc. High mA-low kV conditions, for example, lead to high filament temperatures resulting in among other things a high rate of tungsten deposition on the anode with subsequent spectral contamination. The acceptable ranges of kV and mA are usually defined by power-rating curves supplied with the X-ray tube.

Effect of take-off angle

The effective intensity of the source is dependent both on the receiving slit aperture and the focal spot characteristics of the X-ray tube. The take-off angle of the X-ray tube can be important in the selection of tube conditions owing to the finite depth within the anode at which the characteristic X-radiation is produced. Also because the take-off angle determines the effective width of the X-ray beam it influences the peak and background responses obtained with the diffractometer. Figure 6 shows that the effective width W of the filament shadow on the anode is determined by the angle of view α (i.e., the take-off angle) and the true breadth of the filament shadow B. From the diagrams, it is seen that W is equal to B sinα. The equation indicates that the smaller the value of B the narrower will be W. A small value of α will give a narrow beam and a reasonably high intensity, but this intensity will be very sensitive to the surface roughness of the anode. The optimum size of the receiving slit is typically twice as large as the projected beam width W, because of the broadening of the beam due to axial divergence. As an example, for a fine-focus copper target tube with focal spot dimensions of 12 x 0.5 mm, at a take-off angle of 6° and with Soller collimators giving ±2.5° beam divergence, the beam width at the receiving slit is about 0.2 mm.

Effect of receiving slit width

Changing the size of the receiving slit width affects both the width and intensity of a given line profile. In general, as the intensity increases, the resolution of the lines decreases, and vice versa. This is illustrated in Figure 7 (Parrish, 1965). Here a series of measurements of peak intensity and width have been taken using different receiving slit widths. When the beam width is close to the size of receiving slit, in this instance 0.15°, optimum intensity and resolution are obtained. When the receiving slit is smaller than the beam width, the intensity is reduced in rough proportion to the increase in the slit size with marginal improvement in the

Table 1. Typical X-ray tube focal spot loadings.

Tube type	anode dimensions	loading (kW)	specific loading
Fine	0.5 x 12 mm	2.0	333 W/mm²
Normal	1.0 x 12 mm	2.5	208 W/mm²
Broad	2.0 x 12 mm	3.0	125 W/mm²

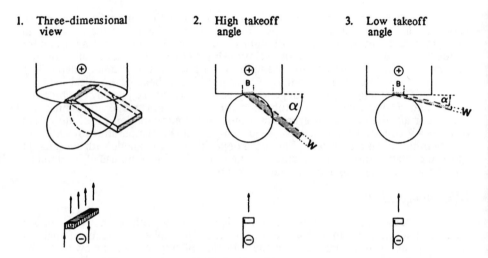

Figure 6. Effect of the take-off angle on the beam width and beam intensity.

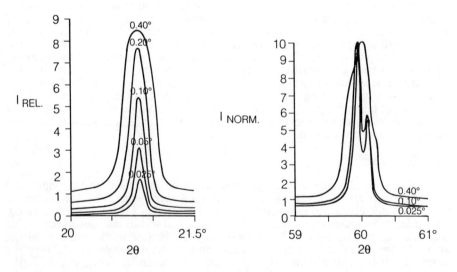

Figure 7. Effect of the receiving slit width on the profile shape.

resolution of the α1/α2 doublet. Thus going from a 0.1° receiving slit to a 0.05° receiving slit reduces the intensity by about a factor of two. When the receiving slit is larger than the beam width, only a marginal increase in intensity is seen and the resolution of the α1/α2 is very poor. This decrease in resolution arises from the inclusion of extra diffracted photons coming from angles slightly smaller or larger than the correct diffraction angles.

Effect of axial divergence

The parafocusing geometry does not give completely perfect, three-dimensional focusing. Referring again to Figure 4, the focusing circle is in the plane of the page, and in these two dimensions, the focusing is reasonably good. However, in the plane normal to the page, the focusing is very poor. This is because an extended X-ray source is used, typically a 12 mm long line focus from a sealed X-ray tube. In order to prevent too much axial divergence of the beam (which would mean that the distances LP and LD might be different, leading to broadening of the diffraction lines), parallel-plate collimators (Soller collimators) are placed in the radiation path, usually before and after the specimen. These collimators typically subtend an angle of ±2.5° and will remove much of the axial divergence. Where a diffracted-beam monochromator is used, it also reduces the axial divergence. For this reason, some instrument configurations employing a diffracted beam monochromator do not use a second Soller collimator.

The use of very narrow collimators also significantly reduces the intensity of radiation (Stoeker and Starbuck, 1965) so different manufacturers approach the problem in different ways. Most manufacturers place fixed Soller collimators between the tube and specimen, and specimen and receiving slit (Klug and Alexander, 1974b). The collimator divergence is then chosen to give an optimum intensity and line shape. These systems typically introduce some distortion of the profile, to the low-angle side at low 2θ values. Some manufacturers allow the user the choice of easily removing the second collimator (between specimen and receiving slit) if intensity is at a premium. Removing the diffracted-beam Soller slits typically doubles the X-ray intensity, increases the background, and introduces additional asymmetry into the diffracted profile. A third alternative is used by at least one manufacturer and this is to use rather long collimators to reduce the axial divergence to around ±1°. This gives a very sharp, symmetrical diffraction line but with the considerable loss of intensity. This last approach is likely to become more popular as the intensity from X-ray tubes increases.

Effect of divergence slit

The function of the divergence slit is to limit the lateral divergence of the X-ray beam so that as much as the sample surface is irradiated as possible, but at the same time avoiding irradiation of the sample support. Figure 8 shows the three conditions which generally predominate at low 2θ values (Jenkins and Squires, 1982). Here we show a specimen of length l mounted in a holder of length L. The beam from the source F passes through the divergence slit DS. The center line of the beam makes an angle θ_H with the sample surface. Under conditions "a", only the specimen is irradiated - this represents the ideal condition. Under conditions "b", both specimen and support are in the beam, which leads to scatter of the beam. As more of the sample support is irradiated, the degree of scatter and the intensity of the background increases. Under conditions "c", both specimen and support are bathed in the X-ray beam, but the length of the sample holder is not enough to intercept the whole beam. Under these conditions, as the angle decreases, less holder intercepts the beam, and the amount of scatter decreases.

The overall effect of this is shown in Figure 9. The upper part of the figure

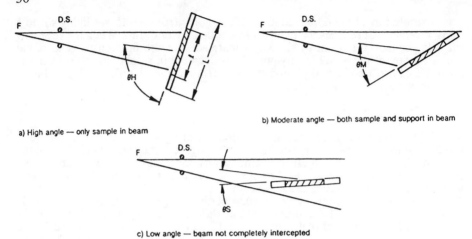

Figure 8. Effect of the angle of the X-ray beam on the specimen surface.

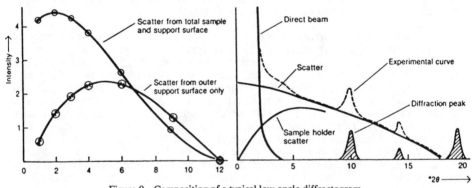

Figure 9. Composition of a typical low-angle diffractogram.

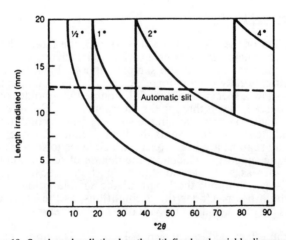

Figure 10. Specimen irradiation lengths with fixed and variable divergence slits.

shows the effect of increasing, then decreasing scatter as the system goes from condition "b" to condition "c". The effect on the actual diffractogram is shown in the lower part of the Figure. Here it will be seen that, under a badly chosen set of conditions (too wide a divergence slit or too small a sample holder length), a small bump appears in the diffractogram at around 4-5° 2θ, which can easily be confused with a broad peak or a basal plane reflection for a clay at about 20 Å.

Choice of fixed divergence slit. Because of the situation described above, the choice of the divergence slit is quite critical, especially at low 2θ values. The majority of commercially available powder diffractometers employ fixed divergence slits. Typically a number of slits are supplied with the instrument, and the operator must choose which one to use. Figure 10 shows plots of length of specimen irradiated as a function of diffraction angle for a typical X-ray powder diffractometer. The irradiation length of the specimen varies approximately with the sine of the angle θ. Figure 10 shows that a 2 cm long specimen will be completely covered by the beam at 18° 2θ when a 1° divergence slit is employed. The angle of complete coverage would be about 9° for a 1/2° divergence slit and so on. In the case of the 1° divergence slit, not only is the specimen completely irradiated below 18° 2θ, but scatter increases from the specimen holder. This is not to say that one could not use a 1° divergence slit below 18° 2θ but simply that care must be exercised at these low angles because backgrounds will be high and ghost peaks may occur as described above. Using the 1° divergence slit at angles higher than 18° 2θ will cause less and less of the specimen area to be irradiated as 2θ is increased. Since the diffracted intensity varies roughly in proportion to the irradiated specimen length, all other things being equal, intensities will decrease at higher Bragg angles.

Use of the variable divergence slit. In an attempt to avoid high-angle intensity loss and to allow working at very low angles without having to put up with high background values, some diffractometers may be equipped with a variable (theta-compensating) slit (Jenkins and Paolini, 1974). The variable divergence slit is simply a pair of molybdenum rods which are mounted on a drum driven by the θ-axis of the goniometer. As the diffraction angle changes, so too does the angle of view subtended by the slit. This gives a constant specimen irradiation length of about 13 mm (see also Fig. 10). A well-aligned variable slit allows one to work at angles very close to the direct X-ray beam - perhaps as low as 1/2° 2θ. A disadvantage of the variable slit system is that the intensities it gives do not agree with those obtained with a fixed divergence slit system and have to be corrected by multiplying by the sine of the diffraction angle.

X-RAY DETECTORS

The function of the detector is to convert the individual X-ray photons into voltage pulses that are counted and/or integrated by the counting equipment, allowing various forms of visual indication of X-ray intensity to be obtained. Detectors used in conventional X-ray powder diffractometers are generally one of four types, gas proportional counters, scintillation counters, Si(Li) detector diodes, and the intrinsic germanium detector. Of these detector systems, the scintillation counter is by far the most commonly employed. In addition to these, there are also some more specialized detectors which are used for unique applications. All X-ray detectors, however, have one thing in common, in that they all depend upon the ability of X-rays to ionize matter. There are three specific properties which are sought in an X-ray detector: quantum counting efficiency, linearity, and proportionality (Jenkins, 1988).

Quantum counting efficiency

Quantum counting efficiency is effectively the efficiency of the detector in collecting radiation incident upon it. Ideally, in the case of X-ray diffraction, one would prefer that the detector be efficient in collecting the diffracted characteristic photons and at the same time be inefficient in collecting the diffracted and scattered short-wavelength radiation, since this greatly contributes to the background, especially at low angles. Gas counters, such as the gas proportional and Geiger counters are effective in this regard since the gas filling of the detector has a much lower absorption for the high-energy, short-wavelength radiation. In the case of the scintillation and Si(Li) detectors, some short-wavelength discrimination is designed into the detector by careful selection of the thickness of the X-ray transducer - in the case of the scintillation counter, this is the phosphor, and in the case of Si(Li) it is the actual thickness of the silicon disk. It must be appreciated that most detectors which have been designed for application in X-ray diffraction have probably been optimized for the measurement of Cu Kα radiation, since this is generally the radiation of choice. Where diffraction measurements are being made with a shorter wavelength (e.g., Mo Kα, $\lambda = 0.71$ Å) or a longer wavelength (e.g., Cr Kα, $\lambda = 2.29$ Å), some loss in quantum counting efficiency will probably result.

Linearity

Figure 11 a shows a single X-ray photon of energy E entering the detector producing a pulse V. Figure 11 b shows a number of photons incident upon the detector at a rate of I photons/sec, producing voltage pulses at a rate of R pulses/sec. The detector is said to be linear when there is a direct proportionality between R and I. Where this is not true we refer to the dead time of the counter (this is somewhat of a misnomer since dead time refers to all losses in the detection system including, in particular, the pulse shaping circuitry (Jenkins et al.,1981a). It is becoming common practice with more modern instrumentation to include dead-time correction circuits in the actual counting chain.

Since the counter is dead for a certain time τ, the measured count rate R_m will always be lower than the true count rate R_t. The following relationship holds approximately true.

$$R_t = \frac{R_m}{1 - R_m \cdot \tau} . \qquad (5)$$

Proportionality

As illustrated in Figure 11 a, the detector is said to be proportional when V is proportional to E. In other words, the size of the output voltage pulse is directly proportional to the energy of the incident X-ray photon. The actual size of the output pulse will also depend upon the gain of any amplifiers which are used in the detector circuit.

The gas proportional counter

When an X-ray photon interacts with an inert gas atom, the atom may be ionized in one of its outer orbitals giving an electron and a positive ion, and this combination is called an ion pair. For example, in the case of xenon which is a typical counter gas, the energy required to remove the electron is about 20.8 eV. In the case of a gas counter detecting Cu Kα radiation, each Kα photon has a total energy of about 8.04 keV, thus each photon can produce 8040/20.8 = 387 ion pairs. The actual number of electrons contributing to the measured signal will be proportional to, but much greater than, this number of primary ion pairs. However, the resolution of the

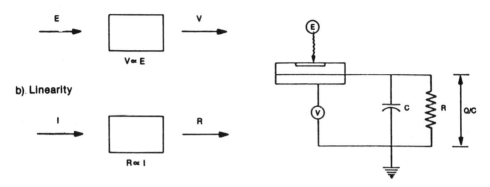

Figure 12. Schematic diagram of the proportional counter.

a) Proportionality

V ∝ E

b). Linearity

R ∝ I

E   energy of a given X-ray photon
V   voltage pulse produced from E
I   flux onto detector (photons/sec)
R   pulse rate produced by detector (counts/sec)

Figure 11. Desired properties of an X-ray detector.

detector will be inversely proportional to the number of primary ion pairs.

The basic configuration of the gas counter is shown in Figure 12. The counter itself consists of a hollow metal tube carrying a thin wire along its radial axis. The wire forms the anode and a high tension of 1.5 to 2 kV is placed across it, the casing of the counter being grounded. Electrons produced by the ionization process described above move towards the anode while the positive ions go to the grounded casing. Under the influence of the positive field close to the anode the electrons are accelerated gaining kinetic energy. Part of this energy can be given up by ionizing after collision with other inert gas atoms which lay in the path of the electron towards the anode. Thus each electron produced in the initial process may give rise to many secondary electrons leading to an effect called gas amplification or gas gain. This gas amplification may have a value between 10^3 and 10^5 depending upon the magnitude of the field (in turn fixed by the voltage of the counter and the radii of the anode wire and the body of the counter). A burst of electrons reaching the anode causes a decrease in the voltage at the condenser C, this decrease also being determined by the value of the resistance R. These pulses are passed through suitable amplifiers to the scaling circuitry.

The scintillation detector

In the scintillation counter, the conversion of the energy of X-ray photons into voltage pulses is a double-stage process. In the first of these processes, the X-ray photon is converted into flashes of blue light by means of a phosphor, or scintillator, which is a substance that has the property to absorb radiation at a certain wavelength, then re-emit radiation at a longer wavelength. For work in the X-ray region it is common to use phosphors of sodium iodide doped with thallium. This phosphor emits light photons with wavelengths of around 4100 Å. In the second stage of the process, the blue light from the phosphor is converted to voltage pulses by means of a photomultiplier. Here the light photons fall onto an antimony/cesium photocathode, each producing a burst of electrons which are then focused onto a chain of typically ten further photosurfaces called dynodes. Each of these dynodes has a successively higher potential and the electrons produced at each dynode are accelerated to the next, such that at the following dynode more electrons can be produced by means of the kinetic energy gained by acceleration. This total effect gives an increase or gain in the signal. After the last dynode, the electrons are collected by the anode and a voltage pulse is formed in a similar way to that already described for the gas counter.

Solid-state detectors

Solid-state detectors were developed towards the end of the 1960's and until recently their major application in analytical instrumentation has been in the areas of X-ray fluorescence spectrometry and electron microscopy. By far the most common of these detectors is the Si(Li) detector diode, although Ge(Li) does offer some advantages for the measurement of short-wavelength radiation. In addition to their use in X-ray fluorescence spectrometry, there has been increasing interest in using these devices for stand-alone X-ray powder diffraction, (Giessen and Gordon, 1968; Drever and Fitzgerald, 1970). More recently, there have been successful recent attempts to replace the scintillation counter by a Si(Li) detector (Bish and Chipera, 1989).

The Si(Li) detector. The Si(Li) detector is a piece of "p-type" silicon which has been drifted with lithium atoms to remove impurities and increase the electrical resistivity. The silicon single crystal is typically several millimeters thick and has an applied reverse bias of 300 to 1,000 volts. A high concentration of lithium at the rear side of the crystal creates an n-type region, over which a gold contact layer is deposited. A Schottky barrier contact is applied at the front side of the crystal, thus producing a p-i-n-type diode. An X-ray photon entering the detector produces a cloud of electron/hole pairs - the number of these pairs being proportional to the energy of the photon divided by the energy required to produce one pair, in the case of silicon, about 3.8 eV. The electrons are swept from the silicon by the potential difference across the diode, and a charge sensitive preamplifier (field effect transistor, FET) is then employed to generate a suitable signal to pass on via the main amplifier to the counting circuit. It is necessary to keep both the detector and the FET under cryogenic conditions to reduce noise and this is achieved by mounting the whole detector/ preamplifier assembly on a liquid-nitrogen-cooled cold finger.

The function of the Si(Li) detector can be compared to a gas counter without gas gain. In other words, whereas the actual number of electrons produced per event in a Si(Li) detector is significantly larger than in the gas counter (which means the energy resolution will be better), the final number of electrons is much less (which means that the signal will be weaker. As far as the direct use of a Si(Li) detector for powder diffraction measurements, work was done in the 1970s to use the energy-dispersive system to collect radiation from a crystalline specimen which has been bombarded with white radiation from an X-ray tube (Drever and Fitzgerald, 1970; Ferrel, 1971; Laine, 1974). Since the individual sets of d-spacings will select certain narrow wavelength portions of the primary spectrum to diffract at the angle of the detector, a correlation can be made between the measured energy of a given band and the d-spacing of the material. Although this technique has found some application in high speed diffractometry, the rather poor relative resolution of the energy-dispersive system (about 200 eV compared to a few tens of eV for the diffractometer) has severely limited this application.

The advantages of the Si(Li) detector include a high quantum counting efficiency and large collection angle. In addition, the ability to select energy (therefore, wavelength) ranges of interest offers some rather unique advantages in powder diffractometry. For many years, the need to run the Si(Li) detector under cryogenic conditions has somewhat limited its usefulness in diffraction applications.

More recently, significant strides in overcoming this problem have been made by using thermoelectric (Peltier) cooling. A major disadvantage of the Si(Li) detector for diffraction applications remains its rather large dead time. Experiments have shown that the dead time effects are quite significant even at relatively low count rates with the introduction of corresponding errors in the relative line intensities (see Chapter 3).

The Ge(Li) detector. Most Si(Li) detectors are designed to be sensitive to the 2-20 keV range. At the short-wavelength end of this region there is considerable loss of quantum counting efficiency because of the rather low absorption cross section of silicon. Germanium has been substituted for silicon in applications involving the detection and measurement of shorter wavelengths and the Ge(Li) detector is similar in its construction and operation to the Si(Li) detector. Early versions of the Ge(Li) detector had two major disadvantages. First, the high mobility of lithium in germanium caused the detector to be destroyed if it was accidentally warmed to room temperature. Second, the detector is characterized by rather dominant escape peaks which cause complications in the measured spectrum and strongly reduce the full energy detection efficiency (Jenkins, et al.,1981b). However, by the late 1970's the Ge(Li) detector was replaced by high purity-germanium (intrinsic germanium), which does not suffer from the shortcomings of Ge(Li), while retaining the advantage of high quantum short-wavelength counting efficiency.

Cadmium telluride. A major disadvantage of the Si(Li) detector is the need to cool with liquid nitrogen. Although some newer developments indicate that Peltier cooling may offer a long-term solution to this problem (Robie and Scalzo, 1985), until recently the extensive use of Si(Li) detector systems has been inhibited by the mechanical inconvenience involved with having to move the whole dewar assembly on the detector mount. For this reason, room-temperature solid-state detectors have been sought. Cadmium telluride is one such detector and has been shown to operate successfully at room temperatures, albeit only for the shorter wavelength radiations. The major concern at the present time in the use of cadmium telluride detectors for X-ray diffraction applications is the high leakage current (and, therefore, higher background noise). This means that it is very difficult to obtain reasonable signal/noise ratios when using Cu Kα radiation unless the detector area is very small. There are some indications that this problem may soon be solved (Roth and Burger, 1986). Even with the current limitations, the cadmium telluride detector does have useful applications when used with Mo Kα radiation or even with Cu Kα where the beam size is limited, for example, in pole-figure devices.

Mercuric iodide. One of the most promising room-temperature, solid-state detection devices to be developed is the mercuric iodide detector. The room temperature energy resolution of these detectors is comparable with cooled Si(Li). A major disadvantage has been the technical problems in growing commercial quantities of the mercuric iodide crystals, and for this reason, there has been little or no application of these devices for X-ray diffraction experiments. Again, recent work has shown that these problems are solvable (Faile et al.,1980) and some success is now being enjoyed in the associated field of X-ray fluorescence spectrometry (Kelliher and Maddox, 1987).

Position-sensitive detectors

Position-sensitive detectors (PSD's)(Gabriel et al.,1978) are finding increasing application in X-ray powder diffraction, mainly because of the advantage they offer of increased speed of data acquisition. In its simplest configuration, the position-sensitive detector is essentially a gas proportional detector in which electron collection and pulse generating electronics are attached to both ends of the anode wire. The anode wire is made to be poorly conducting to slow down the passage of electrons. By measuring the rate at which a pulse develops (the rise time) at each end of the wire, it is possible to correlate these times with the position along the anode wire where the process originated. The use of position-sensitive detectors in X-ray diffraction goes back to the mid 1970's where they were applied mainly to specialized applications such as the measurement of residual stress (James and Cohen, 1975). Within a few years they were finding increasing application in commercial diffractometers (Göbel, 1978; Wölfel, 1983).

Since the position-sensitive detector is able to record data from a range of angles at one time (Gabriel, 1977), it offers special advantages to those cases where speed of data acquisition is critical, for example, in the study of phase transformations. The effective angular resolution of the PSD is equivalent to a few hundredths of a degree 2θ, and although this is somewhat worse than can be obtained using a classical diffractometer, it is more than sufficient for most applications. Recent work (Foster and Wölfel, 1987; Rassineux et al., 1988) has shown that a combination of the PSD and curved-crystal monochromator is a particularly powerful combination for the quantitative analysis of very small (1-5 mg) quantities of material.

TECHNIQUES FOR SELECTION OF MONOCHROMATIC RADIATION

It is useful to review the make-up of a typical diffractogram as illustrated in Figure 13. Here the diffractogram is shown to consist of three major phenomena: diffraction, scatter, and fluorescence. The diffractogram itself is made up of diffraction peaks superimposed upon background. The desired peaks arise from diffraction of the assumed experimental wavelength, whereas the unwanted peaks arise from other (frequently unexpected) wavelengths. The background arises from coherent and incoherent scattering from the specimen, air scatter, and (especially at low values of 2θ) scatter from the sample support. Most of the low-angle background from air scatter can be removed by use of a vacuum or helium path around the specimen. However, such devices are not popular mainly because of the possibility of loss of water of hydration from the specimen under the reduced pressure. When the radiation falling onto the specimen is energetic enough to excite characteristic lines from the elements making up the specimen, and where this fluorescence radiation falls within the acceptance range of the detection/monochromatization system, it too will contribute to the background.

Most routine powder diffractometry is carried out using a sealed X-ray tube, powered from a constant-potential high-voltage generator delivering 1.5 - 2.5 kW. The actual choice of tube conditions will generally be determined by a number of factors including:

    a. type of generator employed,
    b. maximum rating (i.e., mA x kV) of tube,
    c. the choice of monochromatization conditions,
    d. take-off angle of X-ray tube,
    e. optimum kV.

The intensity of a characteristic wavelength $I(\lambda)$ is roughly proportional to the current i and to the over-voltage on the tube raised to the power of about 1.5. The tube over-voltage is the difference between the operating voltage V_o and the critical excitation potential V_c for the line in question. For example, in the case of copper which has Vc equal to 9, at 45 kV, the over-voltage would be 45-9 equal to 36 kV. Design constraints in the X-ray tube generally require an optimum setting of $V_o = 3$

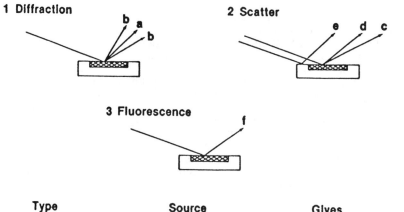

Figure 13. Composition of a diffraction pattern.

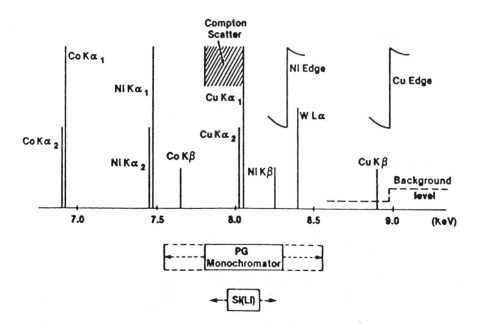

Figure 14. Band pass of the pyrolytic graphite (PG), diffracted-beam monochromator, the Si(Li) detector, and the β filter.

to 5 times V_c. The actual value of the optimum kV depends also on the take-off angle of the beam from the X-ray tube. As the potential on the tube is increased, the electron penetration also increases, as does self absorption of the tube characteristic lines. The total self absorption by the target is a function of the total absorption path length, i.e., the product of the target penetration depth and the sine of the take-off angle. At a take-off angle of 6°, the optimum operating voltage for a copper target tube is 45 kV. At 45 kV, a copper tube will emit a broad band of continuous or white radiation, along with intense characteristic α and β lines. The emitted spectrum may also contain contaminant lines from other elements from the target or tube window. Notable among these contaminant elements is tungsten, which comes from deposition of this element from the tube filament. Radiation striking the specimen or the specimen support may be diffracted, scattered, or produce secondary fluorescence radiation. X-radiation which eventually reaches the detector is further modified by partial monochromatization.

A major contributor of additional weak lines and artifacts in a diffractogram is the polychromatic nature of the source. Probably in excess of 90% of all powder work in the U.S.A. today is done with Cu Kα radiation. Although it is generally the intent to diffract just the Cu Kα1,2 doublet, in fact, the radiation actually being diffracted and detected may be more than simply the Kα doublet. The Cu Kα emission is actually made up of six α lines, but for most practical purposes it can be considered bichromatic (Jenkins, 1989). Figure 14 shows the spectral distribution in the energy region around copper K radiation. The Cu Kα and Kβ doublets are shown along with other lines which could fall within the acceptance range of the monochromatization/detection device. There are three methods that are commonly employed to render the radiation monochromatic (in actual fact, bichromatic because the Kα1/Kα2 doublet is generally employed). These techniques are:

a) Use of a β filter, generally in association with a proportional type detector and pulse-height selector.

b) Use of a diffracted-beam monochromator - typically based on a monocrystal of pyrolytic graphite.

c) Use of a solid-state proportional detector, typically Si(Li).

Figure 14 also shows the relative band-pass regions for these three monochromatization systems. A final point to be considered in the effectiveness of the selection of a monochromatic wavelength is the increasing use being now being made of the synchrotron for powder diffraction measurements (see Chapter 11 of this volume). Although the fraction of diffraction work carried out by this means is still small, the high degree of monochromaticity and wavelength tunability make it for many applications much more useful than conventional X-ray tube source systems.

<u>Use of the β filter</u>

The β filter is a single band-pass device which is mainly used to improve the ratio of Cu Kα to Cu Kβ. It is generally supplemented by some energy discrimination to remove high-energy white radiation also coming from the X-ray tube. The effectiveness of this white radiation removal depends upon the resolution of the detector. For Cu Kα, the resolution of the scintillation detector is about 2,000 eV, and such a resolution would not allow the removal of radiation in the immediate vicinity of the copper K lines. By choosing a filter material with an absorption edge between the Kα doublet and the Kβ doublet of the X-ray tube target element, the transmission ratio of α/β will be much improved. For example, in addition to continuous radiation, a copper target tube emits CuKα radiation with a wavelength of 1.542 Å and CuKβ radiation of 1.392 Å. Nickel has its K absorption edge at 1.488 Å and the mass absorption coefficient of Ni for CuKβ (286 cm^2/g) is much higher than that for CuKα (49.2 cm^2/g). The relative transmission efficiency of Kα and Kβ for a filter of given thickness can be readily calculated by use of the standard absorption equation. For example, a 15 μm thick nickel filter will improve the α/β ratio for copper from 8:1 to 50:1 (Jenkins and de Vries, 1977a).

Unfortunately, however, quite a significant amount of short-wavelength radiation from the continuum will also pass the β filter. For example, in the case of the 15 μm thick nickel filter already mentioned, whereas the percentage transmission immediately to the short-wavelength side of the absorption edge was only 2%, at 1 Å the transmission is 22% and at 0.5 Å about 80%. Thus the spectrum of the filtered radiation is typified by having a fair amount of short-wavelength radiation along with intense Kα and weak Kβ. As the voltage on the X-ray tube is increased, the output of continuum increases and spreads to shorter wavelengths, this in turn leading to an increase in the transmission of unwanted short-wavelength radiation. Hence when the β filter is used on its own, it can never completely remove the continuous radiation unless the voltage employed on the X-ray tube is very low (i.e. < about 25 kV). This is a particular problem when using Debye-Scherrer and Gandolfi cameras where the transmitted white radiation gives rise to low-angle blackening of the film. It should also be noted that where gas proportional counters are employed, the counter itself gives significant gas discrimination against shorter wavelengths. Over recent years, the advent of much higher power X-ray sources, generally means operation of the X-ray tube well in excess of 25 kV in order to give optimum intensity. For this reason, it is common practice to use a β filter in combination with a proportional counter and a pulse-height selector to give optimum separation of the continuum from the Kα.

<u>Use of pulse-height selection</u>

Pulse-height selection is essentially an electronic filtering technique that prevents unwanted radiation being counted. Although this is useful in improving the peak to background ratios of the diffracted lines, it is not very effective for separating radiation of wavelengths similar to that of the characteristic line being used for the measurement. Since the size of a voltage pulse is proportional to the energy of an X-ray photon, where different wavelengths are incident upon a proportional detector, voltage pulses of different sizes will be generated. Figure 15 shows a theoretical case where a diffracted peak is superimposed upon background which arises both from scattered continuum and iron fluorescence. Thus the peak is a composite of wavelengths from Cu Kα (1.5 Å), from Fe Kα fluorescence (1.9 Å), and from short-wavelength continuum (0.75 Å), giving rise to pulses of size 8 mV, 6 mV and 16 mV, respectively. By means of electronic discrimination it is possible to accept only pulses between certain levels, for example the limits defined by the acceptance window (usually called simply the window). In the example shown, pulses from the Fe Kα and the continuum both fall outside the window hence only pulses from Cu Kα are passed on to the scalers. In actual fact, pulses of a certain type, for example the Cu Kα, do not all have exactly the same value since a statistical process is involved. The distribution of pulses is defined by the resolution R of the detector given by

$$R = K/\sqrt{E}, \qquad (6)$$

where E is the energy of the incident X-ray photons in keV. K has a value of about 35 in the case of a xenon proportional counter and about 128 for a scintillation counter. The resolution of the detector is the peak width at half height of the pulse amplitude distribution expressed as a percentage of the average voltage level at which the pulses occur. Pulse-height selection proves invaluable for the removal of

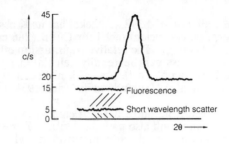

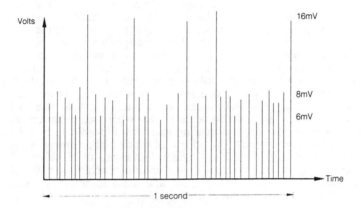

Figure 15. Principle of the pulse-height selector.

such effects as sample fluorescence and background which may arise from shorter wavelengths from the X-ray tube continuum which pass the β filter. However, since the inherent resolution of the detector/pulse-height selector is poor, 15 to 50% depending upon the detector employed and the energy of the measured radiation, it would never be possible to use this combination on its own to completely separate the Kα from the Kβ.

Use of monochromators

The diffracted-beam monochromator consists of a single crystal mounted behind the receiving slit with a detector set at the correct angle to collect the wavelength of interest diffracted by the monochromator crystal. When used with a suitable slit or aperture, the monochromator has a sufficiently narrow band pass to allow rejection of Kβ radiation from the anode. A monochromator can be placed between the source and the specimen (incident beam), or between the specimen and the detector (diffracted beam). The diffracted-beam configuration is the most popular, mainly because it is very effective in removing specimen fluorescence. Since intensity is invariably at a premium in most diffraction experiments, most diffracted-beam monochromators employ a pyrolytic graphite crystal, which has high reflectivity but poor dispersion. The angular dispersion of the pyrolytic graphite crystal is less than 100 eV, but because its mosaic structure allows diffraction over a wide angular range, it has an effective band pass of around 500 to 1,000 eV. The actual band pass value for a given monochromator will depend upon the widths and positions of limiting apertures which are placed somewhere on the focusing circle of the monochromator. It should also be appreciated that a slight missetting of the monochromator can displace its energy acceptance window in either direction. From the foregoing it is clear that even with correctly set-up monochromatizing devices undesirable radiation can pass through the monochromator to the counting circuits. Where the monochromatizing device is incorrectly set up, even more radiation may be passed.

Use of solid-state detectors

The resolution of a detector is inversely proportional to the number of initial ionizations produced within the detector. Solid-state detectors such as the Si(Li) detector have a very low effective ionization energy per primary event and thus have a very good energy resolution. Using the example already given in Figure 10 the actual energy separations for the three wavelengths in question, for the scintillation counter, the gas proportional counter, and the Si(Li) detector would be:

Source	V	Scint	Prop	Si(Li)
Continuum	16 KeV	±10.1	±2.1	±0.4
Cu Kα	8 KeV	±7.2	±1.4	±0.4
Fe Kα	6 KeV	±6.2	±1.2	±0.4

From these data it will be clear that use of the pulse-height selector with the scintillation counter will do little to reduce the effects of Fe K fluorescence, but it would be reasonably effective at reducing any background arising from short wavelength continuum. The proportional counter would be more effective in both areas. In comparison to both scintillation and gas proportional counters, however, the Si(Li) detector is most efficient in completely removing the effects of Fe K fluorescence and continuum.

Figure 14 shows the band pass of the Si(Li) detector along with other monochromatization devices, and it will be seen that the resolution of the Si(Li) detector is such that it can be used to separate Cu Kα and Cu Kβ. An advantage, therefore, to using the Si(Li) detector is that one can easily select either one of these wavelengths. This offers two major possibilities. First, one does not need a β filter or crystal monochromator to select the Kα wavelengths, and second, where advantageous, one can record the diffractogram with the Kβ rather than the Kα radiation. Like Cu Kα, Cu Kβ is also a doublet. However, whereas the energy difference between the Kα1 and Kα2 lines of copper is about 12 eV, the difference between Cu Kβ1 and Cu Kβ3 is only 2 eV. The angular dispersion of a diffractometer dθ/dλ has the form:

$$d\theta/d\lambda = n/2d\cos\theta . \qquad (7)$$

In other words, as the Bragg angle increases the diffractometer is better able to separate wavelength doublets. The absolute value of the angular dispersion varies from about 100 eV at low 2θ values to about 2eV at high 2θ values. Thus, whereas the Kα1/Kα2 doublet typically starts to be resolved at about 50° 2θ, the Kβ1/Kβ3 doublet always appears as a single reflection. Even though the Cu Kβ lines are weaker that the corresponding Cu Kα's, the simplification in the measured pattern sometimes makes this intensity reduction worth while. It should also be noted, however, that the Cu Kβ emission does contain a weak Kβ5 line which is a forbidden 3d -> 1s transition. This line is normally resolved from the Kβ1 Kβ3 doublet and typically has an intensity about 2% of the Kβ1.

One of the most successful recent applications of the Si(Li) detector has been as a replacement for the conventional scintillation detector in powder diffraction instrumentation. While the advantages of wavelength selectability and high quantum counting efficiency are somewhat counteracted by large dead time problems, it is clear that the Si(Li) detector has an important role to play in this area. Bish and

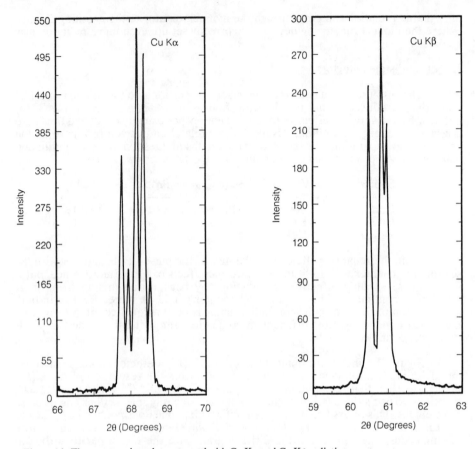

Figure 16. The quartz quintuplet measured with Cu Kα and Cu Kβ radiation.

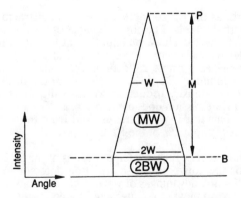

Figure 17. Definition of a peak for a figure-of-merit.

Chipera (1989) recently summarized the advantages and disadvantages of such a replacement and Figure 16, taken from their paper, demonstrates the use of the wavelength selectability of the Si(Li) detector and shows two scans over the α-quartz quintuplet. The first is done using the α1/α2 doublet and shows the classic five lines (two resolved doublets and one overlapped doublet), and the second shows the same d-spacing region with the β1/β3 doublet. Since the energy gap between the copper Kβ lines is only a few eV, the radiation is almost completely monochromatic. Use of β radiation can greatly simplify a diffraction pattern and even though intensities are lower by about a factor of six, the intensity loss can sometimes be justified. Because of the excellent energy resolution offered by the Si(Li) detector, backgrounds are typically lower than in conventional systems, leading to improved signal/noise ratios and improved detection limits.

RESOLUTION VERSUS INTENSITY

Although intensity is important in powder diffraction measurements, the ease of interpretation of the resultant diffractogram is invariably dependent on the resolution of the pattern. A pattern made up of broad lines superimposed on a high, variable background is much more difficult to process than one in which the lines are sharp and well resolved and the background is low and flat. In the setting up of instrumental parameters for a given series of experiments, there are many variables under the control of the operator, such as source conditions, receiving slit width, scan speed, step increment, etc. It is useful to have a figure-of-merit to provide guide-lines in the establishment of optimum conditions. Such figures-of-merit based on counting statistical limitations have long been used in, for example, X-ray fluorescence spectrometry (Jenkins and de Vries, 1977b). However, in X-ray fluorescence the line broadening parameters are determined purely by the fixed collimators and line shapes are easily correlated with diffraction angle. In X-ray diffractometry the situation is much more complicated and conventional figures-of-merit are not directly applicable.

Jenkins and Schreiner addressed this problem in their review of the data from an intensity round robin test (Jenkins and Schreiner, 1989) and suggested an instrument parameter figure-of-merit for X-ray powder diffraction. This figure-of-merit (FOM) has the form:

$$\text{FOM} = M \sqrt{[W/(M + 4B)]} \ . \tag{8}$$

Figure 17 illustrates the case of a simple peak in which the full width at half maximum is given by W, the width at the base as 2W, the average background is B, and the maximum peak intensity above background is given as M. The area under the peak is equal to MW and that under the background is 2BW. It will be clear from this that as W increases (due to, for example, a specimen of small particle size), whereas MW does not change appreciably, 2BW increases significantly. Application of the figure-of-merit given in Equation (8) to a wide range of data from different diffractometer configurations and types has shown that the larger the value of FOM, the statistically better are the net count data.

REFERENCES

Bish, D.L and Chipera, S.J. (1989) Comparison of a solid-state Si detector to a conventional scintillation detector-monochromator system in X-ray powder diffractometry. Powder

Diffraction 4, 137-143.
Bohlin, H., (1920) Anordnung für röntgenkristallographische untersuchungen von kristallpulver. Ann. Phys. 61, 421-439.
Drever, J.I. and Fitzgerald, R.W., (1970) Fluorescence elimination in X-ray diffractometry with solid-state detectors. Mat.Res.Bull. 5, 101-108.
Edmonds, J.E. et al, (1988) in Methods & Practices in Powder Diffraction, Section 13.1.1 - Report on cell parameter refinement. JCPDS-ICDD, Swarthmore, PA 1908.
Edmonds, J.E., (1989) Instrument selection and parameters, Section B in X-ray Powder Diffraction, ACS Audio Course #C-A5. American Chemical Society, Washington DC.
Faile, S.P., Dabrowski, A.J., Huth, G.G. and Iwanczyk, J.S. (1980) Mercuric Iodide platelets for X-ray spectroscopy produced by polymer controlled growth. J.Crystal Growth 50, 752-756.
Ferrel, R.E.Jr., (1971) Applicability of energy dispersive X-ray powder diffractometry to determinative mineralogy. Amer.Mineral. 56, 1822-1831.
Foster, B.A. and Wölfel, E.R. (1988) Automated quantitative multi-phase analysis using a focusing transmission diffractometer in conjunction with a curved position-sensitive detector. Adv. X-ray Anal. 31, 325-330.
Gabriel, A., (1977) Position-sensitive X-ray detector. Rev.Sci. Instrum. 48, 1303-1305.
─────── Dauvergne, F. and Rosenbaum,C., (1978), Linear, circular and two-dimensional position sensitive detectors. Nucl.Instrum. and Methods 152, 191-194.
Giessen, B.C. and Gordon, G.E., (1968) X-ray diffraction: New high speed technique based on X-ray spectrography. Science 159, 973-975.
Göbel, H. (1979) A new method for fast XRPD using a position-sensitive detector. Adv. X-ray Anal. 22, 255-265.
Huang, T.C. and Parrish, W., (1979), Characterization of thin films by X-ray fluorescence and diffraction analysis. Adv. X-ray Anal. 22, 43-63.
James, M.R. and Cohen, J.B. (1975) The application of a position-sensitive X-ray detector to the measurement of residual stresses. NorthWestern University, Dept. of Mat. Sci. and Mat. Res. Center, Technical Report #11, August
Jenkins, R. and Paolini, F.R. (1974) An automated divergence slit for the powder diffractometer. Norelco Reporter 21, 9-16.
─────── and de Vries, J.L. (1977a) An Introduction to X-ray Powder Diffractometry. N.V. Philips:Eindoven, p.14 et seq.
─────── and de Vries, J.L., (1977b) An Introduction to X-ray Spectrometry, 2nd. Ed. Springer-Verlag:New York, Sections 5.8 and 5.9.
─────── Gould, R.W. and Gedcke, D. (1981a) Quantitative X-ray Spectrometry, Dekker:New York, Chapter 4.
─────── Gould, R.W. and Gecke, D. (1981b) Quantitative X-ray Spectrometry, Dekker:New York, Section 4.3.1 Ge(Li) and high purity Ge detectors, p.205 et seq.
─────── and Squires, B. (1982) Problems in the measurement of large d-spacings with the parafocusing diffractometer. Norelco Reporter 29, 20-25.
─────── (1988) X-ray Fluorescence Spectrometry, Wiley/Interscience, p. 59 et seq.
─────── (1989) On the selection of the experimental wavelength in powder diffraction measurements, Adv. X-ray Anal. 32, 551-556
─────── and Schreiner, W.N. (1989) Intensity round robin report. Powder Diffraction 4, 74-100.
Kelliher, W.C. and Maddox, W.G. (1988) X-ray fluorescence analysis of alloy and stainless steels using a mercuric iodide detector. Adv. X-ray Anal. 31, 439-444.
Klug H.P. and Alexander, L.E. (1974a) X-Ray Diffraction Procedures, 2nd Ed., Wiley:New York, 302-303.
─────── and Alexander, L.E. (1974b) X-Ray Diffraction Procedures, 2nd Ed., Wiley:New York, Chapter 5, Profiles and positions of diffraction maxima.
Laine, E, Lähteenmäki,I. and Hämäläinen, M., (1974) Si(Li) semiconductor in angle and energy dispersive X-ray diffractometry J.Phys.E.Sci.Instrum. 7, 951-954.
Le Galley, D.P., (1935) A type of Geiger-Müller counter suitable for the measurement of diffracted X-rays Rev. Sci. Instrum. 6, 279-283.
Parrish W. and Wilson, A.J.C. (1959) Precision measurements of lattice parameters of polycrystalline specimens. International Tables for Crystallography II, 216-234, Kynoch Press, Birmingham.
─────── (1965) X-ray Analysis Papers, Centrex:Eindhoven, Advances in X-ray diffractometry of

clay minerals, p.105 et seq.
———— and Mack, M., (1967), Seemann-Bohlin X-ray diffractometry, I Instrumentation, II Comparison of aberations and intensity with conventional diffractometer. Acta Cryst. 23, 687-692.
———— (1974), Role of diffractometer geometry in the standardization of polycrystalline data. Adv. X-ray Anal. 17, 97-105.
Rassineux,F., Beaufort,D., Bouchet,A., Merceron,T. and Meunier,A. (1988), Use of a linear localization detector for X-ray diffraction of very small quantities of clay minerals. Clays and Clay Minerals. 36, 187-189.
Robie, S.B. and Scalzo, T.R., (1986) A comparison of detection systems for trace phase analysis. Adv. X-ray Anal. 28, 361-365.
Roth. M. and Burger, A., (1986) Improved spectrometer performance of cadmium selenide room temperature gamma-ray detector. IEEE Trans. Nucl. Sci. NS-33, 407-410.
Seemann, H., (1919) Eine fokussierende röntgenspektroskpische anordnung für kristallpulver, Ann. Phys. 59, 455-464.
Stoecker, W.C. and Starbuck, J.W. (1965) Effect of Soller slits on X-ray intensity in a modern diffractometer. Rev.Sci.Instrum. 36, 1593-1598.
Wilson, A.J.C., (1963) Mathematical Theory of X-ray Powder Diffraction, Philips Technical Library: Eindhoven.
Wölfel, E.R. (1983) A novel curved position-sensitive proportional counter for X-ray diffractometry. J. Appl. Cryst. 16, 341-348.

3. EXPERIMENTAL PROCEDURES
Ron Jenkins

INTRODUCTION

Essentially three types of information are derived from a powder diffraction pattern:

a) Angular position of diffraction lines (dependent on geometry and contents of unit cell)

b) Intensities of diffraction lines (dependent mainly on atom type and arrangement, and particle orientation)

c) Shapes of diffraction lines (dependent on instrument broadening, particle dimension and strain)

In this Chapter we are mainly concerned with experiments giving angular and intensity information. Chapters 8 and 9 deal with the various aspects of fitting and use of profile information. There are many instrumental factors which contribute to the shape of a diffraction line profile, some of which were dealt with in Chapter 2. For a further discussion of these factors the reader is referred to survey papers dealing directly with these aspects (e.g., Wilson and Parrish, 1954; Pike, 1957). Like all instrumental methods of analysis, X-ray powder diffractometry is subject to a variety of random and systematic errors. In order for these errors to be controlled, the source of the errors must be understood. One especially critical point in obtaining the three types of information listed above is that it does not necessarily follow that the best set of experimental conditions for obtaining high quality angular data is also the best set of conditions for obtaining intensity data. Optimum conditions for obtaining the best line profile data also require a different set of experimental parameters. Today, most qualitative and quantitative powder diffraction measurements are made using the powder diffractometer, with somewhat less use being made of various camera techniques including the Debye-Scherrer camera (Debye and Scherrer, 1916), the Gandolfi camera (Gandolfi, 1967) or the Guinier camera (Guinier, 1937). Each of these methods is subject to certain errors, the most important of which are listed in Table 1.

The major sources of error encountered in powder diffractometer experiments are as follows:

a) Systematic errors in the experimental positions of the diffraction lines due to inherent aberrations, including the flat-specimen error and axial divergence. These errors are angle dependent and generally manifest themselves as an increasing relative error in d-spacing as the diffraction angle is decreased.

b) Systematic errors in the experimental 2θ maxima due to displacement of the specimen surface from the goniometer focusing circle, and to specimen transparency. Additional errors may accrue due to inadequate alignment or incorrect calibration of the goniometer.

c) Line profile shapes may be asymmetrically distorted at lower 2θ angles because of residual axial divergence allowed by the Soller slits. The amount of distortion increases as the axial divergence of the collimators increase. The

Table 1. Instrumental sources of error.

I Powder Diffractometer

- Specimen displacement
- Instrument misalignment
- Missetting of the 2:1 adjustment
- Error in zero 2θ position
- Specimen transparency
- Flat specimen error
- Axial X-ray beam divergence
- Rate-meter recording
- Peak distortion due to Kα1e and Kα2 wavelengths

II Debye-Scherrer Camera

- Absorption of the X-ray beam by the specimen
- Radius errors - use of incorrect value,
- Eccentricity of the camera circle, etc.
- Eccentricity of the specimen
- X-ray beam divergence
- Film shrinkage
- Peak distortion due to Kα1 and Kα2 wavelengths

III Focusing Cameras

- Camera radius eccentricity
- X-ray beam divergence
- Film shrinkage
- Specimen displacement
- Specimen transparency
- Knife edge calibration

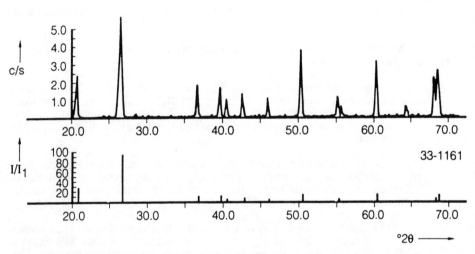

Figure 1. The original and reduced X-ray powder patterns.

collimator characteristics often vary with the design of the diffractometer.

d) The relative intensities in a given pattern are very dependent on particle orientation because the optical arrangement of the goniometer requires that the receiving slit scan at a constant rate through two dimensional space, intersecting diffraction rings on the Ewald sphere (Ewald, 1921). Because the distribution of intensity around these rings is itself orientation dependent, the portion of the ring intercepted by the receiving slit may or may not be a good measure of the average intensity distribution around the ring.

e) Intensities differ depending upon the selection of the divergence slit aperture. If the slit aperture is fixed and the irradiation length of the sample is less than the sample length, the relative intensities are, to a first approximation, independent of the divergence slit aperture. Where a variable divergence slit is used, the observed intensities, relative to those obtained with a fixed-slit, increase at higher angles by as much as a factor of three.

FACTORS INFLUENCING d-SPACING ACCURACY

At this stage of the development of the powder method for qualitative work, it is common practice to use not the full diffractogram as recorded but a reduced pattern, in which profiles are expressed as single unique "d" (or 2θ) values, and intensities are expressed as a percentage of the strongest line in the pattern (see Fig. 1). In the examination of errors in the experimental d-I list to be used, for example, for phase identification, it is necessary to consider errors introduced both in the collection of the original (raw) diffractogram, as well as in the production of the reduced pattern (Schreiner et al., 1982a). In addition to the geometric systematic aberrations and the sample-placement problems discussed in Chapter 2, other influences on the observed d-spacings may be encountered. As an example, phases may sometimes form solid solutions as in the case of sodium and potassium feldspars. When an atom substitutes for another in a given structure, the size difference between the two types of atoms may cause the lattice parameters to increase or decrease, or in some cases, some parameters to increase and others to decrease. Because the d-value is a direct function of the lattice parameters, solid-solution formation may cause significant shifts in the measured d-values. These shifts can be so large as to result in failure to identify a solid solution phase. Although this problem can, in principle, be successfully handled by computer techniques, the quality of the older data in the Powder Diffraction File has inhibited researchers from applying such techniques to any great extent. Some programs are now available on commercial machines which will help in the identification of solid solutions (Schreiner et al., 1982b). In practice the problem is not quite as bad as perhaps it first seems because workers in a specific field quickly become familiar with potential problem areas. A good example of this is the variable c-axis in montmorillonite clays. Although lines can shift by several degrees, workers are generally alert to the problem and rarely fail to identify this particular clay type.

For routine use of the powder diffractometer, probably the biggest source of experimental error is that due to specimen displacement. Also, a major area of uncertainty in a given data set is the quality of the diffractometer alignment, including missetting of the $\theta/2\theta$ adjustment and incorrect alignment of the zero-2θ position. Some peak distortion may occur due to partial separation of the Cu K $\alpha 1/\alpha 2$ doublet over some parts of the 2θ range because the angular dispersion of the diffractometer increases with increase of 2θ (see Eq. 3 in Chapter 2 of this volume). Hence, whereas at low 2θ values the α doublet is unresolved, at high 2θ values it is completely separated. In the mid-angular range the lines are only partially resolved, leading to some distortion of the diffracted line profile. When using Cu Kα radiation, this area of partial resolution occurs between about 30° and 70° 2θ. In order to quantify better the quality of a given set of d-spacings, it is useful to employ a figure-of-merit as shown in Table 2. Here, the figure-of-merit F_N is expressed as the reciprocal of the average value of Δ2θ, corrected for the number of lines observed N_{obs} over the number of lines possible N_{poss}. Figures-of-merit around 80 to 150 are considered high quality and less than 20 rather poor quality.

In spite of the numerous sources of error in the measurement of 2θ values and subsequent conversion to d-spacings, the derived d's are generally sufficient to allow qualitative phase identification. As an example, Table 3 shows data from a recent round-robin test (Jenkins and Schreiner, 1989) in which participants were asked to measure samples of the binary mixture $CaCO_3$/ZnO. Here the multi-person/ multi-laboratory precision is about 0.048°.

Influence of diffractometer misalignment

The three major steps in the mechanical alignment of a powder diffractometer are:

a) setting of the take-off angle of the X-ray tube
b) the setting of the mechanical zero
c) the setting of the 2:1 rotation axes of the goniometer

The effect of mis-setting the take-off angle is simply to change the intensity and line width (and perhaps line shape) of the diffracted profiles and has little effect on the accuracy of the experimental 2θ values. However, mis-setting of either or both the mechanical zero and the 2:1 goniometer setting can introduce significant systematic errors into the observed 2θ values. In addition, mis-setting of the 2:1 can affect both line shapes and intensities. Other mechanical errors (Klug and Alexander, 1974; Jenkins and Schreiner, 1986) such as the accuracy of slit placement, eccentricity of gearing, and temperature sensitivity of components are usually beyond the direct control of the user and will not be discussed here.

The zero error. The mechanical zero of a goniometer is the angle at which a single line bisects the center of the receiving slit, the center of the goniometer rotation axis, and the center of the projected source from the X-ray tube. The correct setting is usually established by use of a single knife edge or slit placed at the center of the goniometer rotation axis. An error of x° (±) in mechanical zero will produce an equivalent error of x° (±) in all observed 2θ values.

Errors in the 2:1 rotation of the goniometer. The function of the goniometer is to rotate the receiving slit at twice the angular speed of the specimen. Should this 2:1 relationship be in error, a cyclic error will be introduced into the observed 2θ value. The 2:1 alignment is generally made with a double knife-edge slit at the specimen position and is typically the last major step made during the alignment process.

Experimental 2θ errors

Counting statistical limitations. The production of X-rays is a random process, and any measurement of N X-ray quanta will have an associated random counting error $\sigma(N)$ equal to $\sqrt{N}$. The number of counts collected is equal to the product of the counting rate R (counts/s) and the count time t (s):

Table 2. Use of the Figure-of-Merit as a measure of data quality.

$$F_N = \frac{1}{|\Delta 2\theta|} \frac{N_{obs}}{N_{poss}}$$

F_N = Figure of merit
$\Delta 2\theta$ = The average error in 2θ
N_{obs} = The number of lines observed
N_{poss} = Number of lines possible

Table 3. Average 2θ precision obtained with ZnO/CaCO3 mixures.

ALL DATA SETS (DEG)

LINE	AVERAGE	SIGMA
Z/100	31.793	0.052
Z/002	34.448	0.041
Z/110	56.603	0.049
Z/103	62.867	0.046
Z/112	67.958	0.048
Z/213	116.269	0.043
C/012	23.072	0.050
C/104	29.428	0.051
C/113	39.433	0.051
C/202	43.169	0.053
C/116	48.519	0.046
C/314	83.759	0.040

AVERAGE 2-THETA PRECISION

ZNO	0.047
CACO3	0.049

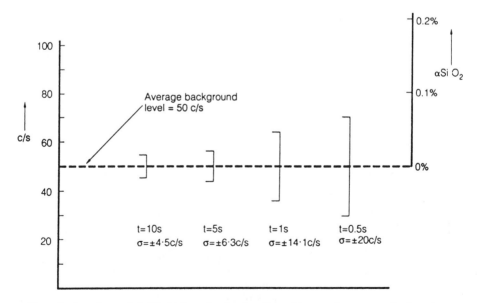

Figure 2. Counting statistical limitations, showing the effect of the counting error on the detection limit.

$$\sigma(N) = \sqrt{N} = \sqrt{(R \times t)} \ . \tag{1}$$

Thus the random counting error is time dependent. The probability of, for example, detecting a signal above background is illustrated in Figure 2. In this example, the average value of the background is 50 c/s and the 2σ (95% probability) errors are shown for t = 10, 5, 1 and 0.5 seconds respectively. With an integration time of, for example, 5 seconds, any count datum greater than 56.3 c/s (i.e., 5.3 c/s above the background) would be statistically significant. To observe a count datum of only 4.5 c/s above background would require a count time of 10 seconds.

In order to express these data in terms of phase concentration, one must convert the counts/second to percent by dividing by the sensitivity of the diffractometer for the appropriate phase in c/s/%. For example, if one were determining the amount of α-SiO$_2$ in an airborne dust sample, and a standard containing 5% of α-SiO$_2$ gave 1550 c/s at the peak-position of the (101) line, with background of 50 c/s, then the c/s/% for α-SiO$_2$ under these conditions would be (1500 - 50)/5 = 300 c/s/%. In the example discussed in Figure 2, this would correspond to a detection limit (2σ) of 0.015% for 10 seconds, 0.021% for 5 seconds, 0.047% for 1 second, and 0.067% for 0.5 seconds!

In quantitative X-ray diffraction, allowance must be made for the fact that diffraction lines may be broadened by factors other than instrumental effects. Strain and small particle sizes are the most important of these non-instrumental factors. It is common practice, therefore, to integrate the total area under the peak to gain a measure of the peak intensity rather than use peak heights. In this instance the net error σ_{net} in counts is given by:

$$\sigma_{net} = \frac{100\sqrt{(N_p - N_b)}}{N_p - N_b} , \tag{2}$$

where Np is the total number of counts accumulated over the selected angular range, and Nb is the total number of counts under the background.

Because an independent measure of both N and t may be required, visual display systems generally incorporate separate scaler and timer units which are coupled. Initiation of a count sequence starts the timer and each pulse entering the scaler trips a series decade display (a scaling switch is often incorporated allowing the counting of every 1/10th, 1/100th, 1/1000th, etc. pulse). When the fixed-count method is employed, once a selected number of counts in the scaler is reached the scaler sends a stop pulse to the timer. Similarly where the fixed-time method is used, the attainment of selected time on the timer causes a stop pulse to be sent to the scaler.

Rate-meter errors. Because the detector circuitry gives information in digital form, some system of integration must be used if a continuous scan over an angular range is required. For this purpose, either a rate meter circuit or a step-scanning system is used. The recorder is generally fitted with a time-interval marker which gives an indication of each half of a degree (2θ). Figure 3 shows a typical analog/digital counting chain where all pulses from the detector are first passed to the pulse-height selector and then allowed to flow in one of two directions. The analog chain makes use of a rate-meter and an x/t recorder. In rate-meter integration, the pulses are integrated for a time related to the RC time constant of the rate-meter. The output of the rate-meter is essentially a voltage level which is then fed to the x axis of the x/t recorder. This makes the x axis an intensity axis. The time (t) axis of the recorder is synchronized to the angular speed

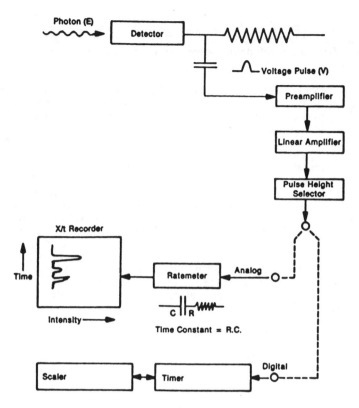

Figure 3. Schematic representation of analog and digital instrumentation commonly used in X-ray powder diffractometers.

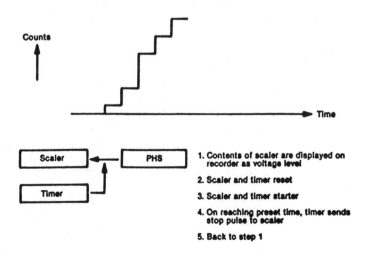

Figure 4. Step-scanning with the timer/scaler.

of the goniometer converting it to a 2θ axis. The digital channel consists simply of a timer scaler. The scaler collects counts as long as the timer is active. Thus the counting rate (R) for a given datum is the count contents (N) of the scaler divided by the time content (t) of the timer.

A rate-meter is essentially a type of smoothing device which gives a measure of the counting rate averaged over a given period of time related to the time constant. In practice, a rate-meter consists of pulse amplifying and shaping circuitry which gives pulses of fixed time and voltage dimensions. These pulses are fed into an integrating circuit consisting of resistor R and condenser C. The time to discharge the condenser to 63% of its maximum charge value is called the time constant RC. Integration of the pulses is thus achieved by a succession of charging and discharging stages, the latter utilizing a value of RC selectable by the operator. In practice, care must be taken in matching the time constant with the scanning speed. Too small a value of RC will lead to a noisy recording, whereas too large a value of RC leads to severe distortion of the line profiles. Of these, too large a value of RC is by far the most dangerous. Because the rate-meter is essentially integrating all X-ray photons impinging on the detector, it is important that the integration time be sufficient to allow a correct measure to be made of photon rate at a given angle, before the receiving slit passes the angle and no longer intercepts the true photon rate. The effect of the use of too large a rate-meter time constant is to distort the diffracted line profile and introduce varying degrees of peak shift, profile asymmetry and peak maximum suppression (Parrish et al., 1964).

A useful general rule for establishing the value of RC is:

$$RC \leq \frac{30 \times \text{Rec Slit width}(°)}{\text{scan speed }(°/\text{min})} . \quad (3)$$

As an example, with a receiving slit width of 0.2° and a scan speed of 2°/min., the time constant should be less than or equal to 3. Too small a value of the time constant will lead to large statistical fluctuations in the recording, and too large a value will cause distortion of the line profiles. Because one is severely restricted in flexibility when working with incremental time-constant values and driving the goniometer with mechanical gearing, almost all modern systems employ a goniometer equipped with a pulsed stepping motor and step-scanning.

Step-scanning errors. Because of the limitations associated with rate-meter scanning, most modern diffractometers employ a system of step-scanning in place of rate-meter scanning. Where step-scanning is required, the goniometer is successively stopped and started by the timer/scaler combination. This allows a certain number of counts to be collected at specific positions of the goniometer where the angular increment is selected by a stepping gear which replaces the usual continuous drive. This is illustrated in Figure 4. Here, a five-stage process involving scaler and timer is used to simulate rate-meter scanning. First, the contents of the scaler are passed through a digital/analog converter and displayed as a voltage level on the recorder. Next the timer and scaler are reset to read zero and at the same time the goniometer is incremented by one step. The scaler and timer are then started. On reaching the preset time, the timer sends a stop pulse to the scaler. Finally the process repeats back to the first step. The step-scan output is then as shown in the upper portion of the figure.

Although the use of step-scanning offers the diffractionist great flexibility, special care must always be taken in setting up the step-scan conditions, especially in those cases where computer treatment of the raw data is anticipated. Typical processing steps include data smoothing, background subtraction, $K\alpha 2$ stripping,

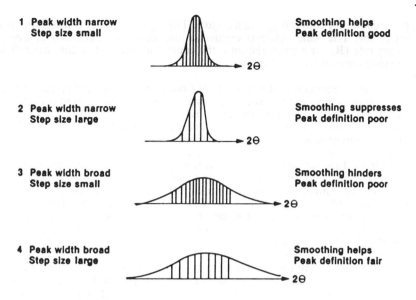

Figure 5. Choice of step size as a function of peak width.

and peak-hunting (Jenkins, 1988a). Each of the steps must be considered in the light of all others. For example, too large a step size followed by a high degree of smoothing will lead to marked suppression of the peak intensity. Conversely, too small a step size and too little smoothing may lead to peak shifts (Jenkins and Nichols, 1989). Figure 5 illustrates this problem in a rather simplified way. It sometimes happens that the lines in a pattern may be broad and, as illustrated in the Figure, the choice of step size and any subsequent data smoothing is predicated on how much of the profile should be sampled. The ideal step size is one which gives between 10 and 20 individual data points above the full width at half maximum of the peak. For well-crystallized materials the peak width is typically (but not always) about 0.1-0.3° 2θ. Thus, applying the generalization above, one would use a step size of 0.02° 2θ. Clearly, as the peak width broadens due to the influence of particle size and/or strain, one should increase the step width accordingly. The situation is clearly different where profile fitting techniques are employed because, in this case, much more of the profile data are being used.

<u>Software-induced errors.</u> More than 90% of all new powder diffractometers now sold in the USA are automated. Many excellent computer programs are provided with these instruments to aid the diffractionist in the reduction of powder data. Unfortunately, there can be pitfalls in the use of some of these programs, especially for the newcomer to the diffraction field. A popular misconception is that automated diffractometers are easier to use than manual systems. What in fact is true, is that automated powder diffractometers can give many hours of unattended and reliable data collection and can greatly assist in the tedious and routine tasks of data analysis provided one is careful to apply the programs in the correct manner.

Software-introduced intensity errors are generally rare and are usually caused by over-smoothing of raw data. One major source of confusion is often the recognition of peaks by a given automated system on a certain specimen, especially where data collected by manual and automated methods are being compared (as is often the case in qualitative phase identification!). The extreme sensitivity of, for example, the second derivative peak-hunting method may give significantly more lines than would be found manually. Additionally, the number of peaks found is very sensitive to the method of defining what a significant signal above background is. Many of these "peaks" may not be true peaks at all but simply statistical variations in the background or on the shoulders of real peaks. As an example, a recent round-robin (Jenkins and Schreiner, 1989) showed that for a well-characterized specimen of α-Al_2O_3, the number of peaks reported by different users, representing a range of instrument types over the same 2θ range, varied from 25 to 53, whereas the calculated number is 42.

Problems in specimen placement

The effect of specimen displacement can be minimized by careful preparation and mounting of the specimen. With care, this displacement can be kept within the range of +/-50 μm, corresponding to a 2θ error of the order of several hundredths of a degree and an equivalent d-spacing error in the range of a few parts per thousand. Where more accurate work is required, for example, in the accurate determination of lattice parameters, it is usual to employ an internal standard to correct for this and other errors.

Although there are many ways of actually preparing the specimen (e.g., Jenkins, 1988b; Chapter 4 of this volume), there are essentially two things which one is trying to achieve. First to obtain a random orientation of particles and second to obtain a smooth flat surface which meets tangentially with the focusing circle of the diffractometer. In the case of large particle-size materials these two aims may well be mutually exclusive.

The magnitude of the specimen displacement error is given by:

$$\Delta 2\theta = -2s\,(\cos\theta/R) \;,\qquad(4)$$

where s is the sample displacement in μm, R the radius of the goniometer circle, and θ one half of the Bragg angle. At moderate to low angles, where $\cos\theta$ approaches unity, the magnitude of the error in 2θ is roughly proportional to (-2s/R). For a goniometer circle radius of 17 cm, the error in 2θ is approximately equal to (-s x 7 x 10^4) degrees, or about -0.01° 2θ for every 15 μm sample displacement. From this it is clear that great care is required when mounting the holder. This is, of course, not so critical when an internal standard is employed.

Many diffractometers are equipped with specimen spinning devices which rotate the specimen in its own plane during measurement. The basic advantage of the specimen spinner is that, particularly in the case of fixed divergence slit systems, it gives a much larger specimen irradiation area (therefore a larger analyzed volume) and so reduces problems of particle statistics. For example, using a 1° divergence slit at 90° 2θ only 8 mm of the total specimen length is irradiated. By spinning the specimen the total volume is increased by about 50%. However, if a platy specimen is strongly oriented in, for example, the (00l) planes such that the platelets are parallel with the specimen surface, rotation will not help. A disadvantage with many commercially available specimen spinners is that they can only be loaded from the top. This tends to increase the orientation effect, thus counteracting the advantage of a greater irradiated volume.

It will be clear from the above that care must be taken in the sample preparation stage to ensure that the volume analyzed is really representative of the whole. Where the particle size is large, a problem may arise simply due to particle statistics. Only crystallites having reflecting planes parallel to the specimen surface can contribute to the diffracted intensity and where this number of crystallites is too small to justify the supposed random distribution, an error in the intensity measurement results. Typically, this error may be as large as 5-10%.

Problems in establishing the true peak position

As previously mentioned, there is a variety of factors which determine the shape of a diffracted line profile, and the more important of these include the axial divergence of the X-ray beam, the particle size and/or micro-strain of specimen, the monochromaticity of the source, and the degree of data smoothing employed in the processing of the raw data. Because of these various influences, the peak shape is typically variable and asymmetric. As an example, the data shown in Table 4 were taken on the (020) line of MoO_3 at 12.8° 2θ and show that the peak widths at 20%, 50%, and 80% of peak maximum vary quite significantly over a range of different instruments. Even so, the relative shapes of the profiles are remarkably consistent (Table 4). Figure 6 illustrates these shape characteristics and gives data for peak widths at 20% and 80% ratioed to the FWHM (widths at 50%) and plotted as a function of FWHM. The range of widths at 20% width is 1.73-2.13, and at 80% width is 0.38-0.58.

Peak distortion of this type can, in turn, lead to problems in the estimation of peak maxima. As illustrated in Figure 7, there are various ways in which a "peak" can be defined. These include: (a) use of $dI/d\theta$ = zero; (b) taking the average of inflexion points; or (c) use of profile centroids. As long as the peak is reasonably symmetrical, each of these methods will give essentially the same result. When the peak is asymmetric, due to specimen or instrument effects, the choice may vary depending on the method. Because the width of a peak may also be dependent upon the state (particle size or strain) of the specimen, it is useful to have available a measure of the peak widths during data evaluation.

<u>Problems with manual methods.</u> Although the human eye is certainly very good at integrating analog data and establishing peak-position, position measurements established by manual methods are not always of the best quality. Most of the problems in this area arise from poor technique on the part of the operator. One such problem is the use of too small a value of the 2θ scale. As an example, with a goniometer speed of 1° 2θ/min and a chart speed of 1 cm/° 2θ, in order to measure a peak value to an accuracy of 0.01°, one must be able to measure the chart to ±0.1 mm - clearly a very optimistic goal. Another common source of systematic error is incorrect alignment of the recorder interval marker pen with the intensity marker pen. In order that data of reasonable quality be obtained, the best solution is to run standard specimens through the whole data collection/ treatment process at frequent intervals (at least once a week) and compare the experimental set of d-I values with the theoretical set of values.

<u>Use of software peak-hunting programs.</u> Essentially two methods are in use at the present time - profile fitting and a combination of smoothing, $K\alpha2$ and background stripping, and a derivative method of peak location. The latter of these options is by far the most common, even though the success of derivative methods has a strong dependence on counting statistical noise. To minimize such errors a combination of raw data smoothing and differentiation of data by the least-squares technique may be employed (Savitsky and Golay, 1964). The second-derivative peak location method is commonly employed (Mallory and Snyder, 1978; Schreiner and Jenkins, 1980; Huang, 1988), and this technique yields information about both the peak width and peak position. Figure 8 lists the typical steps which may be employed in the data treatment process. Following the collection of the digitized data, the data are smoothed and background subtracted. The $\alpha2$ profile may then be stripped (Rachinger, 1948) and a peak-location method applied. This peak location is typically done using a second-derivative method in which the negative space is fitted to a parabola and the minimum calculated. A 2θ calibration may

Table 4. Measured peak shapes for MoO3.

USER	AVERAGE WIDTH (3 RUNS) AT FRACTION OF PEAK MAX			FRACTIONAL WIDTH RATIOED TO WIDTH @ 50%	
	20%	50%	80%	20%	80%
M	0.150	0.087	0.050	1.73	0.58
Q	0.213	0.110	0.057	1.94	0.52
T	0.228	0.111	0.063	2.06	0.57
R	0.207	0.113	0.060	1.82	0.53
P	0.237	0.120	0.067	1.97	0.56
H	0.253	0.133	0.077	1.90	0.58
K	0.270	0.143	0.083	1.88	0.58
L	0.275	0.145	0.055	1.90	0.38
I	0.270	0.150	0.083	1.80	0.56
O	0.273	0.153	0.077	1.78	0.50
D	0.327	0.153	0.080	2.13	0.52
E	0.340	0.173	0.087	1.96	0.50
AVG	0.254	0.133	0.070	1.91	0.53
MAX	0.340	0.173	0.087	2.13	0.58
MIN	0.150	0.087	0.050	1.73	0.38

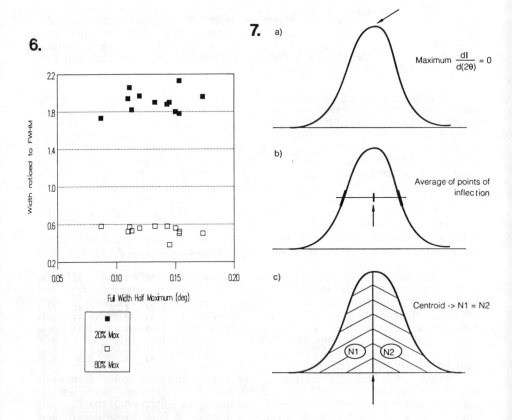

Figure 6. Resolution test on Al$_2$O$_3$ showing the consistency in the shapes of diffraction profiles measured on different types of diffractometers.

Figure 7. Methods of defining the peak-position.

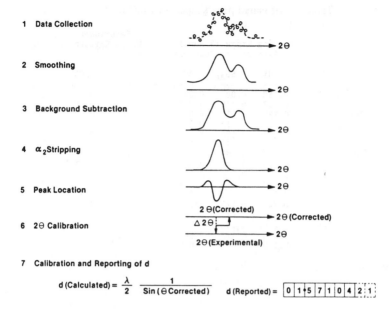

Figure 8. Steps in the treatment of diffraction data.

Table 5. Average counts in the background as a function of tube mA (background taken over 300 contiguous incre-mental steps), compared to the cts/mA obtained on a strong line in the pattern.

User	Cts/mA	Avg Cts
1	4.74	142.3
2	3.00	105.0
3	2.87	100.4
4	2.82	98.6
5	2.68	93.8
6	2.34	93.8
7	1.89	47.2
8	1.38	27.5
9	0.94	37.5
10	0.91	36.3
11	0.47	14.1
12	0.12	4.9

Table 6. Background ratioed to value at 20° 2θ.

			10°	20°	37°	60°
VDS		Q	1.18	1.00	0.63	0.81
Systems		D	1.29	1.00	0.92	1.77
		P	1.33	1.00	0.69	0.78
		R	1.36	1.00	0.87	1.14
		K	1.36	1.00	0.80	0.96
		O	1.37	1.00	0.87	1.13
		H	1.43	1.00	0.58	0.62
FDS		I	1.85	1.00	1.18	1.74
Systems		M	2.01	1.00	0.91	1.03
		T	2.11	1.00	0.80	1.08
		E	2.63	1.00	0.78	1.14
		L	3.25	1.00	0.37	0.35
All		Avg	1.8	1.0	0.8	1.0
Systems		Max	3.2	1.0	1.2	1.8
		Min	1.2	1.0	0.4	0.4

then be applied based either on an internal standard or an external calibration curve. Finally the peak data are converted to a d-spacing and stored to an appropriate significance. Although different data-handling systems may apply these steps in a slightly different order, most software schemes include all of these steps.

Treatment of the background. As long as the background level in a diffraction pattern is reasonably constant over the 2θ range under examination, there are no difficulties in simply subtracting an average background value to give net peak intensities. In general, the average value of the background will increase with the total loading on the X-ray tube. Table 5 compares the average background counts taken on 300 contiguous incremental steps with the c/s per mA obtained on a strong line in the pattern. Each "user" represents a different equipment/operator combination. There is an excellent correlation between the cts/mA and the average background counts taken at the same mA value.

Non-linear background in an X-ray diffraction pattern can, however, cause difficulties in picking out small peaks and defining a true intensity. Variation in the background is mainly due to five factors: scatter from the sample holder (generally seen at low values of 2θ where too wide a divergence slit is chosen), fluorescence from the specimen (controllable to a certain extent by the use of a diffracted-beam monochromator or by pulse-height selection), presence of significant amounts of amorphous material in the specimen, scatter from the specimen mount substrate (seen in "thin" specimens, but controllable by use of "low-background" holders), or air scatter (which has the greatest effect at low 2θ values).

Background values may also be different for fixed and variable divergence slit systems. As an example, Table 6 shows the background count rate values for a MoO_3 sample at a number of different angles. For convenience, all data are ratioed to the line at 20°, and the variable divergence slit data (VDS) are separated from the fixed divergence slit data (FDS). As would be expected, the FDS data show much higher background at low 2θ values due to scatter as the beam exceeds the specimen length.

Treatment of the $\alpha 2$. Most powder diffraction work is carried out using the Cu $K\alpha 1/\alpha 2$ doublet, and one of the greatest experimental inconveniences arising from this choice is the variable angular dispersion of the diffractometer already discussed in Chapter 2 of this volume. Figure 9 shows the angular dispersion as a function of 2θ and indicates where the $\alpha 1/\alpha 2$ lines are resolved at the half-maximum position for different values of the line width. When using manual peak-finding techniques, it is common practice to use the weighted geometric average of the $\alpha 1/\alpha 2$ wavelength as the experimental wavelength, at least until a 2θ value is reached where the $\alpha 1/\alpha 2$ doublet is sufficiently resolved to allow accurate measurement of the $\alpha 1$ line. Where automated methods are used, $\alpha 2$ stripping is generally employed over the whole range of the measured pattern.

Table 7 shows data obtained from 12 different users on a sample of Al_2O_3 in which the valley V between two peaks is ratioed to the height of the strongest line (this gives a rough experimental measure of the "resolution" of the diffractometer). Two data sets are given. In the first the line pair chosen is the $\alpha 1/\alpha 2$ doublet diffracted by the (030) planes of Al_2O_3 at 68° 2θ, and the second is the (324) and (01.14) lines at around 116° 2θ. Also shown is the aperture of the receiving slit. Both data sets show substantially the same effect, namely that as the receiving slit width increases and as 2θ decreases the resolution decreases.

Errors in conversion of 2θ to d

The measured parameters in a diffraction experiment are the absolute line

Table 7. Resolution of line pairs measured on a sample of Al$_2$O$_3$.

USER	68 deg Ref V	(030) α2	116 deg Ref V	(324) 01.14	R.S. Size [mm]
M	0.38	0.44	0.25	0.81	0.05
T	0.47	0.51	0.28	0.88	0.05
L	0.35	0.47	0.29	0.77	0.15
E	0.63	0.52	0.31	0.80	0.15
I	0.53	0.56	0.30	0.89	0.2
K	0.59	0.49	0.33	0.85	0.2
H	0.46	0.49	0.39	0.82	0.2
Q	0.48	0.49	0.40	0.84	0.2
O	0.53	0.52	0.41	0.86	0.2
R	0.52	0.52	0.42	0.85	0.2
P	0.52	0.52	0.43	0.85	0.2
D	0.61	0.53	0.45	0.81	0.3
Avg	0.51	0.50	0.35	0.84	
Sig	0.08	0.03	0.07	0.03	

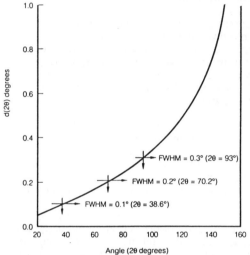

Figure 9. Dispersion of the α1/α2 doublet as a function of 2θ, showing where the doublet is resolved at different values of the FWHM of a peak.

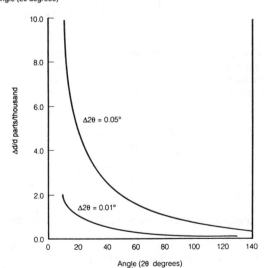

Figure 10. Conversion of 2θ errors to d-spacing errors. Plot showing d-spacing errors versus diffraction angle (2θ) for 2θ errors of 0.05° and 0.01°.

intensities and the angles at which lines occur. In qualitative phase identification the parameters used to characterize a phase are the equivalent inter-planar "d" spacings and the relative intensities. An important point to consider is the effect of an error in 2θ on the d-spacing. The relationship between the two is of the form:

$$\Delta d/d = \Delta\theta \cot\theta \ . \tag{5}$$

This function is plotted in Figure 10. It will be clear from the plot that a fixed error in 2θ has a greater impact on d-spacing accuracy at low angles than it does at high angles.

FACTORS INFLUENCING INTENSITY ACCURACY

There are four major categories of factors which contribute to the measured intensity in a powder pattern. These categories are listed in Table 8 and include: structure-sensitive properties, instrument-sensitive factors, sample-sensitive factors, and measurement- (data reduction) sensitive factors. Each of the categories will be discussed below. Although the integrated peak intensity is the best measure of the diffraction intensity from a given reflection, it is common practice in qualitative phase analysis to use the peak height rather than the integrated peak area. Provided that the broadening of the line profiles due to strain or small particle size is not too great, the error introduced by use of peak heights rather than peak areas is acceptable.

The statement is frequently made in discussions of X-ray powder data that the "d-spacing accuracies are good but intensities are poor". Although it is certainly true that in many cases preferred orientation of particles in the specimen may lead to large systematic errors in the observed intensities giving poor accuracy, it is also true that the precision of intensity data can be very good with careful specimen preparation. In a recent round-robin study (Jenkins and Schreiner, 1989), measurements, including separate data runs, were taken on two samples, Al_2O_3 and MoO_3. Between runs the sample was removed from the holder and reloaded, thus it was possible to test precision of data within each laboratory, including the contribution from specimen preparation. Data showed that typical intensity precision was about 2.5%. The major observations drawn from this round-robin study are:

a) The single-lab precision of raw intensity data in the absence of orientation is essentially limited by counting statistics and implies a significant potential for improved intensity data if other systematic errors can be modeled and removed from measured data.

b) Systematic variations of intensity with 2θ angle are observed by all users. An intensity calibration standard would be of significant help in calibrating instruments to compensate for this effect.

c) Significant intensity differences are apparent among users of variable divergence slit diffractometers. This problem can likely be corrected by proper alignment of the divergence slit.

d) Peak profile shapes were independent of receiving slit.

Two additional effects may also lead to errors in measured intensities. Firstly, the fact that a diffracted line is generally not a single line but an $\alpha 1/\alpha 2$ doublet may cause a worker to read a low-angle unresolved line at a higher intensity than if it were resolved. This causes an intensity distortion which is in turn a function of the angular dispersion of the instrument and which is of course also subject to the vagaries of misalignment. It should be remembered too that this effect also gives an angular displacement to the line thereby also affecting the d-value. Secondly, the statistical nature of photon counting and use of too long a time constant value where count rate-meters are employed can cause intensity values to vary.

Relative and absolute intensities

As shown in Table 8, item 2, there are many instrumental factors which affect the absolute and relative intensities of lines in a given diffraction pattern. Because there is a wide divergence in diffractometer types and configurations, absolute intensities may vary by more than an order of magnitude from instrument to instrument. In diffraction work involving phase identification, this problem is overcome by using relative intensities rather than absolute intensities. The strongest line is designated as 100 relative intensity units, and all other lines are scaled to this value. However, what now become important are factors which may cause changes in the relative intensities in a given pattern. The three main sources of error in this regard are: preferred orientation of particles within a specimen, effect of the divergence slit, and the influence of the dead time of the detector and associated counting circuitry.

Structure-dependent factors

As was discussed in Chapter 1 of this volume, there are a number of structural and physical factors which determine the theoretical set of intensities in a diffraction pattern. These include the atomic scattering factor, the structure factor, the effect of multiplicity, the Lorentz-polarization factor, and the temperature factor. Although we can do little to change these factors in the actual measurement of a pattern (other than perhaps reducing the temperature factor by reducing the temperature of the experiment, or changing the angular range of the experiment by choice of wavelength, thus affecting the magnitude of the polarization factor), where the full structure of the material is known, we can use these factors to calculate a set of relative intensities for a pattern (Smith et al., 1983). Such data are useful in determining the quality of an experimental pattern. As an example, if an experimentally obtained set of relative intensities agree within reasonable limits with a theoretical set of intensities calculated for the same material, the probability is that the experimental pattern is not subject to preferred orientation. In addition, it may be useful to calculate a diffraction pattern for a material where problems of unavailability or instability of the sample may make it difficult to measure the powder pattern.

Instrument- and software-dependent factors

Instrument-dependent factors include source parameters, effect of divergence slit on relative intensities, and the effects of dead time.

Effect of the divergence slit. The effect of the divergence slit on relative intensities has been discussed in detail in Chapter 2 of this volume. Briefly, for a fixed divergence slit, provided that the specimen completely intercepts the diverging beam of radiation coming from the slit, relative intensities will be virtually unaffected by the aperture of the slit. Where variable divergence slit systems are employed, there will be a gain in relative intensity over the fixed slit as the diffraction angle increases. Variable-slit intensity data can be corrected to equivalent fixed slit data by multiplying by $\sin\theta$. Table 9 demonstrates this effect and shows data on a sample of MoO_3 from 18 users, 9 with fixed-slit systems and 9

Table 8. Factors affecting X-ray powder diffraction line intensities.

1) Structure Sensitive

- Atomic scattering factor
- Structure factor
- Polarization
- Multiplicity
- Temperature

2) Instrument Sensitive

a) Absolute intensities

- Source intensity
- Diffractometer efficiency
- Take-off angle of tube
- Receiving slit width
- Axial divergence allowed

b) Relative intensities

- Divergence slit aperture
- Detector dead-time

3) Sample Sensitive

- Absorption
- Crystallite size
- Degree of crystallinity
- Particle orientation

4) Measurement Sensitive

- Method of peak area measurement
- Method of background subtraction
- α_2 stripping or not
- Degree of data smoothing employed

Table 9. Orders of MoO_3 intensities measured with fixed and variable divergence slit systems.

Fixed Divergent Slit Data Sets

USER	Orders of (020)					Orthogonal Directions		
	(020)	(040)	(060)	(080)	(0.10.0)	(020)	(200)	(002)
I	0.33	1.00	0.68	0.03	0.08	0.33	0.04	0.01
E	0.44	1.00	0.64	0.06	0.09	0.44	0.07	0.04
C	0.46	1.00	0.70	0.11	0.10	0.46	0.12	0.12
B	0.48	1.00	0.81	0.17	0.12	0.48	0.15	0.21
G	0.52	1.00	0.52	0.10	0.06	0.52	0.13	0.15
T	0.54	1.00	0.67	0.07	0.07	0.54	0.10	0.07
M	0.56	1.00	0.62	0.09	0.09	0.56	0.10	0.10
S	0.63	1.00	0.58	0.13	0.09	0.63	0.10	0.11
A	1.50	1.00	0.53	0.03	0.06	1.50	0.03	0.01
Avg	0.61	1.00	0.64	0.09	0.08	0.61	0.09	0.09

Variable Divergent Slit Data Sets

USER	(020)	(040)	(060)	(080)	(0.10.0)	(020)	(200)	(002)
J	0.36	1.00	1.08	0.43	0.23	0.36	0.38	0.56
N	0.37	1.00	0.89	0.20	0.15	0.37	0.17	0.19
O	0.37	1.00	0.82	0.13	0.14	0.37	0.12	0.11
D	0.38	1.00	0.92	0.20	0.17	0.38	0.15	0.18
H	0.38	1.00	0.75	0.28	0.17	0.38	0.22	0.31
K	0.41	1.00	0.77	0.08	0.13	0.41	0.08	0.06
R	0.45	1.00	0.71	0.07	0.11	0.45	0.06	0.05
Q	0.46	1.00	0.73	0.17	0.12	0.46	0.16	0.19
P	0.48	1.00	0.75	0.20	0.13	0.48	0.18	0.16
Avg	0.41	1.00	0.83	0.20	0.15	0.41	0.17	0.20

with variable-slit systems. Five orders of the 020 are shown and all data are ratioed to the 040 line. The relative effects of fixed and variable-slits are clearly seen.

Influence of detector dead time. The need for good detector linearity was discussed in Chapter 2. Where the detector is not linear, i.e., the measured count rate from the detector is not directly proportional to the photon rate entering the detector, we refer to the "dead time" of the system (see equation 2-5). Most modern diffractometers are able to handle count rates up to 100,000 c/s without severe dead time constraints, the exception being where the Si(Li) or Ge(Li) detectors are employed. Even though both software and hardware solutions are available, such methods are often not employed in diffraction measurements with the result that relative peak intensities are affected. Because the biggest dead time loss is found with the strongest lines, and all lines are normalized to the strongest line, the effect of dead time is to <u>increase</u> the intensity of the weaker lines in a d-I list.

A simple check can be made to evaluate whether or not dead time is playing an effect in a given set of measurements. Chapter 2 of this volume described how the intensity of a given diffraction line is directly proportional to the current applied to the X-ray tube. Thus, in a given experiment, if the <u>strongest</u> line in the pattern is first measured at the X-ray tube current intended for use with that experiment and then measured a second time at exactly one half that tube current, the measured intensity should also drop by exactly one half. If in practice it gives more than one-half the count rate at half the tube current, it is almost certain that dead time is the cause.

Influence of smoothing. Where automated powder diffractometers are employed, it is fairly common practice to smooth the raw data before peak-hunting methods are applied. Because the production of X-ray quanta is governed by a statistical process, the use of several data points, as opposed to a single data point, can significantly improve the precision of a measurement. A disadvantage with data smoothing is that some suppression of the peak intensity will occur. As an example, for a peak width of $0.20°$ 2θ and a step increment of $0.02°$, a five-point smooth will suppress peak intensity by about 15% and a nine-point smooth by about 25%.

Specimen-dependent factors

The ideal specimen for X-ray powder diffraction analysis is completely homogeneous over the distance diameter of less than 10 μm, has constant particle size of between 1 and 10 μm, and shows no preferred orientation or strain. Table 10 shows typical data from a series of users measuring a mixture of zinc oxide (random particles) and calcite (shows orientation). In each case, six lines were measured for each phase, and data were taken on fixed (FDS) and variable (VDS) divergence slit systems. The effect of the preferred orientation on the precision of the calcite data is clear; the intensity data are about four times worse than the zinc oxide data. Ideally oriented specimens may be difficult to obtain in practice, (see Chapter 4) thus it is important to realize what problems are likely to arise if the specimen deviates from this ideal state. It is useful to differentiate between the sample submitted for analysis and the specimen which is actually placed in the diffractometer or camera. During the process of specimen preparation, changes may occur in the sample due to, for example, loss or gain of carbon dioxide or water, leading to an incorrect analysis of the sample (Jenkins et al., 1986).

The actual volume of a specimen analyzed is rather small - typically in the range of 2-20 mg, depending on the beam penetration. The penetration is generally small at low angles but increases by almost an order of magnitude at high angles. Also, the irradiation area at the surface of the specimen decreases

Table 10. Multi-user/multi-laboratory intensity precision measured on ZnO/CaCo3.

LINE	FDS Data Sets			VDS Data Sets		
	AVG	SIGMA	SIG(%)	AVG	SIGMA	SIG(%)
Z/100	71.0	0.0	0	71.0	0.0	0
Z/002	51.3	2.9	6	53.7	7.1	13
Z/110	42.7	5.1	12	65.8	9.0	14
Z/103	34.0	6.4	19	58.1	4.3	7
Z/112	29.0	4.0	14	51.6	4.0	8
Z/213	10.0	1.4	14	25.3	5.5	22
C/012	6.0	3.7	61	6.9	3.3	48
C/104	77.9	35.4	45	103.8	42.1	41
C/113	13.9	7.9	57	20.7	9.1	44
C/202	11.6	6.7	57	18.6	8.1	44
C/116	14.2	6.6	46	26.5	11.1	42
C/314	2.5	1.9	76	7.0	3.1	44

Average Intensity Precision

ZnO	3.9	13		6.0	13	
CaCO3	10.4	57		12.8	44	

markedly with increase of 2θ unless the divergence slit is changed. For example, if a scan is started at 10° 2θ with a 1/2° divergence slit, complete coverage of a 20 mm long specimen is achieved at the start of the measurement. However, if the scan is allowed to proceed to 150° 2θ with the same slit, at the end of the scan only about 3 mm of the specimen is being irradiated.

Where individual crystallites are very small, < 0.2 μm, the diffraction lines broaden by an amount dependent upon the mean crystallite dimension (D), the diffraction angle (θ), and the wavelength of the radiation (λ). Scherrer (1918) proposed the relationship:

$$D_{(hkl)} = \frac{K\lambda}{\beta \cos\theta}, \qquad (6)$$

where β is the line broadening derived from the measured breadth B corrected for instrumental broadening b. B is (generally) the breadth of the diffraction line at its half-intensity maximum. K is the so-called shape factor which usually takes a value of about 0.9. It will be seen from Equation 6 that below 2,000 Å (0.2 μm) the line breadth begins to become significantly broader than the instrumental broadening. In this instance the assumption cannot be made that the peak height of a diffraction line is proportional to its integrated area.

Measurement-dependent factors

There are a number of measurement-dependent factors that must be considered in the determination of a set of intensities. Because one cannot generally assume that the peak height is an accurate measure of the integrated peak intensity, the choice of the method of measurement is clearly important. Similarly, the treatment of background the degree of data smoothing (if any), and the choice of α2 elimination can all lead to variations within a given set of intensities. Probably the biggest source of systematic error is the choice of α2 elimination. At low values of 2θ, the α1 and α2 are completely overlapped and the net intensity is the sum of α1 (100 relative) and α2 (50 relative). If an α2-elimination procedure is invoked on such a doublet, the effect is to reduce the net intensity by one third. Clearly, the reduction decreases as the angular dispersion increases and the α1 and α2 start to resolve. At high angles, where the α1 and α2 are completely resolved, stripping away the α2 has no effect on the intensity of the α1. Difficulties will arise, therefore, where α2-stripped data are being compared with non-α2 stripped data over a wide range of 2θ angles.

AUTOMATION OF QUALITATIVE PHASE ANALYSIS

Qualitative X-ray diffraction analysis involves matching the diffraction pattern from the unknown material with patterns from single-phase reference materials. The basic concepts incorporated in the original method of retrieval of potential pattern matches proposed by Hanawalt and his coworkers 50 years ago (Hanawalt et al., 1938) are still in use today. The Powder Diffraction File (PDF) is a collection of single-phase X-ray powder diffraction patterns in the form of tables of the interplanar spacings (d) and relative intensities (I/I_1) characteristic of the compound (Jenkins and Smith, 1987). Each individual data set in the file contains as a minimum a list of d-I pairs, chemical formula, name, a unique identification (PDF) number, and a reference to the primary source. In addition to this information, supplemental data may be added including Miller indices for all lines, unit cell data, physical constants, experimental details, and other comments. The PDF has been used for almost five decades (Hanawalt, 1983) and, as of Set 39, contains about 50,000 patterns. The PDF is distributed and maintained by the International Centre for Diffraction Data, located in Swarthmore, Pennsylvania, USA.

There has been a major effort over the past several years to meet the ever increasing demand for the higher quality data needed because of improved instrumentation and better techniques (Jenkins et al., 1987). One of the most significant advances in recent years has been the development of fully automated diffractometer systems, or "APD's" (Rex, 1966; Baker et al., 1967, Jenkins et al., 1971; Segmüller, 1971; Mallory and Snyder, 1978; Norelco Reporter, 1983). The introduction of the computer for data collection, treatment, and processing has improved the quality of measured d-spacings, leading to an ongoing need for improvement in the quality of reference patterns. As an example, the modern automated powder diffractometer offers the user the possibility of producing d-spacing accuracies of about 1 part per 1,000 for all but the larger d-values. This quality of data corresponds to an average angular error of 0.01 to 0.03° 2θ.

Computer data-bases

Although the full file, now referred to as the 'PDF-2' file, has been available on magnetic tape since 1964, the limited disk storage available on most commercial APD systems has prevented its use in that environment. In excess of 120 Mbytes are required for the storage of the full PDF-2 file, and most APD systems developed in the late 1970's and early 1980's had limited storage capabilities (Jenkins and Holomany, 1987). In order to provide users with a useful subset of the full data base, in 1977 the International Centre introduced the PDF-1 data base comprising the d-I lists along with the PDF number, quality mark, the chemical formula name, mineral name, and $I/I_{corundum}$ value. This data base, which requires only about 5 Mbytes, has provided and continues to provide an extremely useful product, both for individual users and for third-party suppliers of custom-made data bases for proprietary search/match computer programs. At this time, seventeen different organizations, ranging from automated powder diffractometer suppliers to software entrepreneurs, offer software for searching custom versions of the PDF-1 data base.

Role of the personal computer

Table 11. Types of standards used in X-ray powder diffraction.

a) d-spacing standards

   Primary: Si (SRM640b), fluorophlogopite (SRM675)

   Secondary: W, Ag, Quartz, diamond, etc

(b) Intensity standards

   Quantitative analysis: α-Al_2O_3 (SRM676)
   α and β silicon nitride (SRM656)

   Intensity: Al_2O_3, CeO_2, Cr_2O_3, TiO_2(R) and ZnO (SRM674)

   Respirable quartz: α-SiO_2 (SRM1878)
   Tridymite (SRM658)

   Note: all SRM materials are available from:

(c) Line profile standards

   line broadening calibration: LaB_6

   The Office of Standard Reference materials
   Room B311 Chemistry Building
   National Institute of Standards and Technology
   Gaithersburg, MD 20899

Early attempts to automate the collection and processing of powder diffraction data go back to the mid-1960's. Today, a large fraction of all powder diffractometers have at least some degree of automation. Chapter 7 lists the wide range of software for the treatment of diffraction data which is now available. The role of the personal computer in this rapid recent growth is significant. The cost of an adequate PC system is now well within the budget of most laboratories, and one can expect the use and application of these systems to increase dramatically over the next several years. The recent availability of low-cost, high-storage-capacity Compact Disk, Read-Only Memory (CD-ROM) systems (Levine, 1987) will be discussed in Chapter 8 of this volume. Use of such devices now allows easy and inexpensive access to a variety of different diffraction data bases.

USE OF STANDARDS

Table 11 lists the various types of standards which are typically employed in X-ray powder diffraction. In diffraction work essentially there are three main uses for standards (Dragoo, 1986): (a) d-spacing standards; (b) intensity standards; and (c) line-profile standards. Within each of these categories the standards may be considered primary or secondary, depending on the degree of characterization which they have been given. As an example, silicon and fluorophlogopite are considered to be primary d-spacing standards, and tungsten, silver, quartz, diamond, etc., are considered to be secondary standards (Wong-Ng and Hubbard, 1987). Other types of well characterized materials may be used for special purposes such as instrument alignment.

Use of standards to establish quality of alignment

One of the most important uses of standards is for checking the alignment of a diffractometer. The various processes involved in the alignment of a powder diffractometer were discussed in Chapter 2 of this volume. Once an alignment has been established, it is necessary to check first that the alignment is correct and at frequent intervals to ensure that the integrity of the alignment is maintained. The first of these steps might be quite complex and takes several hours to perform. The second of the steps needs to be based on a rapid, simple procedure which can be completed in a few minutes. In both cases, however, it is necessary that the tests be very reproducible without the introduction of day to day errors, for instance, in sample mounting. To aid in this area, most equipment manufacturers supply a mounted, permanent alignment reference with the diffractometer. Today, this is typically a specimen of fine-grained quartz (novaculite) cemented into a holder and surface ground to ensure that there is no specimen displacement. Other materials which have been used for this purpose include silicon and gold. Figure 5 in Chapter 2 of this volume shows a typical calibration curve obtained with a quartz alignment standard.

Use of external 2θ-alignment standards

The type of instrument alignment standard described above can also be used as an external alignment standard to establish a calibration curve for a given instrument. By this means, all data can be corrected to the external standard. One important point to note, however, is that use of the external standard will not correct for specimen-displacement error or transparency. It is also not good practice to use an external standard to compensate for poor alignment of the diffractometer.

Use of internal 2θ-alignment standards

Where very accurate d-spacing data are required, it is good practice to use an internal standard, such as silicon. Because the silicon powder is intimately mixed with the specimen under analysis, any specimen displacement is immediately compensated for. Where many of the important lines in the specimen occur at low angles ($<20°$ 2θ), it is useful to employ fluorophlogopite as the internal d-spacing standard. Main disadvantages with the use of internal standards are the additional time required for specimen preparation, the addition of more lines to the experimental pattern, and contamination of the specimen.

Use of standards to model systematic relative intensity errors

Because of the wide range of factors which can influence the relative intensities in a given diffraction pattern, and because of the need to provide reliable digitized pattern data sets which will be usable in years to come, tests are now being made (JCPDS-ICDD, 1989) to "model' intensities on different equipment configurations using well-characterized specimens. By storing such a standard data set with each experimental digitized pattern, one can reduce each pattern to a common intensity scale. This procedure is somewhat akin to the use of the Reference Intensity Ratio technique discussed in Chapter 5 of this volume.

REFERENCES

Baker, T.W., George, J.D., Bellamy, B.A. and Causer, R., (1967) Fully automated high precision X-ray diffraction. Adv. X-ray Anal. 11, 359-375.

Dragoo, A.L., (1986), Standard reference materials for powder diffraction, Part I - overview of current and future standard materials. Powder Diff. 1, 294-298.

Debye, P. and Scherrer, P., (1916) Interferenzen an regellos orientierten teilchen im rontgenlight. Physik. Z. 17, 277-283.

Ewald, P.P., (1921) Das reziproke gitter in der strukturtheorie. Z. Krist. (A) 56, 148-150.

Gandolfi, G., (1967) Discussion upon methods to obtain X-ray powder patterns from a single crystal. Mineral Petrogr. Acta 13, 67-74.

Guinier, A., (1937) Arrangement for obtaining intense diffraction diagrams of crystalline powders with monochromatic radiation. C.R. Acad. Sci. Paris 204, 1115-1116.

Hanawalt, J.D., Rinn, H.W. and Frevel, L., (1938) Chemical Analysis by X-ray Diffraction - Classification and Use of X-ray Diffraction Patterns. Ind. Eng. Chem., Anal. Ed. 10, 457-512.

——— Hanawalt, J.D., (1983) History of the Powder Diffraction File in Crystallography in North America - Apparatus and Methods, Chapter 2, pp.215-219, American Cryst. Assoc.

Huang, T.C., (1988) Precision peak determination in X-ray powder diffraction. Aust.J.Phys. 41, 201-212.

JCPDS-ICDD (1989) Intensity modelling round-robin, established by the JCPDS-ICDD, Swarthmore, PA 19081

Jenkins, R., Haas, D, and Paolini, F.R., (1971) A new concept in automated X-ray powder diffractometry. Norelco Reporter 18, 1-16.

——— Jenkins, R., et al., (1986) Sample preparation methods in X-ray powder diffractometry - report of JCPDS Task Group. Powder Diff. 1, 51-63.

——— and Schreiner, W.N., (1986) Considerations in the design of goniometers for use in X-ray powder diffractometers. Powder Diff. 1, 305-319.

——— Holomany, M. and Wong-Ng, W., (1987) On the Need to for Users the Powder Diffraction File to Update Regularly. Powder Diff. 2, 84-87.

——— and Smith, D., (1987) The Powder Diffraction File. IUCr Data Base Commission Report, August, 1987.

——— and Holomany, M., (1987) PC-PDF - A search/display system utilizing the CD-ROM and the complete Powder Diffraction File. Powder Diff. 2, 215-219.

——— (1988a) Profile data aquisition for the JCPDS-ICDD database. Austr.J.Phys. 41, 145-153.

——— Ed., (1988b) Methods and Practices in X-ray Powder Diffraction, International Centre for Diffraction Data: Swarthore, PA., Section 4 - Specimen preparation methods.

——— and Nichols, M., (1989) Problems in the derivation of d-values from experimental digital XRD patterns. Adv. X-ray Anal. 33, in press

——— and Schreiner, W.N., (1989) Intensity round-robin report. Powder Diffraction 4, 74-100.

Klug,H.P. and Alexander, L.E., (1974) X-ray Diffraction Procedures, Wiley:New York, Chapter 5 - diffractometric powder technique.

Levine, R., (1987) Optical Storage. DEC Professional February 1987, p.30.

Mallory, C.L and Snyder, R.L., (1978)The control and processing of data from an automated powder diffractometer. Adv. X-ray Anal. 22, 121-131.

Norelco Reporter (1983) - Special issue on Automated Powder Diffractometry. Volume 30, No. 1X, February.

Parrish, W., Taylor, J. and Mack, M, (1964) Dependence of lattice parameters on various angular measures of diffractometer line profiles. Adv. X-ray Anal. 7, 66-85.

Pike, E.R., (1957) Counter Diffractometer-The effect of Vertical Divergence on the Displacement and Breadth of Powder Diffraction Lines. J.Sci.Instrum. 34, 355-363. and ibid (1959) 36, 52-53.

Rachinger, W.A., (1948) A correction for the $\alpha 1/\alpha 2$ doublet in the measurement of widths of X-ray diffraction lines. J.Sci.Instrum. 25, 254-259.

Rex, R.W., (1966) Numerical control X-ray powder diffractometry. Adv. X-ray Anal. 10, 366-373.

Savitsky, A. amd Golay, M.J.E., (1964) Smoothing and differentiation of data by simplified least squares procedures. Anal. Chem. 36, 1627-1639.

Scherrer, P., (1918) Bestimmung der grosse und der inneren struktur von kolloidteilchen mittels rontgenstrahlung. Nachr. Ges. Wiss. Gottingen. September, 98-100.

Schreiner, W.N. and Jenkins, R., (1980) A second derivative algorithm for identification of peaks in powder diffraction patterns. Adv. X-ray Anal. 23, 287-293.

——— Surdukowski, C., Jenkins, R., and Villamizar, C., (1982a) "Systematic and Random Powder Diffractometer Errors Relevant to Phase Identification". Norelco Reporter 29, 42-52.

——— Surdukowski, C. and Jenkins, R., (1982b) An approach to the isostructural/isotypical, and solid solution problems in multiphase X-ray analysis. J. Appl. Cryst. 15, 605-610.

Segmüller, A., (1971) Automated X-ray diffraction laboratory system. Adv. X-ray Anal. 15, 114-122.

Smith, D.K., Nichols, M.C. and Zolensky, M.E., (1983) A Fortran IV program for calculating X-ray powder patterns - Version 10. The Pennsylvania State University, Department of Geosciences, March.

Wilson, A.J.C. and Parrish,W., (1954) A Theoretical Investigation of Geometrical Factors Affecting

Line Profiles in Counter Diffractometry. Acta Cryst. 7, 622-624.

Wong-Ng, W. and Hubbard, C.R., (1987) Standard reference materials for X-ray diffraction, Part II - calibration using d-spacing standards. Powder Diff. 2, 242-248.

4. Sample Preparation for X-ray Diffraction

D.L. Bish & R.C. Reynolds, Jr.

INTRODUCTION

The preparation of samples is one of the critical steps in many analytical methods, and particularly for X-ray powder diffraction (XRD) analysis. Depending on the nature of the powder XRD study and the desired results, sample preparation can be of minimal importance or it can be crucial. If a qualitative analysis of a simple, familiar material is desired, the material can be quickly ground in a hand mortar and pestle and packed into a cavity-type sample holder, paying no attention to orientation or particle-size effects. On the other hand analyses of complex materials involving search/match procedures, quantitative phase analysis, lattice-parameter refinement, or Rietveld refinement of crystal structures require much more careful attention to the method and execution of sample preparation and mounting. The two most important results desired in these more complex cases are accurate and precise line positions and intensities. An analysis as simple as a manual or computer search/match on an unknown will be much more successful if accurate data are obtained. Minimizing the systematic sample-related effects can be as important as accounting for instrumental systematic errors (Chapter 3) in the collection of accurate and precise data.

COMMON PROBLEMS IN SAMPLE PREPARATION

Due to the nature of geologic materials and differences among modern X-ray diffractometers, a variety of sample preparation problems is commonly encountered. Some of these concern the preparation of the powdered material and others arise during mounting of the powders in a suitable holder.

<u>Grinding problems</u>

Because diffractionists encounter materials having hardnesses ranging from 1 to perhaps 9 on the Mohs scale, it is not surprising that one of the most frequent problems in sample preparation is related to the quality of the powdered material. Probably the most common fault is the use of samples that are insufficiently ground, and in fact, many publications incorrectly recommend the use of powder that has passed through a 325 mesh sieve (44 µm). As discussed below, materials of this size are much too coarse for most quantitative powder diffraction experiments for a variety of reasons, including extinction, particle statistics, and microabsorption effects. Using powder that is too coarse can give rise to inaccurate and imprecise intensities, and one will often obtain spurious or unusually intense and sharp reflections from material that is poorly ground.

On the other hand, overgrinding materials, particularly those that are very soft, can present other problems. In our experience this is seldom a problem with geological materials and appropriate grinding procedures. If overgrinding is encountered, one may change the method of grinding or periodically sieve the ground sample during grinding to remove material of sufficiently small size (being sure to combine the complete sample after grinding of multicomponent mixtures). The eventual effect of grinding a material excessively is to broaden some or all reflections (anisotropic grinding effects) and to produce small amounts of amorphous surface layers (e.g., Nakamura, 1989). Materials will usually begin to show some particle-size related broadening when individual crystallites are smaller than about 1000 Å. Strain-related broadening, however, can be evident in some materials having crystallites much larger than this (see Chapter 9). Although the breadth or full-width-at-half-maximum (FWHM) of observed reflections increases with decreasing crystallite size, the integrated intensities remain the same. It is important to realize, however, that the position of a broadened reflection may be shifted from the position of the unbroadened reflection, particularly for low-angle reflections, due to variations of the Lorentz-polarization factor and possibly the structure factor across the broad reflection (see Chapters 1 and 7; Reynolds, 1968; Ross, 1968). In addition, broadening due to deformation faulting shifts reflections in a predictable manner (e.g., Warren, 1969). Although not common with geological samples, numerous authors have shown that overgrinding can also induce phase transformations and solid-state reactions. Jenkins et al. (1986) summarized some of the reactions that have been observed, including the transformation of calcite to aragonite, of wurtzite to sphalerite, of kaolinite to mullite, and of vaterite to calcite. We have found that such reactions seldom if ever occur if correct grinding procedures are used, including grinding samples under a liquid such as acetone or alcohol.

A final aspect of sample grinding that may be important in some multicomponent samples is differentiation of particle sizes due to differential grinding. For example, a sedimentary rock containing quartz, feldspars, a mica, and a smectite will probably yield a distribution of particle sizes after grinding, the softer materials producing smaller particles than the hard materials. This size distribution results both from the greater resistance of the hard minerals to grinding by the grinding device and grinding of softer materials by the harder materials in the sample. In addition, platy materials such as micas are poorly ground by some methods. This effect can be minimized by sieving during grinding and recombining the complete ground sample, or it can be ignored in many cases as long as the harder materials are sufficiently ground. In most cases, wet grinding of such mixtures produces appropriate particle sizes of the hard minerals while not overgrinding the softer, generally finer-grained materials.

Sample mounting

Sample area. Most samples to be examined by powder diffraction methods require mounting of the (powdered) sample in an appropriate holder, and this step can give rise to systematic errors affecting both reflection positions and intensities. Although sample-mounting errors are not always obvious, they generally produce the most common errors in peak position and intensity encountered in X-ray powder diffraction analysis. Perhaps the most common mounting error is the use of holders with sample areas of insufficient size. Because the irradiated area increases with decreasing diffraction angle, 2θ, it is important to ensure that the X-ray beam is fully within the sample at the lowest angle of interest (see Parrish et al., 1966). This is particularly important in examinations of samples containing layer silicates, such as chlorites, micas, and smectites, that have low-angle reflections. It is important to know accurately the size and shape of the irradiated area at a variety of angles and with a variety of divergence slits for the diffractometer used. This information can be determined using a fluorescent screen in place of a sample or it can be calculated by simple trigonometry using the proper values for the goniometer radius and the angular divergence of the primary or beam slit. In addition, if the sample area is too short to contain the beam at the lowest angle of interest, intensities from reflections occurring at angles where the beam is not fully within the sample will be relatively weak compared with higher-angle reflections where the X-ray beam is within the sample. This effect is most important when accurate intensities are required. With this information, the appropriate size requirements for samples can be determined. It is also important to ensure that the width of the sample is completely within the beam. Using sample areas that are too small can produce diffraction from the sample holder and give rise to spurious diffraction peaks or increased background.

Sample thickness. The X-ray powder diffraction method generally assumes that one has an <u>infinitely</u> thick powder sample. This means that the sample must be thick enough so that essentially all of the incident X-ray beam interacts with the sample and does not pass through it. The important parameters in calculating this thickness are the path length and the linear absorption coefficient, μ. Because of the variable angle of incidence of the X-ray beam on the sample, the minimum required thickness varies with diffraction angle. In addition, the minimum thickness is dependent on packing density of the sample and varies with type of sample and X-ray radiation. The relationship between the fraction of X-ray intensity transmitted, sample thickness, and linear absorption coefficient is

$$I_x = I_0 e^{-\mu x}, \qquad (1)$$

where I_x is the intensity of the transmitted X-ray beam after passing through a thickness x, I_0 is the intensity of the incident beam, and μ is the linear absorption coefficient of the sample (Cullity, 1978). Recasting Equation 1,

$$I_x/I_0 = 1 - e^{-2\mu x \csc\theta}, \qquad (2)$$

where θ is half the diffraction angle, allows calculation of the percentage of diffracted intensity at any angle 2θ that originates from a sample of a given thickness x (Brown and Brindley, 1980, p. 309). The value "2" in Equation 2 appears because the total path length is equal to the sum of the distances in and out of the powder. Following Cullity (1978, p. 135), if we decide that we want the intensity on the back side of a sample to be 1/1000 of the diffracted intensity on the front side, then

$$e^{2\mu t/\sin\theta} = 1000 \qquad (3a)$$

and

$$t = 3.45/\mu \sin\theta. \qquad (3b)$$

From these equations, one can appreciate that a significant thickness (> 50 μm) of a typical silicate sample such as quartz ($\mu = 91$ cm^{-1}) is needed to satisfy the requirement of infinite thickness at diffraction angles of 40° to 60° 2θ with Cu $K\alpha$ radiation. It is important to consider the fact that diffraction occurs from a measurable thickness of sample for most geologic materials, that is these samples have significant transparency. Whenever a cavity mount is used in powder diffraction, errors can be introduced by having a sample with a finite thickness, even though the sample may be effectively infinitely thick. For Cu $K\alpha$ radiation and typical silicate minerals in cavity mounts, errors result from penetration of the X-ray beam into and diffraction from the bulk of the sample. These transparency effects can give rise to potentially significant shifts in the positions of diffraction maxima and changes in the widths of the observed profiles (e.g., Klug and Alexander, 1974, p. 302; Chapter 3). This effect is greatest for thick samples with small linear absorption coefficients. For this reason, if one is interested in accurate peak positions, a thin sample on a non-diffracting substrate (see below) is recommended over a cavity mount, although this method may give rise to finite sample-displacement errors depending on its execution. Usually if accurate intensities are required, such thin samples are not used.

Sample packing. Attention to sample packing is important because diffraction occurs from a volume of sample, not just the surface. Loose sample packing can give rise to transparency effects in addition to those related intrinsically to the sample. Systematic errors related to sample packing and thickness are relatively unimportant for materials having high linear absorption coefficients, such as hematite (Fe_2O_3) with Cu $K\alpha$ radiation, since the incident X-ray beam penetrates little more than a single layer of crystallites on the surface of the sample. However, materials with high absorption coefficients have other inherent problems that will be addressed below.

Sample surface. Ideally the surface of a mount should be flat, with no roughness or curvature, and not tilted in any direction. Any roughness or curvature has the potential to produce systematic deviations in the positions and breadths of observed reflections related to sample-height-displacement and flat-specimen errors (Klug and Alexander, 1974, p. 302). Unless specifically desired, for example in a study of preferred orientation, sample tilting should be avoided because it changes the 2:1 angular relationship between the receiving slit and the sample surface and can give rise to systematic errors in intensity and peak breadth (de Wolff et al., 1959).

Sample-height displacement. Sample-displacement errors are among the most common in powder diffractometry even though modern diffractometers have sample-mounting devices that allow relatively reproducible placement of the sample on the axis of goniometer rotation. The effects of incorrect sample height are commonly manifested as systematic errors in line positions. The magnitude of the error in peak position, $\Delta 2\theta$ in radians, is

$$\Delta 2\theta = -2S(\cos\theta)/R , \qquad (4)$$

where S is the sample displacement from the focusing plane in mm, θ is half the diffraction angle in radians, and R is the goniometer radius in mm. Given a goniometer radius of 200.5 mm, a sample displacement of 50 μm will yield an error of 0.027° 2θ at 20° 2θ. The direction of this error is such that a displacement of the sample surface downward on a vertical goniometer (away from the center of the focusing circle) will yield a negative error in 2θ, that is the peak positions will be displaced to lower 2θ. In Equation 4, the positive direction of S is away from the center of the focusing circle.

Some texts eliminate the minus sign in Equation 4, in which case the positive displacement direction is reversed. In practice, the sample displacement error is one of the largest systematic errors affecting peak position and should always be considered when one desires accurate peak positions. It is possible to correct for this error either by inclusion of an internal standard or through analytical processing of diffraction data covering a large range of 2θ. In fact, several lattice-parameter least-squares refinement computer programs have incorporated a correction for sample-displacement error. Likewise, several Rietveld refinement programs include a sample-displacement correction term. Use of either of these correction procedures can significantly improve the accuracy of lattice parameters obtained from diffraction data.

METHODS OF GRINDING AND DISAGGREGATION

There is a variety of methods and equipment available to reduce the particle sizes of different types of samples, ranging from a simple hand mortar and pestle or percussion mortar to automatic grinding machines that may cost as much as $10,000. As mentioned above, there are several things to consider when grinding samples, such as not adversely affecting the materials being ground, e.g., by overgrinding or contaminating the sample, and obtaining a relatively homogeneous particle size. Although most methods of grinding introduce low levels of contamination detectable by chemical methods, contamination is not generally a problem for diffraction measurements. Commercially available hand mortars and pestles made of agate, mullite, or corundum are acceptable for general grinding. On the other hand, porcelain or iron mortars are generally too soft for grinding mineral samples and can introduce considerable contamination. Grinders made of tungsten carbide are usually excellent for reducing particle sizes, although we have found some tungsten carbide contamination in powders after grinding silicate samples in a shatterbox. One of the most important requirements in grinding silicate samples is the use of a liquid lubricant such as alcohol or acetone. Lengthy grinding of a liquid-saturated sample will result in the desired reduction in particle size with little or no structural damage to the sample. Several automatic grinders, such as ball mills or shatterboxes, that do not easily accommodate liquids, often produce significant amounts of damage to the sample. The most significant drawback to hand grinding in a conventional mortar and pestle is the difficulty in obtaining sufficiently small particle sizes (see below). We have obtained very good results using an automatic "mortar and pestle" made by Retsch and marketed in the US by Brinkmann. This unit consists of a mortar with a cylindrical cavity and a cylindrical pestle with a step on the bottom and a cam ground on the side. During operation, both the mortar and pestle rotate, usually at different speeds. Mortars are available in agate and tungsten carbide (and others). This device can routinely produce powders with average particle sizes ~3 μm, as measured by a commercial particle-size analyzer, after grinding under acetone for 12 min. Contrary to reports in the mineralogical literature on the preparation of <10-μm powders, we have not been able to prepare homogeneous powders of this particle size using a hand mortar and pestle without considerable effort. Typical literature reports of the production of particle sizes below 10 μm do not specify if and how particle sizes were measured, and thus the quoted particle sizes are suspect. Another device that is able produce very fine particle sizes easily is the McCrone micronizing mill, reportedly yielding material averaging about 10 μm.

Sieving is an effective method of isolating particles of the appropriate size. It is quite common in powder diffraction analysis to use material passed through a 325-mesh (44-μm) sieve in the belief that this particle size is sufficiently small. As described below, this particle size is significantly too large for anything but qualitative diffraction work. It is difficult manually to sieve powders much finer than 400 mesh (38 μm), although automatic devices are available (e.g., sonic sifter) that perform well down to micron sizes.

Some materials, notably micaceous minerals, are not amenable to grinding by conventional methods. Numerous authors have obtained good results using a metal file on such materials, and S. W. Bailey (personal communication) has determined that filing micas and chlorites produces much less layer-stacking disorder than attempting to grind these minerals by conventional methods. The common laboratory or kitchen blender (Waring) has also been used effectively to reduce the particle size of micaceous materials, although the final particle size achieved is usually relatively large. It is possible that the newer high-speed laboratory homogenizers such as those marketed by Brinkmann may prove superior to conventional blenders at reducing the particle size of platy or fibrous materials. We have obtained good results with fibrous materials such as chrysotile using a device commonly used to grind biologic materials known as a Thomas-Wiley mill. As with the laboratory blender, this device produces material of relatively large particle size, but it is effective at cutting fibrous materials into a powder that can be ground further by other methods.

Ultrasonic vibration is a method often used to disaggregate loosely consolidated materials such as some sedimentary and volcanic rocks. Both ultrasonic baths and high-power probes are available. The high-power ultrasonic probes are quite effective at disaggregating even slightly consolidated materials and are particularly useful for removing clay minerals from rocks. However, the user should be aware that these devices can significantly heat the suspension, an effect that may be undesirable with some samples. We have found that the ultrasonic baths are relatively ineffective at disaggregating most materials due to their low power.

PARTICLE-SIZE REQUIREMENTS

Accurate X-ray diffraction intensities from a powder sample require that the grain size be small, at least 10 μm and preferably smaller. By "grains", we mean crystallite size. For intrinsically fine-grained rocks, such as shales or basalts, a particle produced by grinding may consist of an aggregate of crystals that are much smaller. For coarse-grained rocks that are ground in the laboratory, grain size is usually equal to crystal size. The problems encountered with coarse grain size, including extinction, particle statistics, and microabsorption, are discussed below. The particle size of a powder can also influence the degree of preferred orientation exhibited by the sample, and this factor will be discussed in the section on preparing random-powder mounts.

Extinction

The intensity of X-radiation diffracted by typical powder samples is traditionally interpreted using the theory of X-ray diffraction for ideally <u>imperfect</u> crystals that have numerous lattice imperfections and small mosaic blocks (as opposed to ideally perfect crystals). However, in some cases the intensities measured from real crystals are measurably less than would be predicted for ideally imperfect crystals. This effect is described by the term extinction, and both primary and secondary extinction can occur in real crystals. Extinction is an important process in thick and nearly perfect crystals, but it is not a common problem with the kinds of fine powders normally dealt with in powder diffraction techniques.

To envision primary extinction, consider a perfect crystal fragment oriented at a suitable Bragg angle to diffract strongly. Upon diffraction, the primary beam experiences a phase shift of $\pi/2$ (James, 1965). The diffracted beams can be re-diffracted and therefore will be exactly π out of phase with the primary beam (having experienced an additional $\pi/2$ phase shift upon re-diffraction). The re-diffracted beams, now travelling in the direction of the incident beam, will thus destructively interfere with and reduce the intensity of the primary beam. This effect is most important for reflections with very large structure factors and/or for crystals with large mosaic blocks. For reflections with small structure factors, the destructive interference of the diffracted beam with the primary beam is insignificant compared with the intensity of the primary beam. In addition, destructive interference will occur only between primary and diffracted beams within a given mosaic block, and thus primary extinction will be insignificant for crystals with small mosaic blocks. When primary extinction occurs, the crystal appears to have an unusually large linear absorption coefficient when measured at a crystal angle that coincides with that of a strong reflection. This effect is due to the significant reduction in intensity of the primary beam, and the strongest reflections will be weaker than expected.

Even for ideally imperfect crystals with small mosaic blocks, another type of extinction known as secondary extinction occurs. Whether or not diffraction is taking place, the intensity of an X-ray beam is attenuated when passing through a material via absorption of the X-rays by an amount $\mu I_o t$ due to ordinary absorption and conversion into thermal energy. However, when the crystal is in a position to diffract, the incident X-ray beam power is reduced by an additional amount equal to the power of the diffracted beam. Therefore intense diffraction from successive lattice planes extracts energy from the incident beam with the result that the remaining transmitted energy is weakened, causing lower lattice planes to receive and thus diffract less than would be predicted by the normal absorption coefficient. The process is repeated throughout the crystal fragment until the incident energy is "used up" at some depth that corresponds to infinite thickness. This process can also occur during re-diffraction of diffracted radiation. However, because we are considering an ideally imperfect crystal, primary extinction will not be important because the phase shifts from different mosaic blocks will have no fixed relations and thus there can be no destructive interference. At orientations corresponding to weaker diffraction peaks, less energy is diverted by successive diffraction events with the result that the incident beam penetrates further and the effective thickness is greater.

The integrated intensity of diffraction is proportional to the volume of the crystalline substance irradiated, and because the effective thicknesses for weak and strong reflections differ, the irradiated volumes and hence diffraction intensities differ. A general version of the intensity equation is given by

$$I(hkl) = \frac{KLp(\theta)|F(hkl)|^2}{V^2\rho\mu^*}, \quad (5)$$

where K is a constant, Lp is the Lorentz-polarization factor, F is the structure factor, V is the volume of the unit cell, ρ is the density, and μ^* is the mass absorption coefficient measured for a <u>non-diffracting</u> orientation of the crystalline substance. If extinction is present, then μ^* is not constant but increases proportional to $|F(hkl)|^2$, and this is the source of the intensity discrepancies. The magnitude of this effect can be considerable, and the effective linear absorption coefficient can be many times larger than the normal value of μ for strong reflections (Woolfson, 1970).

Correction for extinction effects is difficult because the factors involved are usually unknown. Fortunately, extinction is rarely a problem in fine powders, and if they have been ground, the introduction of small amounts of strain helps minimize the long sequences of precise lattice-plane alignment required to produce these undesirable effects. If the mosaic blocks have a range of orientations and there is a finite amount of lattice strain, then the blocks do not all diffract simultaneously, minimizing secondary extinction. Finally, it is noteworthy that detector dead time can yield effects closely resembling extinction, in which the strongest reflections are weaker than expected. Excellent discussions of primary and secondary extinction are given by James (1965) and Woolfson (1970, p. 182-189) to which the interested reader should refer.

Particle statistics

Suppose that two megascopic crystals of different minerals are oriented at random on a substrate and analyzed in a flat-specimen diffractometer. The chances are that there will be no measurable diffraction at all, because no lattice planes from either are oriented suitably for diffraction. If the crystals are finely ground so that a very large number of them are randomly oriented, then somewhere in the powder many crystal fragments are suitably aligned to produce diffraction peaks as the diffractometer detector scans over the appropriate angles of 2θ. The intensity of the diffraction series from each mineral is proportional to the relative abundance of the two minerals present. Between the extremes of the examples cited, the relative intensities of the diffraction lines from each mineral depart from the theoretical intensity distribution because, by chance, some orientations are missing, and/or some are over-represented because of the finite number of crystals and hence the finite number of possible orientations. This condition is easily detected by a lack of repeatability of the powder diffraction pattern. Generally, the orientations of the crystal fragments cannot be repeated by dumping the sample and repacking it, so successive experiments of this type show large variations in the intensities of all of the peaks on the powder pattern. Accurate relative intensities require the averaging of the diffraction phenomena from millions to billions of individual crystal fragments, and there are several ways to achieve these particle statistics.

For a fixed volume of powder specimen, a suitable number of crystallites can be obtained by grinding to very small particle sizes. The volume of sample irradiated is given approximately by

$$V \sim [W] \left[\frac{R_0 \tan\gamma}{\sin\theta}\right] \cdot \left[\frac{3.2\sin\theta}{\theta}\right] \,. \tag{6}$$

V is the volume of sample irradiated in cm^3, and W is the width of the incident beam in cm. The second term describes the length of the sample illuminated, and the variables here are R_0, the goniometer radius in cm, and γ, the angular aperture of the divergence slit in degrees. The third term gives the integrated thickness that contributes to diffraction (Klug and Alexander, 1974), and in Equation 6 μ is the linear absorption coefficient. Consider the quartz 113 reflection at 26.65° 2θ for Cu $K\alpha$ radiation. Assume a goniometer radius of 20 cm, a one-degree divergence slit, and a beam width of 2 cm. The linear absorption coefficient of quartz is 96.5 cm^{-1}. Solution of Equation 6 shows that the volume irradiated is equal to 0.023 cm^3. We ignore the pore space in the sample for simplicity and assume a perfect packing of cube-shaped crystallites of length 10 μm = 1 x 10^{-3} cm = 1 x 10^{-9} cm^3, so the approximate number of crystallites that are bathed in the incident radiation is 0.023 / 1 x 10^{-9} = 23,000,000. The correct number is less than this because the grains at the bottom of the sample contribute much less to the diffraction pattern than those near the surface, but it still serves as a good reference point because experience has shown that a sample described by these parameters provides acceptable (±2 %) accuracy for most diffraction work. Note that increasing the grain lengths to 20 μm decreases the number of "active" crystals eight-fold.

The accuracy of intensity data can be improved by repacking and reanalyzing the sample a number of times and averaging the results. The standard deviation for the mean intensity of an individual peak is equal to $1/\sqrt{n}$ times the standard deviation of the set of experimental measurements about the mean, where n is the number of measurements in the average. Another approach is to use a wider divergence slit and thus increase the area irradiated, although the irradiated length must be consistent with the length of the sample and hence the amount of sample available. Decreasing the wavelength of the incident beam decreases the absorption coefficient and increases the volume irradiated (Eq. 6), but this is not very practical because the next easily available target material is Mo and its radiation is too penetrating to be of much use in reflection diffractometry. A step in the wrong direction is the use of Cr radiation, because this increases the absorption of, for example, quartz, by about three-fold, decreasing by a similar factor the number of crystallites that contribute to the diffraction pattern. Another effective means of improving particle statistics is to use a rotating sample holder. Several authors, notably de Wolff et al. (1959) and Parrish and Huang (1983) demonstrated the effects of rotating samples of different particle sizes, and these papers are recommended to the interested reader. This discussion clearly shows that the most practical method of achieving good particle statistics is to reduce the particle size of the powder as much as possible, consistent with present-day grinding technology and minimal crystal-structure damage.

The effects of particle size on the reproducibility of diffraction intensities are demonstrated by a very informative experiment described by Klug and Alexander (1974, p. 365-368). They used <325-mesh quartz powder, Cu $K\alpha$ radiation, and studied the variability of the intensity of the quartz 113 reflection. Ten replicate analyses were performed on each of four particle size ranges. The mean percentage deviations in peak intensity were 18.2, 10.1, 2.1, and 1.2 for, respectively, the 15 - 20-μm, 5 - 50-μm, 5 - 15-μm, and < 5-μm fractions. Clearly, <325-mesh powders are not sufficiently fine for anything but qualitative measurements. The 5 - 15-μm data are a good compromise between reliability and particle size, suggesting that 10-μm powders are suitable for applications in which a few percent error can be tolerated in intensities from intermediate-absorbing silicate minerals irradiated by Cu radiation. If better data are required, averaging multiple analyses probably is the best choice, because it quickly becomes impractical to implement further and further reductions in particle size. Tedious experimental procedures to minimize further particle statistics errors may not be worth the effort because of the dominance of other sources of error caused by, for example, microabsorption, preferred orientation, and non-representative mineral-intensity standards. It is important to note that although good results can be obtained using 5 - 15-μm powders of typical silicate minerals, the required sizes for powders of more highly absorbing materials, such as Fe oxides, are much smaller. For the analysis of minerals with good cleavage(s), problems of preferred orientation are likely to cause the most significant source of intensity errors.

<u>Microabsorption</u>

Microabsorption is an effect that causes two or more phases in a mixture to contribute diffraction intensities that are related to both their relative proportions and their mass absorption coefficients (see Brindley, 1945; de Wolff, 1947; Wilchinsky, 1951; Gonzalez, 1987; Hermann and Ermrich, 1987). The mass absorption coefficient is defined as the linear absorption coefficient divided by the density. Quantitative analysis cannot be accomplished when microabsorption is significant. Microabsorption occurs in coarse powders if the constituents have very different mass absorption coefficients, and it causes the underestimation of the highly absorbing constituents. The effect is decreased as particle size is decreased, and the powder mixture behaves ideally if the particle size is less than a critical value that is dictated by the phase with the highest absorption coefficient. Absorption coefficients depend on X-ray wavelength as well as chemical composition, so a powder suitable for quantitative analysis with one wavelength may not be suitable for another.

For silicate mineralogists, the most serious manifestations of microabsorption arise from the use of Cu $K\alpha$ radiation with minerals rich in Fe. Consider a powder that contains large grains of two different minerals, one with high and one with low mass absorption coefficients (μ^*). Equation 7 relates diffraction intensity to μ^* and g; the latter is defined here as the mass of material, in g per cm^2 of sample, traversed by an impinging beam. If $g\mu^*$ is large for single grains, I is proportional to $1/\mu^*$ for each because the exponent term of Equation 7 approaches zero. A mixture of large grains produces diffraction intensities that are proportional to g_j / μ_j^* for each of the j phases present.

$$I = \left[\frac{I_0 \sin\theta}{2\mu^*}\right] 1 - e^{-2\mu^* g/\sin\theta} \quad . \tag{7}$$

Suppose now that two constituents are very fine grained, and Equation 7 is applied to the relative intensities each produces for an increment of sample that consists of a pair of the two kinds of grains that are successively penetrated by the incident X-ray beam. If $g\mu^*$ is very small for each, then by the limit theorem, $\lim_{x \to 0} (1 - e^{-x}) = x$, Equation 7 can be written

$$I = \left[\frac{I_0 \sin\theta}{2\mu^*}\right] \cdot \left[\frac{2g\mu^*}{\sin\theta}\right] = I_0 g . \qquad (8)$$

The mass absorption coefficient cancels, and the relative intensities produced by the two phases are proportional to their relative masses, that is, the relative intensities are proportional to the weight proportions of the minerals present. The quantity $\sin\theta$ remains, but for a long sample, the length irradiated is proportional to $1/\sin\theta$, and the volume irradiated contains no angle-dependent quantity. The <u>absolute</u> intensities are proportional to $1 / \overline{\mu}^*$ where $\overline{\mu}^*$ is the weighted-mean sample mass absorption coefficient. Brindley (1945) noted that for ideal absorption behavior, the particle size must be small enough so that each individual grain absorbs only about one percent or so of the X-ray energy impingent upon it.

Table 1. Effects of microabsorption on apparent concentrations of minerals mixed in equal proportions with quartz ($\mu^* = 36.4$ cm^{2}g^{-1}).

Mineral	μ^*(cm^{2}g^{-1}, Cu $K\alpha$)[1]	Apparent Concentration
K-Feldspar	50.6	50
Calcite	73.4	47
Siderite	151.5	39
Pyrite	190.0	32

[1]Data from Brindley, 1980

Table 1 shows calculated results for various minerals that are assumed mixed with quartz in equal proportions. The calculations were made by a Monte Carlo method in which thousands of integrations were made for X-ray paths through randomly generated permutations of the two kinds of grains. For no microabsorption effects, the "Apparent Concentration" values should be 50 %. They are diminished below 50 % to the extent that their absorption coefficients are greater than μ^* for quartz. Cu $K\alpha$ radiation and a uniform 10-µm particle size were assumed. The results show that accurate analysis cannot be made for Fe-rich phases in common rocks using Cu $K\alpha$ radiation for grain sizes of 10 µm and greater. Iron or cobalt radiation would be much better choices because their $K\alpha$ lines lie to the low-µ side of the Fe absorption edge, and for these radiations, the mass absorption contrasts are greatly reduced. Microabsorption is rarely a problem for the clay minerals because of their intrinsically small particle size. Calculations indicate that for kaolinite-glauconite and kaolinite-chlorite mixtures, grain thicknesses of 1 µm produce errors of only a few percent in randomly oriented mixtures. An unanticipated and unwelcome result of these calculations lies in the finding that for oriented micaceous or clay mineral aggregates, microabsorption becomes markedly 2θ-dependent because at low diffraction angles the apparent thickness of the grains is increased due to the low angle of incidence of the incident and diffracted beams. Partially compensating for this is the fact that in oriented aggregates, the minimum dimension of the grain is presented to the incident beam. One might think that microabsorption effects can be calculated and then used to correct the results for quantitative analysis. The analysis presented above, however, shows that to do this, you need to know the particle size distribution for each phase in the sample, and the preferred orientations of each phase if the powder is not randomly oriented. The best recourse is to use a different kind of primary X-radiation.

RANDOMLY ORIENTED POWDER PREPARATION

One of the long-standing problems in X-ray powder diffraction analysis is the elimination of preferred orientation of sample particles. Unless an oriented aggregate or deposit is needed for special purposes, e.g., for analysis of basal reflections from clay minerals, one almost always strives for a completely random orientation of crystallites. This is true whether for qualitative phase identification, quantitative analysis of multicomponent mixtures, or collection of data for structural investigations. Numerous methods have been proposed to minimize preferred orientation and these are summarized in several publications (e.g., Klug and Alexander, 1974, p. 364-376; Smith and Barrett, 1979). Procedures for producing a randomly oriented powder mount for diffractometry include side- or back-packing a powder into a cavity mount, mixing a powder sample with a filler material (usually amorphous), mixing a powder with a viscous material, dispersion in a binder, and more exotic methods such as spray drying, tubular aerosol suspension, or liquid-phase spherical agglomeration (LPSA). Most of these procedures are relatively simple and produce sample mounts with varying degrees of randomness. Some of the more involved methods such as spray drying and LPSA require special equipment and are not well documented in the geological literature. These different preparation procedures are briefly summarized here, and the more exotic methods are described in detail.

In addition to sample preparation methods, several instrumental techniques are reported to minimize preferred orientation. These include the use of a sample spinner and slightly rocking the sample (by about 1°) about the θ axis of the diffractometer. The former improves particle statistics but does little to eliminate preferred orientation of crystallites parallel to the sample surface (e.g., Parrish and Huang, 1983). The latter will probably be of little help in eliminating all but small degrees of preferred orientation. In addition, de Wolff et al. (1959) commented that the rocking method, primarily used to improved particle statistics, must be used with care because it changes the 2:1 angular relationship between the receiving slit and the specimen surface. Even small changes in the 2:1 relationship will cause line broadening and a reduction in peak intensity. Systematic intensity errors may result because this effect increases with decreasing 2θ. Although it is usually best to minimize orientation during sample preparation, preferred orientation can be qualitatively evaluated using a conventional θ-2θ diffractometer by rocking the sample about the θ axis of the diffractometer (e.g., Yukino and Uno, 1986). Because this is a very specialized area of powder diffraction, more closely related to texture studies of materials, measurement of orientation will not be addressed in this chapter. Applications of these methods include studies of textures in metamorphic rocks, of orientation in sample mounts of clay minerals (Reynolds, 1986), and of mineral orientation in natural bone samples (Bacon et al., 1979).

Side- and back-packed mounts

Methods involving packing of prepared powders into the sides or backs of cavity mounts are simple and often yield comparatively good results considering the effort and equipment required. With these methods, the front or top of the cavity mount is covered with a microscopically rough solid material, such as a frosted glass slide or cardboard, and the mount is filled either from the back or the side. After loading, the opening is covered and the rough material on the front is carefully removed. Depending on the particle size and shape of the material being mounted, these methods can yield mounts with far less preferred orientation than conventional front-packed mounts. We have found that back-packed mounts almost completely eliminate preferred orientation in finely powdered samples that are not extremely orienting and are relatively transparent to Cu $K\alpha$ X-rays, e.g., kaolinite (Bish and Von Dreele, 1989). However, such mounting methods are not very effective at eliminating preferred orientation for materials that are notorious for orienting (e.g., chlorite, Walker and Bish, 1989; albite and sanidine, Bish, unpublished). One of the prerequisites for preparing an orientation-free sample mount is the use of material of sufficiently fine particle size. Materials with good cleavage will orient much more effectively when in the size range greater than about 30 µm than they will when ground to <5 µm.

Solid filler materials

Both crystalline and amorphous filler materials have been used with varying degrees of success in an effort to surround the particles of interest in a homogeneous, non-orienting medium. Amorphous materials used as fillers include powdered glass, starch, gum arabic, cork, gelatin, and amorphous boron (Smith and Barrett, 1979). The disadvantages to this method are that the sample is contaminated and the amorphous component gives rise to scattering in the diffraction pattern, significantly increasing background. In addition, admixture with one of these materials of low linear absorption coefficient can significantly increase the transparency of the sample. Crystalline materials have also been used to minimize preferred orientation. Corundum, which is available commercially in very fine particle size (< 1 µm), has been used with some success. The advantage of using a material such as corundum is that it can act as an internal d-spacing and intensity standard in addition to minimizing preferred orientation. The disadvantages are that the sample is contaminated, and corundum has diffraction lines that may interfere with those of the sample. In a survey of several methods for reducing orientation, Calvert et al. (1983) found that mixing an orienting sample (MoO_3) with ~50% silica gel and side-mounting dramatically reduced but did not eliminate preferred orientation.

Use of a viscous mounting material or dispersion in a binder

A very common method of quickly preparing a random-powder mount is to sprinkle a finely ground powder onto a slide or solid plate coated with a viscous material such as vaseline, silicone grease, collodion, or machine oil. The powder is best applied by sprinkling through a fine sieve, preferably 400 mesh (38 µm) or smaller. This method has several disadvantages and is recommended only as a rapid way to produce a random mount of a small amount of powder. Because the mount is usually very thin and the powder is imbedded in a material with a low atomic number, the resultant sample mount usually will not be infinitely thick and may be relatively transparent to X-rays, yielding unreliable intensities. Thus the reason for choosing such a mount, to obtain accurate intensities, may be defeated. All of these viscous mounting media also give rise to significant amorphous scattering. Examples of the diffraction patterns resulting from some common mounting media, including double-sticky tape, high-vacuum silicone grease, vaseline, and household 3-in-1 oil are given in Figure 1. It is clear from this figure that the type and amount of viscous mounting medium should be considered carefully. Finally, unless the mounting medium is placed on a special recessed slide, use of this method will produce a significant specimen displacement, yielding potentially serious errors in peak positions. With consideration of specimen displacement, such sample mounts can provide accurate peak positions unaffected by sample transparency.

A related mounting method is the use of hardening binders. Once set, the resultant mass may be either sectioned or crushed. Bloss et al. (1967) minimized orientation by lightly spraying a dispersed powder with a clear acrylic and mounting the resultant clumps of powder. These methods suffer from most of the problems inherent with other methods involving addition of a foreign substance to the sample, including contamination, amorphous scattering, and sample-transparency effects exacerbated by the addition of a low-absorbing matrix to the sample powder.

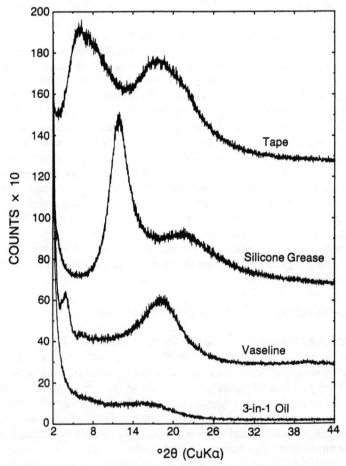

Figure 1. Diffraction patterns resulting from common mounting media (on a cut quartz single-crystal plate). Patterns offset for clarity.

Special methods for minimizing preferred orientation

Whereas the above methods for reducing preferred orientation are relatively straightforward, the techniques described here are either more involved or require special equipment. The tubular aerosol suspension method derives from sampling methods used in the atmospheric sciences, and the spray drying and liquid-phase spherical agglomeration methods evolved from methods commonly used in ceramic sciences. These methods have a better potential than the simple methods for eliminating preferred orientation.

Tubular aerosol suspension. This method of sample preparation has been developed primarily by Briant Davis and co-workers at the South Dakota School of Mines and Technology and has been applied mainly to problems in quantitative analysis. The method uses what Davis (1986) calls a tubular aerosol suspension chamber (TASC) which consists of a buret suspension chamber with a fluidized bed of fine (< 1 mm) glass beads in the well. Beneath the buret is a trap and a connection to the pressure side of a rotary vacuum pump. At the top of the buret is a 4.25- or 4.7-cm Whatman GF/C filter with a connection to the vacuum side of the pump. The TASC operates by creating a convection cell in the buret, ejecting the sample upwards into the glass fiber filter. This method appears to form samples with little preferred orientation if filter loading is light. However, in these cases, amorphous scattering is contributed by the glass filter. In addition, for materials of low mass absorption coefficient and/or for light filter loading, the filter mounts are not effectively infinitely thick. For quantitative measurements on such materials, a transparency and matrix absorption correction must be made using the known sample loading and the mass absorption coefficient of the sample and matrix (Davis and Johnson, 1982; Davis, 1986). The transparency problem diminishes as the filter loading increases, but this may lead to measurable preferred orientation. In addition, the method may not be ideal for multiphase mixtures such as rock samples, in which one has a distribution of particles of different sizes and densities. For these materials, the method appears to have the potential for forming significantly segregated sample mounts.

It is likely that this method will provide sample mounts with little or no preferred orientation and no transparency problems for materials of high mass absorption coefficient (greater than ~200 cm^2/g). The special equipment and corrections necessary for typical silicate minerals make this method less attractive than simpler methods for minimizing preferred orientation. Post (personal communication) successfully used a modification of this method in which he sieved pure samples of Mn-oxides directly onto the glass fiber filters. Rietveld refinements using data obtained with these samples show that preferred orientation is drastically reduced. However, it is likely that this modification will be useful only with materials of high linear absorption coefficients for which sample transparency will not be a problem.

Liquid-phase spherical agglomeration. Calvert and Sirianni (1980) described a method for minimizing preferred orientation which they called spherical agglomeration or liquid-phase spherical agglomeration. In this method, a finely powdered sample is suspended in a liquid such as water and a second liquid which wets the sample and is immiscible with water is added. With agitation by, for example, a laboratory shaker or blender, the suspended solid separates into spherical bodies. Calvert and Sirianni (1980) used varsol as the binding, immiscible liquid, and tested the method with good results on brucite, talc, graphite, and palygorskite. By controlling the conditions, they were able to prepare spheres ranging from 50 μm to 1 mm in diameter. Obviously, with a method such as this, care must be taken to reduce the particle size of the sample to considerably less (< 10 μm) than the size of the resultant spheres. Calvert and Sirianni did not describe how they removed the spheres from the liquid.

Spray drying. This common industrial technique was used over 30 years ago to minimize preferred orientation (Flörke and Saalfeld, 1955) and has been used sporadically since then, primarily with clay minerals (e.g., Jonas and Kuykendall, 1966; Hughes and Bohor, 1970). There has been renewed interest in spray drying in the last ten years, mainly through the efforts of Prof. R. Snyder of Alfred University and his students (Smith et al., 1979a,b). The method consists of atomization of a slurry of the solid of interest mixed with a liquid, usually water, and a small amount of an organic binder into a high-temperature gas zone. The liquid rapidly evaporates from the droplets, leaving spherical and often hollow agglomerates. The binder provides the structural integrity necessary for packing. Smith et al. (1979a,b) constructed a spray dryer in order to evaluate the method as a means of eliminating preferred orientation. Their spray dryer is quite large, and the drying chamber alone is over 1 m high. To our knowledge, there are two commercial spray dryers on the market today (Yamato and Brinkmann-Buchi). Neither unit appears to be ideal for producing spray-dried agglomerates from small samples. Blake Industries is reportedly ready to market a spray dryer manufactured by Glatt Air Techniques, Inc., and designed by Prof. Snyder specifically for preparing X-ray powder diffraction samples. Only prototype instruments have been built, but the instrument appears to be well suited for working with small samples. The primary drawback to this unit may be the price (over $15,000).

In their evaluation of the method, Smith et al. (1979a) examined samples of muscovite, wollastonite, pyrite, siderite, and synthetic corundum. Materials such as muscovite and wollastonite are particularly notorious for orienting in conventional cavity mounts. They compared results obtained from spray-dried and non-spray-dried samples and used a quantitative agreement index to assess deviations from calculated diffraction patterns. In every case, data for the spray-dried sample agreed remarkably well with the calculated pattern. The method appeared to accommodate the platy muscovite, the rod-like wollastonite, and the rhombohedral siderite crystallites equally well. Smith et al. (1979a) concluded that spray drying is an efficient means of minimizing and perhaps eliminating preferred orientation.

As an additional evaluation of spray drying, Smith et al. (1979b) used the method in quantitative analyses of Devonian shales. Their samples contained minerals prone to orientation, such as chlorite, kaolinite, illite, and feldspar. Results of their analyses of several prepared mixtures of minerals with known proportions were very good, with absolute errors between 2 and 4%. However, their results with a pure sample of chlorite again emphasized the necessity of having very finely powder samples; <325 mesh powder is usually not small enough for spray drying. Smith et al. (1979b) concluded that they had eliminated the problem of preferred orientation and that the principal source of error in the analyses of natural shales was the differences between a particular mineral in the shale and the material used as a standard. But the accurate quantitative analysis of samples containing these orienting materials is noteworthy and illustrates the potential of the spray-drying method.

Although the method is remarkably effective, it does not seem to be well suited to small samples (< 1 g). In addition, in order to produce uniform spheres of ~50 μm in size, the particles in powders should ideally be < 5 μm in size, considerably smaller than the average diffractionist normally uses. As mentioned, a significant disadvantage of this method is the special, expensive equipment required and the time needed to prepare a sample.

In an evaluation of methods used to reduce preferred orientation, Calvert et al. (1983) compared side packing of samples mixed with an amorphous diluent, spray drying, and LPSA. Side-packed and spray-dried samples were mounted in cavity holders and examined on a conventional diffractometer (Cu $K\alpha$ radiation), and the LPSA samples were mounted in capillaries and examined on a diffractometer with Debye-Scherrer geometry (Mo $K\alpha$ radiation). For the LPSA, samples were suspended in a 1:1 mixture of hexane and octane and agglomerated by shaking with H_2O in a blender. Results showed that all three methods produced results for the MoO_3 far superior to those obtained with a normal packed mount, but only LPSA gave good results for fluorophlogopite. Calvert et al. (1983) speculated that the spray-dried sample of fluorophlogopite did not consist of perfect spherical agglomerates (exhibited some preferred orientation) due to a large crystallite size. For LPSA as for spray drying, the sample particle size should be < 5 μm to prepare agglomerates of appropriate size (< 50 μm). The improved results seen with the LPSA sample may be due solely to the instrument geometry used with these samples (Debye-Scherrer, samples in capillaries); no direct comparison of samples mounted identically was made because the agglomerates resulting from LPSA do not have a binder, are structurally weak, and thus cannot be packed into cavity mounts. It therefore appears that LPSA is best suited for sample mounts not requiring packing, e.g., capillaries.

PREPARING ORIENTED CLAY MINERAL AGGREGATES

The identification and quantitative analysis of the clay minerals is almost always based on aggregates that are oriented so that the basal or $00l$ diffraction series is enhanced. These minerals have structures that are very similar in the X-Y directions, but the unit-cell parameters along Z are diagnostic. The relatively small Z dimensions of clay mineral crystallites mean that peaks are broad and peak heights are small, with the result that the peaks are difficult to resolve from the background unless integrated intensities are large. This condition requires that as many crystallites as possible be oriented with their X-Y planes parallel to the sample surface for reflection diffractometry, which is the most widely used technique. High degrees of preferred orientation are maximized by the following conditions.

(1) The crystallites must have a good platy morphology.

(2) The crystallites must be separated from each other in aqueous suspension, that is, the sample must be well dispersed in water (peptized, or deflocculated) and individual

grains must be disaggregated into individual crystallites. Cements must be chemically removed.

(3) The sample should contain no non-platy minerals such as quartz, for these interfere with the oriented stacking of crystallites that is produced by the sample-preparation process.

(4) The sample substrate should be smooth so that the first layers of the clay film do not produce an irregular surface for the layers that are deposited later.

The crystallite morphology is not a controllable variable. This is what causes different specimens of the same mineral type to produce very different diffraction intensities even though the sample preparation methods are kept scrupulously constant. Two different kaolinite minerals, prepared for analysis identically, produced basal diffraction intensities that differed by 60-fold (Reynolds, unpublished data). TEM photographs of the two are consistent with the diffraction data. The mineral that produced the weak $00l$ pattern has almost no discernible platy morphology. The crystals or crystal aggregates resemble wadded balls of paper. The kaolinite in the well-oriented aggregate looks much like the shingles on a roof, and measurements of its preferred orientation show a 4 % standard deviation of the tilt angles about the mean orientation that parallels the sample surface.

Aqueous dispersion of clay minerals and the chemical removal of cements such as carbonates, iron oxides, and organic matter is a specialized subject that is beyond the scope of this work. Workers in the field commonly refer to the many techniques developed by M. L. Jackson and his coworkers, and summarized by Jackson (1969). The important point to make here, however, is that some clay minerals are easily damaged by aggressive chemical treatments, so such techniques should be used only when necessary, and then, with the full realization that the sample may have been altered in ways that compromise certain types of interpretations. Examples of these include alteration of layer charge and hence expandability in smectite or vermiculite-containing mixed-layered clay minerals by the use of strong oxidizing or reducing agents, and partial dissolution of Fe-rich chlorites by acid treatment.

Clay mineralogists almost always separate the clay fraction (e.g., < 4 µm or <2 µm, equivalent spherical diameter, or finer) for analysis, and the main reason is to exclude minerals such as quartz that interfere with the preparation of oriented aggregates. This procedure is useless if the sample has been thoroughly ground in a ball mill or similar device, for then, the coarse-grained non-clay minerals have been reduced in particle size to the clay-size fraction where they are inseparable. Rocks must be disaggregated, not ground, and this can be accomplished by crushing and then, in aqueous suspension, by long-term (days) agitation in water, high-intensity ultrasonic vibration, or dispersion by an industrial-grade Waring blender. At the conclusion of such procedures, the sample is washed repeatedly by centrifugation to remove any salts present that may prevent dispersion of the colloidal (clay) fraction. Dispersion is aided by the addition of small amounts of alkali phosphates which buffer the pH to relatively high levels and presumably reverse the edge charge of silicate layers by the adsorption of the phosphate ion.

The sample preparation method used depends on the application. Unfortunately, methods that produce the best preferred orientation also are most likely to yield a sample mount with pronounced particle-size segregation with depth through the clay film, and if different minerals are concentrated in different particle-size ranges (i.e., different depths in the clay film), the sample is useless for quantitative analysis. But the intense basal diffraction intensities from such sample mounts make them a best choice for crystal structure analysis of pure minerals, which requires accurate intensity data for a long series of basal reflections. The different sample preparation methods require different levels of operator skill and some are more time consuming than others. Many different methods are in use that involve a wide variety of substrates. A representative few are described here, and these have been selected for the purposes of reconnaissance, quantitative analysis, and crystal structure analysis.

Reconnaissance

The glass slide method is the most commonly used reconnaissance method. An aqueous suspension of clay is pipetted by medicine dropper onto a clean glass microscope slide that has been labelled and placed in an oven set at approximately 90°C. The suspension dries into a clay film that remains (usually) attached to the glass substrate. The method takes about 1 h for drying and is easy to do. It produces severe particle-size segregation effects with the finest material at the sample surface, causing the peaks from the fine particle-size constituents to be enhanced in the diffraction pattern. In addition, the clay film is invariably too thin to give accurate intensities at moderate to high diffraction angles.

Quantitative analysis

Quantitative analysis requires that the sample be long enough so that the incident-beam spread does not exceed the sample length at the lowest diffraction angles used. The sample must be thick enough at the highest diffraction angles recorded, or some method must be employed to correct for sample thickness (see Moore and Reynolds, 1989). In addition, the different mineral types must be uniformly distributed throughout the clay film.

The Millipore filter-transfer method (Drever, 1973) produces samples that have only fair crystallite orientation but gives accurate intensities because there is little or no particle-size segregation present. The clay suspension is drawn through a filter by means of a vacuum system. Remaining suspension is decanted off and the filter is placed face down on a glass slide. After partial drying, the filter is stripped off and the specimen is dried. The method requires considerable operator skill, but when that is acquired, it is easy to accomplish and requires far less time than any of the other methods discussed here. Filtration time should not exceed 3 or 4 minutes or particle size segregation can become a problem. If longer times are needed because of slow filtration rates, implement some device that stirs the suspension slowly during the filtration period. The method is difficult to use for samples rich in smectite, because the expanded form of that mineral quickly clogs the filter and makes it impractical to obtain clay films of sufficient thickness for quantitative analysis. The initial flow rates through the filter are high because the filter has not yet been clogged with the sample. As long as the movement of the liquid through the filter cake is fast compared to the settling rate of the coarsest particles, there can be no significant particle size segregation. These conditions apply to the first-formed portion of the sample, and that is the portion nearest to the filter face. Inverting the filter cake presents the most particle-size-representative portion of the sample to the X-ray beam, and the least representative (enriched in the finest sizes) portion is the bottom of the clay film which does not contribute much to the diffraction pattern if the sample is reasonably thick.

The smear method is a good option for quantitative analysis because its proper application allows no particle-size segregation, and it produces thick preparations for any kind of clay mineral. Orientations are likely to be poor, however, and thus the samples may not be suitable for the measurement of peak intensities from minerals that are present in small concentrations. Water- or ethylene-glycol-solvated smear samples can be analyzed in the moist or wet condition (Barshad, 1960), and this can be a real advantage in maintaining constant salt and water-activity conditions, and in obtaining data from samples that tend to crack apart and curl upon thorough drying. A useful option for such samples is to smear them onto ceramic tile. The capillary-sized pores in the tile draw liquids from the smear, leaving a moist but coherent clay film that is suitable for X-ray analysis.

A clay suspension is concentrated and collected by high-speed centrifugation. Liquid is drained and the sedimented aggregate should have the consistency of a thick paste. This is smeared by spatula onto glass or some other substrate and dried. Tien (1974) described a simple apparatus for the rapid and repeatable preparation of clay films of uniform thicknesses. Any of the smear methods might profit by polishing the surface of a partially dried preparation with a spatula in order to enhance preferred orientation.

Crystal structure analysis

The highest degrees of preferred orientation are produced by centrifuge methods, because the large g-forces involved press the filter cake and dewater it to produce minimum porosity and maximum preferred orientation. The centrifuged porous-plate method (Kinter and Diamond, 1965) produces thick aggregates that have very high degrees of preferred orientation and therefore produce excellent diffraction patterns. Indeed, the best crystal morphologies produce integrated diffraction intensities that are comparable to those from single crystals. Many clay mineral aggregates give measurable and useful $00l$ intensities to diffraction angles as high as $130°\ 2\theta$ (Cu $K\alpha$).

With this method, a clay suspension is centrifuged onto or through an unglazed ceramic tile or plate to produce the oriented clay mineral aggregate. Any remaining liquid is drained and the specimen mount is dried at room or elevated temperature. Ethylene-glycol solvation is accomplished by the drip or vapor method. A particular advantage of the method is that the ceramic tile serves as a glycol reservoir, allowing long (days) periods of diffraction analysis on a sample that maintains a uniform solvation state. The ceramic tile withstands heat treatments to 800°C or more, if they are required.

The method is somewhat difficult to master and it is time consuming. It requires special centrifuge-cup holders for the ceramic tiles (see Kinter and Diamond, 1965), unglazed ceramic tiles which are difficult to locate, considerable material (200-600 mg), and the technique suffers from particle-size segregation effects that make it useless for quantitative analysis. But in the writers' experience, no other method of sample preparation approaches it in producing infinitely thick, long specimens that allow the collection of high-quality diffraction data from an extensive 00l diffraction series--the kind of data that are indispensable for modeling the structures of mixed-layered clay minerals, or for deductions of one-dimensional atomic structures by Fourier syntheses.

These and other methods of sample preparation have been evaluated by Gibbs (1965). Beginning workers in clay mineralogy will profit by a study of that work. There is still a finite number of laboratories that continue to use sample preparation methods that are entirely unsuitable for quantitative analysis, yet their results are reported with confidence. Demonstrations of precision do not guarantee accuracy.

SPECIAL SAMPLE PREPARATION TECHNIQUES

<u>Mounting small samples</u>

Although camera methods for studying very small samples (μg to mg) have existed for many years, techniques for examining such small samples with diffractometry are not as familiar to the average user. Diffractometric methods do not require recording film and a darkroom and they provide the ability to measure intensities quickly and accurately (after correction for transparency). Frevel and Roth (1982), Baldock and Parker (1973), and Zevin and Zevin (1987) summarized the methods and literature on X-ray diffractometry of low-mass samples, and most of the discussion here is from Zevin and Zevin (1987). Davis and Cho (1977) and Moore and Reynolds (1989) described the transparency corrections necessary when using samples that are not infinitely thick. If only qualitative measurements are required, these transparency corrections are not required. Zevin and Zevin focused their attention on low-mass samples which can be powdered and mounted in or on a suitable holder. In a derivation of limit of detection, they showed that the minimum amount of quartz that can be detected using a conventional diffractometer is below 1 μg! Other types of small samples mentioned by them were thin films and small inclusions which cannot be extracted from their matrix. The former include both filter-mounted airborne dust samples and conventional electronic thin films and encompass a large amount of diffractometry literature which will not be repeated here. The latter samples are usually examined with microbeam techniques, such as the Rigaku microdiffractometer. This very specialized area also will not be discussed here.

In examination of very small samples, there are several important parameters that must be optimized. These include maximizing intensity, decreasing the background count rate, and ensuring that acceptable particle statistics are obtained. Generally there is little one can do to increase intensity for a given phase other than to increase the incident power or increase the diffracting sample volume. In order to determine a balance between high diffracted intensity and low background, Zevin and Zevin (1987) derived a relationship between peak intensity and the irradiated surface area and mass of the sample, the sample mass absorption coefficient, and diffraction angle. For optimum peak intensity-background count rates, they formulated a relationship where S is controlled so that

$$S = 10G\mu^*/\sin\theta , \qquad (9)$$

where S is the irradiated surface area of the sample (cm^2), G is the sample mass, μ^* is the sample mass absorption coefficient, and θ is one-half the diffraction angle (degrees 2θ). Use of <u>irradiated</u> sample areas determined from this equation should maximize diffracted intensity and minimize background count rates. Spreading a sample over a greater area than specified by Equation 9 will yield a nearly linear increase in background intensity and almost no increase in diffracted peak intensity. Decreasing the sample area (i.e., increasing sample thickness) increases absorption and decreases the diffracted intensity. For the example given by Zevin and Zevin (1987), the optimum sample area for 10 µg of quartz with Cu $K\alpha$ radiation is ~1.5 mm^2.

In addition to optimizing the sample area, there are several other steps than can be taken to obtain accurate intensities and decrease the background count rate. As pointed out by many authors (e.g., Klug and Alexander, 1974), one of the most important factors in obtaining accurate and reproducible intensities is the attainment of good particle statistics. In the examination of very small amounts of sample, particle statistics become even more important, and samples must generally be reduced in particle size to below 5 µm. de Wolff et al. (1959) described the relationship between the relative root mean square (rms) deviation in intensity, σ, and a number of instrumental and sample parameters:

$$\sigma = 4R[(\sin\theta)/(m\ w\ h\ N_{eff})]^{1/2} , \qquad (10)$$

where R = goniometer radius, m = multiplicity factor, w = $\varepsilon R + [(w_f+w_s)/2]$ (for integrated intensities), ε = the rocking angle of crystallites, w_f = the apparent width of the focal spot, w_s = width of the receiving slit, h = $(h_f+h_s)/2$, h_f = the length of the focal spot, h_s = the length of the receiving slit, and N_{eff} = the effective number of irradiated crystallites. N_{eff} is related to the volume concentration of the diffracting phase, the area of the primary beam at the axis of rotation, the linear absorption coefficient of the sample, and the effective crystallite volume. From these relationships, it is obvious that as one reduces the size of the irradiated area and the sample volume, σ will increase. The best way to reduce σ while keeping the irradiated area the same as the sample area is to reduce the particle size. Zevin and Zevin (1987) concluded that grinding a 10-µg sample to 2 µm will give particle statistics adequate for qualitative identifications.

In order to keep background counts to a minimum, it is important to collimate the primary beam so that only the sample area is illuminated and background counts are not generated from substrate outside the sample area. In practice this may be difficult, due to the small illuminated areas required with small samples. The nature of the substrate also is crucial to reducing background count rates. Typical holders, such as those made of plastic, glass, or stainless steel should be avoided because of the contribution these holders make to the background, either from amorphous scattering or fluorescence. Figure 2 shows diffraction patterns from a glass slide and a plastic mount supplied by an

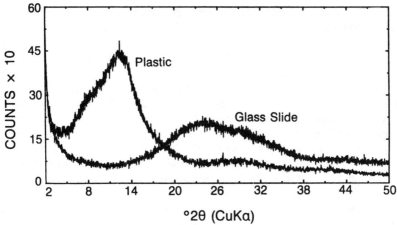

Figure 2. Diffraction patterns from a glass and a plastic mount.

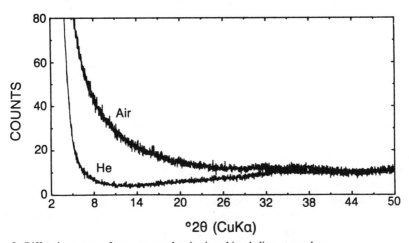

Figure 3. Diffraction patterns from a quartz plate in air and in a helium atmosphere.

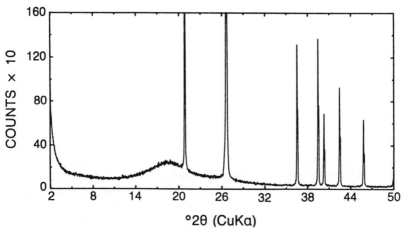

Figure 4. Diffraction patterns from a polycrystalline quartz specimen mounted in a cavity and covered by a Kapton film.

instrument manufacturer. Both of these substrates clearly give rise to high background counts and should be avoided for anything but infinitely thick samples.

It has been recognized for some time that a single-crystal substrate, cut so that the angle between the surface and any diffracting crystallographic plane is at least several degrees, makes an ideal substrate for small or transparent samples. Buerger and Kennedy (1958) described the use of BT-cut quartz single crystals as low-background substrates. BT-cut quartz crystals are cut at an angle of 49° from the c axis, sub-parallel to the (101) face, and they yield the weak 203 reflection at 68.16° $2\theta_{Cu}$. Quartz crystals cut 6° off the c axis toward [110] are available commercially through the Gem Dugout, State College, Pennsylvania. Such single-crystal "zero"-background plates give a relatively flat but non-zero background; in fact they give rise to significant background count rates, primarily due to air scatter at low angles and incoherent (Compton) scattering and thermal-diffuse scattering at higher angles. It is possible to verify experimentally that materials of higher atomic number have significantly lower background count rates due to a decrease in incoherent scattering. Thus, it is probable that appropriate single-crystal substrates of higher atomic number than quartz have a more nearly "zero" background. Figure 3 compares diffraction patterns obtained from a Gem Dugout quartz plate in air and in a helium-containing sample cell. This figure illustrates the considerable reduction in background intensity achieved with the quartz plate compared with the glass slide (~12 counts/s versus ~190 counts/s with the glass plate at 26° 2θ), and it also emphasizes the significant amount of air scatter at low angles. Examining a sample in a vacuum has a similar effect on the low-angle scattering and in some cases enables one to see clearly reflections that were not visible when examined in air. The subject of examining samples in non-ambient atmospheres is related to experiments in which air-sensitive materials are to be examined. Such experiments should preferably be done in an environmental cell. One should avoid covering a powder sample with any surface film to minimize interaction of the sample with air. Figure 4 shows the diffraction pattern of a polycrystalline quartz sample mounted in a cavity and covered by a thin sheet of Kapton, a polymer similar to Mylar. Because the polymer film is directly on the powder sample and is in a position to diffract X-rays, it gives rise to significant amorphous scattering, and absorption by the film causes powder diffraction intensities to decrease with diminishing 2θ.

Gude and Hathaway (1961) described an alternate method for mounting small samples. They mounted small amounts of powder on or between thin membranes of material such as collodion and supported the membrane on an aluminum holder. Although they were not able to compare their results with those obtained using BT-cut quartz plates, diffraction patterns of the bare collodion mounts had lower background count rates than those from an AT-cut quartz plate. AT-cut quartz crystals are cut 35° 15' from the c axis and yield the 101 reflection at 26.67° $2\theta_{Cu}$.

Both of these sample-mounting methods potentially share the problem of mounting the small powder sample to the substrate. Powder may stick to the collodion mount, but often a sticky mounting medium must be used with the single-crystal substrates. Small amounts of powder should preferably be deposited on the quartz plates from a slurry of the sample in a volatile liquid such as water or acetone. If this is not possible, the choice of mounting medium should be considered carefully. Figure 1 illustrates the high backgrounds resulting from a variety of mounting media. Even small amounts of common oils or greases yield high, nonuniform backgrounds. Clearly, the choice of mounting media is as crucial as the choice of substrate for small, transparent samples.

Another method for obtaining diffraction patterns from small amounts of material was discussed by Rassineux et al. (1988). Rather than rely on optimizing sample mounting parameters, these authors used a position-sensitive detector (PSD) to improve dramatically the counting statistics. The PSD detected an angular range of 14° 2θ so that useful portions of diffraction patterns could be obtained without scanning the detector. The obvious advantage of this detector is that diffraction data can be obtained for a 14° 2θ interval in 60 min comparable to data that would require 700 h using 0.02° 2θ increments in the step-scan mode (700 steps of 60 min count each). Rassineux et al. (1988) illustrated a very good diffraction pattern obtained from only 1 µg of corrensite in 60 min. Foster and Wolfel (1988) used a similar method to perform quantitative analyses of 1- to 5-mg samples with a transmission diffractometer. The primary disadvantage to the use of a PSD in diffractometry is the high initial cost of a PSD system, and these detectors generally require more routine maintenance than a conventional scintillation detector. In addition, a PSD system can suffer from the presence of fluorescence and $K\beta$ radiation; an incident-beam monochromator can eliminate only the latter. The conversion of a diffractometer to the PSD mode may make the instrument unsuitable for the analysis of highly oriented clay aggregates. Visualize a random powder specimen that is not rotated during the diffraction experiment. Crystallites that produce, say, the 001 and 002 reflections from a mica in the powder are different crystals that have different but suitable orientations for the two reflections. Thus the diffraction geometry is correct for a random powder. But for an oriented aggregate that normally is rotated at half of the angular rate of the detector, the same crystallites produce all of the 00l series. If the sample is not rotated but remains sensible fixed as it is for the PSD application, only one crystal orientation and corresponding 00l reflection can be optimal and the other 00l reflections will be weak or absent.

This potential undesirable characteristic of the PSD geometry is minimized as the preferred orientation becomes more nearly random. To date, however, we know of no published systematic and quantitative evaluation of this effect. Despite these potential shortcomings, a combination of the methodology recommended by Zevin and Zevin (1987) and a PSD appears great to analyze microscopic samples by diffractometry.

REFERENCES

Bacon, G. E., Bacon, P. J., and Griffiths, R. K. (1979) The orientation of apatite crystals in bone. J. Appl. Crystallogr. 12, 99-103.

Baldock, P. J. and Parker, A. (1973) X-ray powder diffractometry of small (20mg to 1µg) samples using standard equipment. J. Appl. Crystallogr. 6, 153-157.

Barshad, I. (1960) X-ray analysis of soil colloids by a modified salted paste method. Clays & Clay Minerals, Seventh National Conf., Earl Ingerson, ed., Pergamon Press, London, 350-364.

Bish, D. L. and Von Dreele, R. B. (1989) Rietveld refinement of non-hydrogen atomic positions in kaolinite. Clays & Clay Minerals 37, 289-296.

Bloss, F. D., Frenzel, G., and Robinson, P. D. (1967) Reducing orientation in diffractometer samples. Am. Mineral. 52, 1243-.

Brindley, G. W. (1945) The effect of grain or particle size on X-ray reflections from mixed powders and alloys considered in relation to the quantitative determination of crystalline substances by X-ray methods. Phil. Mag. 36, 347-369.

Brindley, G. W. (1980) Quantitative X-ray mineral analysis of clays, Ch. 7 in Crystal Structures of Clay minerals and their X-Ray Identification (G. W. Brindley and G. Brown, Ed.) Monograph No. 5, Mineralogical Soc., London.

Brown, G. and Brindley, G. W. (1980) X-ray diffraction procedures for clay mineral identification, Ch. 5 in Crystal Structures of Clay minerals and their X-Ray Identification (G. W. Brindley and G. Brown, Ed.) Monograph No. 5, Mineralogical Soc., London.

Buerger, M. J. and Kennedy, G. C. (1958) An improved specimen holder for the focusing-type X-ray spectrometer. Am. Mineral. 43, 756-757.

Calvert, L. D. and Sirianni, A. F. (1980) A technique for controlling preferred orientation in powder diffraction samples. J. Appl. Cryst. 13, 462.

Calvert, L. D., Sirianni, A. F., Gainsford, G. J., and Hubbard, C. R. (1983) A comparison of methods for reducing preferred orientation. Adv. X-ray Anal. 26, 105-110.

Cullity, B. D. (1978) Elements of X-ray Diffraction: Addison-Wesley, Reading, 555 pp.

Davis, B. L. (1986) A tubular aerosol suspension chamber for the preparation of powder samples for X-ray diffraction analysis. Powd. Diff. 1, 240-243.

Davis, B. L. and Cho, N.-K. (1977) Theory and application of X-ray diffraction compound analysis to high-volume filter samples. Atmosph. Environ. 11, 73-85.

Davis, B. L. and Johnson, L. R. (1982) Sample preparation and methodology for X-ray quantitative analysis of thin aerosol layers deposited on glass fiber and membrane filters. Adv. X-ray Anal. 25, 295-300.

de Wolff, P. M. (1947) A theory of X-ray absorption in mixed powders. Physica 13, 62-78.

de Wolff, P. M., Taylor, J. M., and Parrish, W. (1959) Experimental study of effect of crystallite size statistics on X-ray diffractometer intensities. J. Appl. Phys. 30, 63-69.

Drever, J. I. (1973) The preparation of oriented clay mineral specimens for X-ray diffraction analysis by a filter-membrane peel technique. Am. Mineral. 58, 553-554.

Flörke, O. W. and Saalfeld, H. (1955) Ein verfahren zur herstellung texturfreier Rontgen-Pulverpraparate. Z. Krist. 106, 460-466.

Foster, B. A., and Wolfel, E. R. (1988) Automated quantitative multiphase analysis using a focusing transmission diffractometer in conjunction with a curved position sensitive detector. Adv. X-ray Anal. 31, 325-330.

Frevel, L. K. and Roth, W. C. (1982) Semimicro assay of crystalline phases by X-ray powder diffractometry. Anal. Chem. 54, 677-682.

Gibbs, R. J. (1965) Error due to segregation in quantitative clay mineral X-ray diffraction mounting techniques. Am. Mineral. 50, 741-751.

Gonzalez, C. R. (1987) General theory of the effect of granularity in powder X-ray diffraction. Acta Crystallogr. A43, 769-774.

Gude, A. J., 3rd, and Hathaway, J. C. (1961) A diffractometer mount for small samples. Am. Mineral. 46, 993-998.

Hermann, H. and Ermrich, M. (1987) Microabsorption of X-ray intensity in randomly packed powder specimens. Acta Crystallogr. A43, 401-405.

Hughes, R. and Bohor, B. (1970) Random clay powders prepared by spray drying. Am. Mineral. 55, 1780-1786.

Jackson, M. L. (1969) Soil Chemical Analysis-Advanced Course: Published by the author, Madison, Wisconsin, 53705, 895 pp.

James, R. W. (1965) The Optical Principles of the Diffraction of X-Rays: Cornell Univ. Press, Ithaca, New York, 664 pp.

Jenkins, R., Fawcett, T. G., Smith, D. K., Visser, J. W., Morris, M. C., and Frevel, L. K. (1986) JCPDS - International Centre for Diffraction Data sample preparation methods in X-ray powder diffraction. Powd. Diff. 1, #2, 51-63.

Jonas, E. C. and Kuykendall, J. R. (1966) Preparation of montmorillonites for random powder diffraction. Clay Minerals 6, 232-235.

Kinter, E. G. and Diamond, S. (1956) A new method for preparation and treatment of oriented-aggregate specimens of soil clays for X-ray diffraction analysis. Soil Sci. 81, 111-120.

Klug, H. P. and Alexander, L. E. (1974) X-ray Diffraction Procedures for Polycrystalline and Amorphous Materials: Wiley, New York, 966 pp.

Moore, D. and Reynolds, R. C. (1989) X-ray Diffraction and the Identification and Analysis of Clay Minerals: Oxford University Press, Oxford, 332 pp.

Nakamura, T. Sameshima, K., Okunaga, K., Sugiura, Y., and Sato, J. (1989) Determination of amorphous phase in quartz powder by X-ray powder diffractometry. Powd. Diffraction 4, 9-13.

Parrish, W. and Huang, T. C. (1983) Accuracy and precision of intensities in X-ray polycrystalline diffraction. Adv. X-ray Anal. 26, 35-44.
Parrish, W., Mack, M, and Taylor, J. (1966) Determination of apertures in the focusing plane of X-ray powder diffractometers. J. Sci. Instrum. 43, 623-628.
Rassineux, F., Beaufort, D., Bouchet, A., Merceron, T., and Meunier, A. (1988) Use of a linear localization detector for X-ray diffraction of very small quantities of clay minerals. Clays & Clay Minerals 36, 187-189.
Reynolds, R. C. (1968) Effect of particle size on apparent lattice spacings. Acta Crystallogr. A24, 319-320.
Reynolds, R. C. (1986) The Lorentz factor and preferred orientation in oriented clay aggregates. Clays & Clay Minerals 34, 359-367.
Ross, M. (1968) X-ray diffraction effects by non-ideal crystals of biotite, muscovite, montmorillonite, mixed-layer clays, graphite, and periclase. Z. Krist. 126, 80-97.
Smith, D. K. and Barrett, C. S. (1979) Special handling problems in X-ray diffractometry. Adv. X-ray Anal. 22, 1-12.
Smith, S. T., Snyder, R. L., and Brownell, W. E. (1979a) Minimization of preferred orientation in powders by spray drying. Adv. X-ray Anal. 22, 77-87.
Smith, S. T., Snyder, R. L., and Brownell, W. E. (1979b) Quantitative phase analysis of Devonian shales by computer controlled X-ray diffraction of spray dried samples. Adv. X-ray Anal. 22, 181-191.
Tien, Pei-Lin (1974) A simple device for smearing clay-on-glass slides for quantitative X-ray diffraction studies. Clays & Clay Minerals 22, 367-368.
Walker, J. R. and Bish, D. L. (1989) Rietveld refinement of IIb chlorite. in Prog. and Absts., 26th Annual Meeting of the Clay Minerals Society, Sacramento, California.
Warren, B. E. (1969) X-ray Diffraction: Addison-Wesley, Reading, 381 pp.
Wilchinsky, Z. W. (1951) Effect of crystal, grain, and particle size on X-ray power diffracted from powders. Acta Crystallogr. 4, 1-9.
Woolfson, M. M. (1970) An Introduction to X-ray Crystallography: Cambridge Univ. Press, London, 380 pp.
Yukino, K. and Uno, R. (1986) "ε-scanning"--A method of evaluating the dimensional and orientational distribution of crystallites by X-ray powder diffractometer. Jap. J. Appl. Phys. 25, 661-666.
Zevin, L. S., and Zevin, I. M. (1987) X-ray diffractometry of low-mass samples. Powd. Diff. 2, 78-81.

5. Quantitative Analysis

R.L. Snyder & D.L. Bish

Quantitative phase analysis by X-ray powder diffraction (XRD)[1] dates back to 1925 when Navias (1925) quantitatively determined the amount of mullite in fired ceramics, and to 1936 when Clark and Reynolds (1936) reported an internal standard method for mine dust analysis by film techniques. The development of the Geiger-counter diffractometer in 1945 by Parrish led to increased possibilities, and in 1948 Alexander and Klug (1948) presented the theoretical background for the absorption effects on diffraction intensities from a flat brickette of powder. Since then there have been numerous methods developed based on their basic equations. Methods applicable to a wide range of phases and samples include the method of standard additions (also called the addition of analyte or spiking method) (Lennox, 1957), the absorption diffraction method (Alexander and Klug, 1948; Smith et al., 1979b), and the internal-standard method (Klug and Alexander, 1974). Formalisms have been established to permit inclusion of overlapping lines (Copeland and Bragg, 1958) and chemical constraints (Garbauskas and Goehner, 1982) in the analysis.

Specific mineralogical examples of quantitative phase analysis by XRD are not numerous in the literature, but several successful applications have been published. Brindley (1980) provided an excellent summary of quantitative analysis of clay minerals using X-ray powder diffraction, and the methods and caveats described therein can be equally well applied to any mineral system. Quite often, quantitative analyses have been directed at determining the amounts of only one or two phases, such as the analyses of Parker (1978) for analcime in pumice. Carter et al. (1987) used either a an internal standard or the measurement of the mass absorption coefficient to determine the amounts of quartz and cristobalite in bentonites. Their results appeared excellent, with lower limits of detection of 0.01% for quartz and 0.03% for cristobalite and absolute errors less than 0.35% for both phases. In another analysis for only two phases, Boski and Herbillon (1988) determined the amounts of hematite and goethite in bauxites using a combination of XRD and heating methods.

Comprehensive whole-rock analyses analogous to modal analyses were performed by Davis and Walawender (1982) who compared XRD reference intensity ratio analyses to optical results. They obtained good agreement between the two methods for most rocks analyzed, demonstrating that it is possible to obtain accurate modal analyses quickly by XRD. Pawloski (1985) applied the Chung (1974b) external-standard method to volcanic tuffs from the Nevada Test Site, and she obtained reproducible results for the fine-grained rocks. Maniar and Cooke (1987) performed modal analysis of natural and artificial granitic rocks using the quartz present in all samples as an internal standard. They obtained rapid modal analyses with average standard deviations of 7%. Clearly, although quantitative analysis by

---

[1] Sections of this chapter were abstracted from "NBS*QUANT84: A System for Quantitative Analysis by Automated X-ray Powder Diffraction", NBS Special Publication, (1984) by Robert L. Snyder and Camden R. Hubbard (Oak Ridge National Laboratory) and an article entitled "Reference Intensity Ratio - Measurement and Use in Quantitative XRD", by C. R. Hubbard and R. L. Snyder published in the Journal of Powder Diffraction 3, 74 (1988).

XRD has not been widely used to obtain modal analysis of rocks, the methods have the potential for quickly and accurately providing mineralogical information.

Quantitative analysis using any method is a difficult undertaking. It requires careful calibration of the instrument using carefully prepared standards and many repetitions during the set up-phase to establish the required technique. In general quantitative analysis of a new phase system will require a minimum of a few days and often a week of setup time. After the technique and standards have been established, a well-designed computer algorithm such as the NBS*QUANT84 system (Snyder et al., 1981; Snyder and Hubbard, 1984) can produce routine analyses in as little as 15 minutes to an hour per sample, depending on the precision desired.

Perhaps it is the difficult nature of quantitative analysis that has limited the number of carefully done literature examples. This in turn has resulted in a concentration of effort on a specific analysis technique rather than on understanding the sample and instrument-dependent limitations of quantitative phase analysis by X-ray powder diffraction.

The purpose of this chapter is to summarize the basis for quantitative analysis by the most popular methods including the most recent developments using Reference Intensity Ratios and the Rietveld refinement. An emphasis will be placed on developments which have occurred since the publication of the "the bible" by Klug and Alexander (1974).

THE INTERNAL-STANDARD METHOD OF QUANTITATIVE ANALYSIS

The internal-standard method is the most general of any of the methods for quantitative phase analysis. In addition this method lends itself most easily to generalization into the RIR (Reference Intensity Ratio) or Chung Method (1974a,b) and even further can be generalized into a system of linear equations which allow for the use of overlapped lines and chemical analysis constraints. For this reason we will place our greatest emphasis on this method, simply outlining the principals of the other techniques.

Background

The intensity of a diffraction line i from a phase α from a flat plate is given by

$$I_{i\alpha} = \frac{K_{i\alpha} X_\alpha}{\rho_\alpha (\frac{\mu}{\rho})_m}, \qquad (1)$$

where:

- X_α = weight fraction of phase α
- ρ_α = density of phase α
- $(\frac{\mu}{\rho})_m$ = mass absorption coefficient of the mixture

- $K_{i,\alpha}$, is a constant for a given crystal structure α, diffraction line i and set of experimental conditions. This constant is modeled exactly as:

$$K_{i\alpha} = \frac{I_o \lambda^3 e^4}{32\pi r m_e^2 c^4} \frac{M}{2V_\alpha^2} |F_{i\alpha}|^2 \left(\frac{1+\cos^2 2\theta}{\sin^2\theta \cos\theta}\right), \qquad (2)$$

where:

- I_o = incident-beam intensity
- r = radius of the diffractometer (i.e., sample to detector distance)
- λ = wavelength of the X-radiation
- c = speed of light
- e and m_e = charge and mass of an electron
- M = Multiplicity for reflection i
- V_α = volume of the unit cell of phase α
- $\left(\frac{1+\cos^2 2\theta}{\sin^2\theta \cos\theta}\right)$ = the Lorentz and polarization corrections for a diffractometer.
- $F_{i\alpha}$ = the structure factor for reflection i which relates the intensity to the crystal structure.

$$F_{hk\ell} = \sum_{j=1}^{\#of\ atoms\ in\ cell} f_j \exp 2\pi i(hx_j + ky_j + \ell z_j),$$

where:

* $hk\ell$ = Miller indices of line i of phase α
* x_j, y_j, z_j = fractional coordinates for atom j of phase
* f_j = the atomic scattering factor for atom j:

$$f_j = f_{corr} \exp -B_j \sin^2(\frac{\theta}{\lambda})^2,$$

where:

· f_{corr} = The atomic scattering factor corrected for anomalous dispersion namely

$$|f_{corr}|^2 = (f_o + \Delta f')^2 + (\Delta f'')^2.$$

· B_j = Debye-Waller temperature factor for atom i in phase α
· θ = Bragg diffraction angle

When more than two components are present in an unknown (whether they are identified phases or not) and when the $(\frac{\mu}{\rho})$ of the unknown cannot be computed (i.e., the chemical composition is not known) then we resort to the most general quantitative analysis technique–the internal-standard method. This method applies only to powders, in that a known weight percent of a standard material must be homogeneously mixed with the sample of unknown phase composition.

The internal-standard method is based on elimination of the matrix absorption effects factor $(\frac{\mu}{\rho})_m$ by computing the ratio

$$\frac{I_{i\alpha}}{I_{js}},$$

the intensity for line i of phase α divided by the intensity for line j of the internal standard s. This ratio of the intensities comes from dividing two equations of type 1 and gives Equation 3, linear in the weight fraction of phase α:

$$\frac{I_{i\alpha}}{I_{js}} = K' \frac{X_\alpha}{X_s}. \quad (3)$$

Equation 3 permits the analysis of samples with addition of an arbitrarily selected, but known, amount of internal standard. To establish a calibration curve, $X_s \frac{I_{i\alpha}}{I_{js}}$ is plotted versus X_α. The slope of this straight line is K'. Note that X_α represents the weight fraction of phase α in the mixture of original sample after it was diluted by the addition of the internal standard. The weight fraction of phase α in the original sample before dilution, X_α, is given by

$$X'_\alpha = \frac{X_\alpha}{1 - X_s}. \quad (4)$$

For routine analyses the weight fraction of the internal standard is specified in the procedure. For example, a laboratory quantitative XRD procedure may call for adding 0.1 gm of internal standard to 0.9 gm of sample. In such cases X_s and $(1 - X_s)$ are constants of the procedure and Equation 3 can be simplified to

$$\frac{I_{i\alpha}}{I_{js}} = K'' X_\alpha. \quad (5)$$

Use of a pre-established calibration curve defined by the slope K'' yields the weight fraction of phase α in the original mixture. Note that K'' is a function of i, j, α, and s for a constant weight fraction of standard.

Systematic errors which vary with X_s would not be detected in a calibration using Equation 5. Both micro-absorption (Cline and Snyder, 1987) and preferred orientation (Cline and Snyder, 1985) can be dependent on X_s. Thus, for improved reliability, use of several mixtures of the internal standard with a calibration sample, each with a different X_s, is recommended to determine K'. The reduction of preferred-orientation effects can also be achieved by measuring several diffraction lines from phase α and phase s.

The generalized Reference Intensity Ratio

The use of multiple lines from each phase can be accommodated by independently determining K' (or K'') for each analyte line-internal standard line pair. If three analyte lines and two standard lines were to be measured, a total of 6 calibration constants would be required. A more general approach which avoids introduction of multiple calibration constants, is to use the relative intensities I^{rel} and the Reference Intensity Ratio, RIR. This more general equation is

$$\frac{I_{i\alpha}}{I_{js}} \frac{I^{rel}_{js}}{I^{rel}_{i\alpha}} \frac{X_s}{X_\alpha} = K = RIR_{\alpha,s}. \tag{6}$$

In Equation 6 (Hubbard and Snyder, 1988) K is a function of α and s but not i, j, or X_s.

In the example above (where 3x2 lines were measured) the RIR and five I^{rel} values are required. However, since one line of the analyte and one of the internal standard can be assigned $I^{rel} = 100$, a total of only 4 calibration constants is required. An added benefit is that the I^{rel} constants are readily understood by the user and are frequently used in other applications. It should be noted, however, that the relative intensities in the Powder Diffraction File (PDF) are seldom accurate enough for this application. The user should accurately determine the relative intensities in question or, more simply, allow calculation of them from the calibration sample measurements. Once this calibration constant K is determined it need not be redetermined before each analysis since variations in the incident-beam intensity cancel in the $\frac{I_{i\alpha}}{I_{js}}$ ratio just as $\left(\frac{\mu}{\rho}\right)_m$ is eliminated.

The slope K of the calibration curve defined by Equation 6 is theoretically a universal constant relating the scattering power of phase α to that of phase s. K is known as the Reference Intensity Ratio or RIR (Chung, 1974a,b; Hubbard et al. (1976); Hubbard and Snyder, 1988). To specify which phase and which internal standard are referred to, a pair of subscripts (α, s) are used. Requirements on measurement procedures to obtain accurate RIR values are:

- Freedom from preferred orientation, extinction, and microabsorption,
- Constant irradiated *volume* of sample independent of scattering angle,
- Correction for monochromator polarization effects,
- Use of integrated intensities.

Of course the calibration sample (α) must be well characterized chemically and represent the chemistry of the analyte. Non-interfering second phases, including amorphous phases, may be present; however, the concentration of α and s must be known exactly. If calibration and analytical data are to be collected on a single instrument, then the requirements for measuring the K value are reduced to:

- Reproducible orientation, extinction (i.e., particle size) (Cline and Snyder, 1987), and microabsorption effects for calibration standard and unknown
- Reproducible irradiated volume of sample for each 2θ
- Measurement of peak or integrated intensities in a reproducible way.

It is clear from the general definition of the Reference Intensity Ratio given in Equation 8 that this value can be measured for any substance using any convenient diffraction line of both phase α and the internal standard when the two substances are mixed in a known weight ratio. The RIR, when accurately measured, is a true constant and allows comparison of the absolute diffraction line intensities of one material to another. It also enables quantitative phase analysis in a number of convenient and useful ways.

$I/I_{corundum}$

A particularly important case is when corundum is chosen as the reference material and integrated intensities are used. In this case the RIR is referred to as "I over I corundum" or more simply, I/I_c. The definition of I/I_c was first proposed by Visser and deWolff (1964). It was proposed as a way of comparing the reflectivity of phases by urging users to agree to use corundum as a common comparison reference. The above equations for the Reference Intensity Ratio simply generalize the I/I_c definition. The PDF contains I/I_c values for over 2500 phases which are published in the Search Manuals and on the PDF cards when they are known. Nearly all have been determined using peak-height ratios as an approximation for integrated-intensity ratios or are calculated using reported atomic position and thermal parameters. Although the experimental values are subject to measurement and sample-related errors (Cline and Snyder, 1983; Gehringer et al., 1983) they are useful in obtaining a quick estimate of the concentration of the phases in an unknown. However, unless the relative intensity and the RIRs are carefully determined by the user, the analysis based on tabulated I/I_c should be referred to as semi-quantitative.

Measurement of I/I_c

By definition $(I/I_c)_\alpha$ is the ratio of the integrated intensities of the strongest line $(I_{hk\ell}^{rel} = 100)$ of phase α to the strongest line of corundum for a 1:1 mixture by weight using Cu Kα radiation. With this definition it is easy to measure I/I_c. One makes the 1:1 mixture, measures the integrated intensities, and directly calculates I/I_c. For even quicker results the ratio of peak-height intensities can be used to approximate the ratio of integrated intensities. However the peak-height approximation should only be used when the grain size is small and all the line breadths are approximately equal.

The simple 2-line procedure for measuring I/I_c is quick but suffers from several drawbacks. Preferred orientation commonly affects the observed intensities unless careful sample preparation methods are employed. Other problems include extinction, inhomogeneity of mixing, and variable crystallinity of the sample due to its synthesis and history. All of these effects conspire to make the published values of I/I_c subject to substantial error. For greater accuracy Hubbard and Smith (1976) recommended using multiple lines from both the sample (α) and the reference phase (corundum, in this case) and using multiple sample mounts. Such an approach often reveals when preferred orientation is present and provides realistic measurement of the reproducibility in the measurement of I/I_c. Although there is a method for assessing the quantitative extent of preferred orientation normal to the sample surface (Snyder and Carr, 1974), it is never recommended to use a computational procedure in place of proper experimental technique. In general spray drying the samples (Smith et al., 1979a,b; Cline and Snyder, 1985) is the best way to eliminate this serious source of error. In the spray-drying process the crystallites of the powder sample are agglomerated into spheres. The crystallites spread on the surface of these spheres and ideally expose all crystallographic surfaces to the X-ray beam as described in Chapter 4.

Quantitative analysis with I/I_c–the RIR or Chung method

Analysis when a known amount of corundum is added to an unknown. In this case corundum is the internal standard. If we rearrange Equation 6, we get

$$X_\alpha = \frac{I_{i\alpha}}{I_{jc}} \frac{I^{rel}_{jc}}{I^{rel}_{i\alpha}} \frac{X_c}{RIR_{\alpha,c}} = \frac{I_{i\alpha}}{I_{jc}} \frac{I^{rel}_{jc}}{I^{rel}_{i\alpha}} \frac{X_c}{\left(\frac{I}{I_c}\right)_\alpha}. \tag{7}$$

The use of an $(I/I_c)_\alpha$ value from the PDF allows us to avoid preparing our own standards to determine this calibration constant. When the quantitative analysis of one or more phases with known (I/I_c)s is desired, we simply have to add a known amount of corundum to the sample and use Equation 7. Analysis is possible for X_α using any combination of diffraction lines from phase α and corundum. This use of RIR_α constants (i.e., I/I_c values) and possibly I^{rel} values from the PDF should be considered as semi-quantitative due to the unspecified errors in the constants. However, if I^{rel} and RIR values are accurately known and other systematic errors are absent, Equation 7 can yield accurate quantitative results.

Analysis when a known amount of an internal standard is added to a sample. From Equation 6 it is readily shown that

$$RIR_{\alpha,s} = \frac{RIR_{\alpha,c}}{RIR_{s,c}} = \frac{\left(\frac{I}{I_c}\right)_\alpha}{\left(\frac{I}{I_c}\right)_s}. \tag{8}$$

Hence, we can combine I/I_c values for phases α and s to obtain the Reference Intensity Ratio for phase α relative to phase s. Taking s to be an internal standard and substituting Equation 8 into Equation 7 and rearranging we have

$$X_\alpha = \frac{I_{i\alpha}}{I_{js}} \frac{I^{rel}_{js}}{I^{rel}_{i\alpha}} \frac{RIR_{s,c}}{RIR_{\alpha,c}} X_s. \tag{9}$$

This equation is quite general and allows for the analysis of any crystalline phase with a known RIR in an unknown mixture by the addition of a known amount of any internal standard. However, if all four of the required constants ($I^{rel}_{i\alpha}, I^{rel}_{js}, RIR_{s,c}$ and $RIR_{\alpha,c}$), each of which may contain significant error, are taken from the literature the results should be considered as only semi-quantitative.

Equation 9 is valid even for complex mixtures which contain unidentified phases, amorphous phases, or identified phases with unknown $RIR'{}^s$. The ratio of the weight fractions of any two phases whose $RIR'{}^s$ are known may always be computed by choosing one as X_α and the other as X_s. That is

$$\frac{X_\alpha}{X_\beta} = \frac{I_{i\alpha}}{I_{j\beta}} \frac{I^{rel}_{j\beta}}{I^{rel}_{i\alpha}} \frac{RIR_{\beta,c}}{RIR_{\alpha,c}}. \tag{10}$$

<u>Analysis of a mixture of identified phases without the addition of an internal standard–The RIR or Chung method.</u> Chung (1974a,b; 1975) was the first to point out that if all the phases in a mixture are identified and the RIR value is known for each phase, then an additional equation holds:

$$\sum_{k=1}^{n} x_k = 1, \tag{11}$$

where n is the number of phases in the mixture. Equation 11 permits analysis without adding any standard to the unknown specimen. This is seen by arbitrarily choosing one of the phases in the mixture as s and then using Equation 10 to solve for the ratio of the weight fraction of each phase to the weight fraction of s namely

$$\frac{X_\alpha}{X_s}, \frac{X_\beta}{X_s}, etc.$$

For n phases in the mixture this produces n-1 ratios. The relationship given in Equation 11,

$$X_s + X_\alpha + X_\beta + ... = 1,$$

allows the writing of the nth equation. This system of n linear equations can be solved to produce the weight fractions of the n unknown phases from the following equation:

$$X_\alpha = \frac{I_{i\alpha}}{RIR_\alpha I^{rel}_{i\alpha}} \left[\frac{1}{\sum_{j=1}^{\#of\ phases} \frac{I_{j\alpha}}{RIR_j I^{rel}_{j\alpha}}} \right]. \tag{12}$$

It is important to note that the presence of any amorphous or unidentified crystalline phase invalidates the use of Equation 12. Of course, any sample containing unidentified phases cannot be analyzed using the Chung method in that the required $RIR'{}^s$, will not be known.

When the RIR values are known from another source, for example published I/I_c values, it may be tempting to use them along with the I^{rel} values on the PDF card to perform a completely standardless quantitative analysis using the Chung Equation 11. One should resist this temptation! For any reasonable accuracy it is strongly recommended that before using the RIR or Chung Method, pure specimens of the phases to be determined be obtained and an internal standard added to each of them. These samples should be carefully measured to determine accurate values for the $I^{rel's}$ and $RIR's$ and only these values should be used in the analysis.

Constrained XRD phase analysis–generalized internal-standard method

Although this chapter is devoted to quantitative analysis by X-ray powder diffraction, it is worthwhile to mention briefly the methods that use a combination of X-ray diffraction results and other observations, such as chemical analyses, cation exchange capacities, or thermal analyses. These methods, using linear programming, can be particularly valuable when other such observations are routinely available and when the systems being analyzed are chemically diverse. Chemical data are particularly appropriate for use with this method because they are often available and are usually accurate. These techniques are distinct from those that only use chemistry as an aid to identification (e.g., Garbauskas and Goehner, 1982).

Suppose that a sample containing only quartz (SiO_2), corundum (Al_2O_3), and hematite (Fe_2O_3) was analyzed by X-ray powder diffraction, yielding results of 30% quartz, 50% corundum, and 20% hematite. Obviously, such a mixture could equally well be analyzed chemically, giving accurate mineralogical results from the total chemical composition. Similarly, if X-ray diffraction data indicated that the sample contained calcite, quartz, and kaolinite, the absolute amounts of calcite and kaolinite could be obtained from their respective weight losses upon heating, and quartz could be determined by difference.

A quantitative analysis combining both the X-ray diffraction and chemical or thermal results with a knowledge of the composition of the individual phases can yield results of higher precision and accuracy than is generally possible with only one kind of observation. The analysis becomes more complex when several phases in a mixture have similar compositions and/or potential compositional variability, but it is possible with appropriate constraints during analysis to place limits on the actual compositions of the constituent phases.

The most general formulation of these ideas has been described by Copeland and Bragg (1958). Their internal standard equations for multicomponent quantitative analysis with possible line superposition and chemical constraints take the form of a system of simultaneous linear equations. Each equation is of the form

$$\frac{I_i}{I_n} = (\frac{C_{i1}}{C_{ns}})\frac{X_1}{X_s} + (\frac{C_{i2}}{C_{ns}})\frac{X_2}{X_s} + \ldots + (\frac{C_{ik}}{C_{ns}})\frac{X_k}{X_s} + \epsilon, \tag{13}$$

where

- I_i is the intensity of the ith line from the mixture of P phases ($1 < i \leq n$ and $1 \leq k \leq P$),

- I_n is the intensity of a *resolved* line of the internal standard s,
- X_k is the weight fraction of phase k in the mixture of sample plus internal standard,
- C_{ik} is a constant including the mass absorption coefficient of the mixture and the diffraction intensity of the ith line of phase k,
- ϵ is a least-squares error term.

Line overlap is allowed for by having each intensity ratio $(\frac{I_i}{I_n})$ have contributions from multiple lines from multiple phases. As many terms involving C_{ik} contributions to the intensity ratio as needed are included in each linear equation. From X_k and X_s, the weight fractions X_k of the original sample can be computed by

$$X_k = \frac{X_k}{1 - X_s}. \qquad (14)$$

Thus, to allow for mixtures of the sample having differing concentrations of internal standard, the matrix coefficients are:

$$\left(\frac{C_{ik}}{C_{ns}}\right)\frac{1}{X_s}\frac{1}{1 - X_s}. \qquad (15)$$

All these matrix coefficients must be known for one to determine the X values. The $\frac{C_{ik}}{C_{ns}}$ ratios are given by the product of the ratio of the corresponding relative intensities of the pure phases and the Reference Intensity Ratio (Hubbard et al., 1976; Hubbard and Snyder, 1988). That is

$$\left(\frac{C_{ik}}{C_{ns}}\right) = \frac{I_{ik}^{rel}}{I_{ns}^{rel}} RIR_{ks}. \qquad (16)$$

From this expression for $\frac{C_{ik}}{C_{ns}}$, the Copeland and Bragg matrix elements can readily be obtained from I^{rel} and RIR tabulations, calculations, and/or experimental measurements. If phase k has zero intensity at pattern line i then $I_{ik}^{rel} = 0$ and thus $\frac{C_{ik}}{C_{ns}} = 0$. In general $m > P$ and the system of equations can be solved by least-squares techniques where the sum of ϵ_i^2 is minimized.

Quantitative elemental data, obtained from X-ray fluorescence analysis for example, can be added to the system of equations in diffraction intensity (Garbauskas and Goehner, 1982) without increasing the number of unknowns. Coupling both the elemental and diffraction data will result in more accurate quantitative analysis of multicomponent mixtures. More importantly, estimates of the standard deviation of the results will be more accurate. The equation relating weight fraction in each phase to the overall weight fraction, expressed as a percentage for element A, is

$$(\%A)_{mixture} = (\%A)_1 X_1 + (\%A)_2 X_2 + \ldots + (\%A)_k X_k + \epsilon, \qquad (17)$$

where (%A) is the weight percent of element A in phase i. Such linear equations in $X_1, X_2, \ldots$ can be combined with Equation 13. The expanded system of equations is again solved by least-squares techniques. Both the line overlap and chemical analysis features of the Copeland-Bragg formalism have been included in the NBS*QUANT84 system (Snyder and Hubbard, 1984).

Other examples of linear programming and a combination of X-ray diffraction data and other observations have been described by several authors. One of the first applications (Hussey, 1972) used linear programming to aid in quantitative clay mineral analysis. He used the weight loss from 300° to 950° C, the Ca/Mg cation-exchange capacity (CEC), and the K/NH$_4$ CEC as observed properties. His method was described further by Johnson et al. (1985). Their analyses required that the number of individual properties (i.e., analyses for individual elements, weight losses, CEC$'^s$) equal the number of components being analyzed and that the minimum number of samples must equal the number of properties measured. In a typical analysis, one sets limits on the ranges of component properties, e.g., defining the maximum chemical variability a particular phase can exhibit. Using this methodology, they were able to calculate component proportions and component property values for soil clays and sediments. Obviously this is potentially a very valuable method because it can provide not only quantitative phase analyses but information on various properties of the individual components. Such property information is difficult if not impossible to obtain in other ways for fine-grained materials.

Braun (1986) used chemical information alone for a mixture of montmorillonite, plagioclase feldspar, quartz, and opal-cristobalite to estimate mineral abundances. As would be expected, he was able to determine the amounts of montmorillonite and feldspar but only the sum of quartz and opal-cristobalite since chemical data were the only observations used. The use of other data allowing discrimination between the isochemical phases quartz and opal-cristobalite, such as X-ray diffraction data, would facilitate a complete phase analysis. The complete presentation of the theory behind this method is beyond the scope of this chapter, and the interested reader is referred to the papers mentioned above and similar papers by Hodgson and Dudeney (1984) and Renault (1987).

THE INTERNAL-STANDARD METHOD – AN EXAMPLE

Since the internal-standard method is the most general and by far the most commonly used quantitative phase analysis procedure, we will outline an example quantitative analysis using it. In the example we will take full advantage of the equations previously described and use multiple lines in our analysis. A procedure that has been developed for collecting such data is the AUTO algorithm (Snyder et al., 1981,1982). The processing of these "intelligent" data is carried out by the NBS*QUANT84 analysis system (Snyder and Hubbard, 1984). These algorithms are implemented (or in the process of being implemented) by all of the major X-ray diffraction equipment manufacturers. Thus, giving our example in the conceptual framework of this system will serve two purposes; to give the details of an analysis and introduce the concepts of this widely used analysis system.

However, rather than introduce the method simultaneously with the line overlap and chemical constraints which can be incorporated, we will work through an example of the procedure as it is most often used. Hubbard et al. (1983) gave an example of a complex fly-ash analysis in which line-overlap constraints and Copeland-Bragg formalism are used.

Figure 1 illustrates the steps used in carrying out an internal standard analysis. Assume that we wish to analyze a sample containing only quartz and corundum in unknown concentrations. It of course does not matter to the internal-standard method that only these two phases are present. The presence of other phases would not affect the results. We will analyze each of the steps, and the considerations at each, that must be carried out to perform the analysis. An outline of the method is as follows:

Internal standard selection

In order to carry out an internal-standard analysis we must first choose an internal standard which has diffraction lines which do not interfere with the lines of the phases to be determined. All analyses should begin by collecting powder diffraction patterns for the unknown and for pure samples of the phases to be analyzed. In Figure 2 we see that the position of the peaks for Al_2O_3 and SiO_2 do not overlap the peaks of Si, and thus Si may be chosen as the internal standard.

Diffraction line selection

Next we must establish which diffraction lines are to be used for each phase, including the internal standard. We want to avoid peaks that overlap each other and would like to select the strongest peaks of each phase to minimize count time. In general it is best to choose at least three peaks from each phase to be used in the analysis. If a suitable set of unoverlapped peaks for a phase cannot be located, one might be forced to use peaks that are partially resolved and measure peak height intensities rather than determine integrated intensities. Integrated peak intensities are collected with step-motor driven diffractometers as illustrated in Figure 3.

In complex multi-phase mineralogical samples it is quite common to find peaks not resolved to background. In this case care should be taken in choosing the location of the points at which to measure background. It is not required that a high and low background be measured. In cases of severe peak crowding one may prefer to measure the average background at a single point rather than averaging over N_{bkg} points as is done by AUTO. Another serious concern in measuring peak heights is the specimen-displacement error. The displacement of the effective diffracting surface of a specimen from the focusing circle of a diffractometer, whether by physical mispositioning of the specimen or due to sample transparency, causes a peak to shift its observed location. When peak heights are being measured at a fixed, predetermined

113

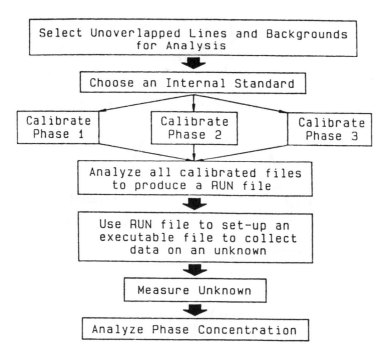

Figure 1. Flow chart of an Internal-Standard Analysis.

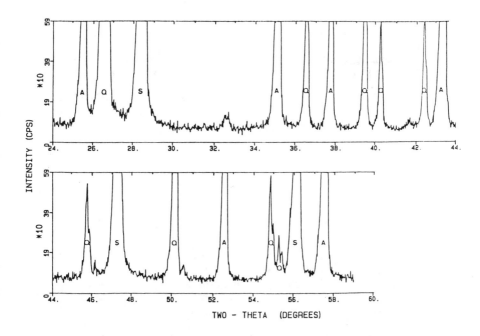

Figure 2. Scans of the reference standards and internal standard.

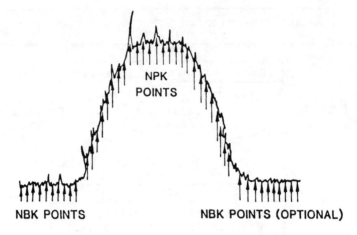

Figure 3. The strategy for collecting integrated intensity data.

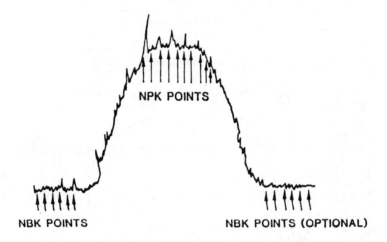

Figure 4. The strategy for collecting peak-height data.

Table 1. 2θ values (Cu Kα) for background and peaks of Al_2O_3 and Si.

Phase	$hk\ell$	Low Angle Background	Peak Location	High Angle Background
Si	111	27.46	28.42	29.52
Al_2O_3	104	34.30	35.12	35.91
Al_2O_3	110	37.00	37.75	38.30
Al_2O_3	113	42.77	43.32	43.88
Si	311	46.43	47.27	48.32

2θ position, sample displacement may cause the actual peak to shift away from the measured position, lowering the observed intensity by a substantial amount. For this reason, when peak heights are being used a peak-location optimization must be performed before the intensity is measured. The AUTO algorithm performs this optimization using the Savitsky-Golay procedure and then measures not one but N_{pk} points at the top of the peak and averages these values as shown in Figure 4.

Table 1 shows the background and peak positions chosen for the calibration of Al_2O_3 with Si as the internal standard. These values were obtained from an analysis of Figure 2.

Calibration of the Al_2O_3 standard

To calibrate the Al_2O_3 we must carry out the following steps:

1. Obtain a pure sample of Al_2O_3 and grind it to a particle size between 1 and 10 microns. This fine particle size is essential to improve particle counting statistics (Klug and Alexander, 1974) and to eliminate extinction effects (Cline and Snyder, 1985).

2. Obtain a pure sample of the internal standard, Si, and grind it to a particle size between 1 and 10 microns.

3. Mix a known amount of internal standard into the calibration phase so that each of the two phase's 100%-intensity lines have about equal intensity. For the Al_2O_3 calibration with Si, approximately equal amounts were weighed out and mixed together giving an Si concentration of 49.5% in the two-phase calibration mixture.

4. Prepare the Al_2O_3–Si mixture so as to minimize preferred orientation. In general it is recommended that the powder be spray dried. If, as in this case, orientation effects have been shown to be minimal, then a simple side-drift mount into the sample holder may be sufficient.

5. Mount the sample on the diffractometer and set up a run to measure the lines already chosen and shown in Table 1.

Calibration of the SiO_2 standard

The next step is to calibrate the SiO_2 phase following exactly the same procedure described above. An analysis of Figure 2 yields the lines shown in Table 2. For the sample used in this example the concentration of Si was 20% in the two-phase calibration mixture.

Data collection and analysis

The AUTO data-collection algorithm requires all of the data for a calibration

Table 2. 2θ values for background and peaks of SiO_2 and Si (in degrees for Cu Kα radiation).

Phase	$hk\ell$	Low Angle Background	Peak Location	High Angle Background
Si	111	27.46	28.42	29.52
SiO_2	110	35.90	36.51	37.00
SiO_2	102	38.70	39.43	39.86
SiO_2	200	41.85	42.42	42.77
Si	331	46.43	47.27	48.32

standard to be entered onto a special kind of file called a "RUN" file. A RUN file contains all of the information needed to collect the required quantitative analysis data (i.e., background locations, scan parameters, etc.). The required information on each line is entered onto a RUN file for each of the calibration samples and the AUTO algorithm reads this set of directions and carries out the measurements writing the output to a "RAW" file. For our example the Al_2O_3 and the SiO_2 calibrations were written to separate files. The NBS*QUANT84 system processes these files computing the average values for I^{rel} and RIR from each diffraction line measured and outputting them along with all of the information read from the RUN files to a new RUN file which will direct the analysis of unknowns. The printed output from the calibration of our example files is shown in Figure 5.

The calibration and RUN file creation output shows the initial table of intensities measured for each line in the first calibrant to be processed (SiO_2). Then relative intensities and lastly $RIR's$ are computed and averaged. The process is repeated for each of the phases to be calibrated and at the end, the information output to the new RUN file is tabulated. The RUN file produced in the example differs from the RAW file in that information for all analyte phases is present rather than just information pertaining to a single calibration phase.

It is recommended that each standard be measured at least three times (even though our example violates this advice in order to conserve space), and that the resulting series of files should be processed to compute relative intensities of the lines selected for measurement averaged over each of the data files collected.

Running an unknown

The reason for the rather elaborate procedures required in calibrating standards was to prepare a RUN file which can be used repeatedly to analyze unknowns in daily routine analysis with a minimum of operator intervention. The unknown is prepared in exactly the same manner (ground, internal standard added, spray dried) as described above for the calibration samples. A 50:50 mixture of Al_2O_3 and SiO_2 was prepared with Si as the internal standard and no attempt was made to control orientation. The RUN file prepared from single scans of the calibrants shown above was used to collect a single RAW data file. The output from this

NBS*QUANT88 - RUNFIL

Calculates Irel and/or RIR and/or I/Io and/or Quantitative Analysis RUN files

Written by R. L. Snyder and C. R. Hubbard

Run on 26-MAY-88 at 08:51:32

Title: SIO2 WITH .50000 WT FRACTION SI AS INT STD. SNYDER RUN File: SISIO2
File: SISIO2 collected on: 25-MAY-88 17:34:55 # Repetitions=1 # of Peaks = 5

Data collected on unit 1-A with a CU Target with a wavelength of 1.54056 A
Divergence Slit of type 1   Dead Time = 0.00

Negative numbers in the following table imply quantities which are meaningless for the current run, or to be calculated later

ln #	fl #	Two Bkglo	Theta Peak	Region Bkghi	Delt Ang	Phase Name	h	k	l	Rel Int	Sig Rel	Intstd Conc	Ref Conc	Reference Intensity	Sig of Refint
1	1	27.46	28.42	29.52	.010	SI	1	1	1	-1.	-1.0	0.500	-1.000	-1.0000	-1.0000
2	1	35.90	36.51	37.00	.010	SIO2	1	1	0	-1.	-1.0	-1.000	1.000	-1.0000	-1.0000
3	1	38.70	39.43	39.86	.010	SIO2	1	0	2	-1.	-1.0	-1.000	1.000	-1.0000	-1.0000
4	1	41.85	42.42	42.77	.010	SIO2	2	0	0	-1.	-1.0	-1.000	1.000	-1.0000	-1.0000
5	1	46.43	47.27	48.32	.010	SI	3	1	1	-1.	-1.0	0.500	-1.000	-1.0000	-1.0000

cnt Typ	Peak Int.	Sig PkI	Intg Area	Sig Area	Req Err	Obs Err	Bkg. Time	Bkg Pts	Peak Time	Pk. Pts	Max Tim
Int	7958	239	79614	193	0.5	0.2	14.39	10	1.67	5	10.
Int	1402	212	39917	191	0.5	0.5	14.11	10	3.53	5	10.
Int	1066	142	28650	153	0.5	0.5	15.00	10	3.13	5	10.
Int	880	119	29709	173	0.5	0.6	14.64	10	4.21	5	10.
Int	2957	518	46457	153	0.5	0.3	13.90	10	1.91	5	10.

Relative Intensity Analysis:
The following line(s) will be used in scaling
Ln Phase h k l Method
 2 SIO2   1 1 0 INTEGRT
 1 SI   1 1 1 INTEGRT

Meas Analyte Line Irel=100 Line   Relative Standard Root Mean Number of
 # Phase h k l   Name h k l   Intensity Deviation Square Dev. Measurements
 1 SIO2 1 1 0   SIO2 1 1 0   100.000   74.090

Meas Analyte Line Irel=100 Line   Relative Standard Root Mean Number of
 # Phase h k l   Name h k l   Intensity Deviation Square Dev. Measurements
 1 SIO2 1 0 2   SIO2 1 1 0   71.774   60.239

Meas Analyte Line Irel=100 Line   Relative Standard Root Mean Number of
 # Phase h k l   Name h k l   Intensity Deviation Square Dev. Measurements
 1 SIO2 2 0 0   SIO2 1 1 0   74.426   58.025

Meas Analyte Line Irel=100 Line   Relative Standard Root Mean Number of
 # Phase h k l   Name h k l   Intensity Deviation Square Dev. Measurements
 1 SI   1 1 1   SI   1 1 1   100.000   73.306

Meas Analyte Line Irel=100 Line   Relative Standard Root Mean Number of
 # Phase h k l   Name h k l   Intensity Deviation Square Dev. Measurements
 1 SI   3 1 1   SI   1 1 1   58.353   48.704

Summary of relative intensities
 Phase h k l Method   Intensity e.s.d. RMS Dev #
 SIO2   1 1 0 INTEGRT 100.000   74.090 0.000   1
 SIO2   1 0 2 INTEGRT 71.774   60.239 0.000   1
 SIO2   2 0 0 INTEGRT 74.426   58.025 0.000   1

 Phase h k l Method   Intensity e.s.d. RMS Dev #
 SI   1 1 1 INTEGRT 100.000   73.306 0.000   1
 SI   3 1 1 INTEGRT 58.353   48.704 0.000   1

RIR Calibration for quantitative analysis by the internal standard method
***** Line by line RIR Analysis *****
Meas Analyte Line Standard Line   RIR   Standard   Root Mean Num Deg of
 # Phase h k l   Name h k l   Constant Deviation Square Dev. Obs Freedom
 1 SIO2 1 1 0   SI   1 1 1   0.5014   0.6400

Figure 5. Runfile Creation Output

```
Meas Analyte Line Standard Line    RIR  Standard Root Mean Num Deg of
 #   Phase  h k l Name    h k l  Constant Deviation Square Dev. Obs Freedom
 1   SIO2   1 0 2  SI     1 1 1   0.5014   0.6984

Meas Analyte Line Standard Line    RIR  Standard Root Mean Num Deg of
 #   Phase  h k l Name    h k l  Constant Deviation Square Dev. Obs Freedom
 1   SIO2   2 0 0  SI     1 1 1   0.5014   0.6627

Meas Analyte Line Standard Line    RIR  Standard Root Mean Num Deg of
 #   Phase  h k l Name    h k l  Constant Deviation Square Dev. Obs Freedom
 1   SIO2   1 1 0  SI     3 1 1   0.5014   0.6998

Meas Analyte Line Standard Line    RIR  Standard Root Mean Num Deg of
 #   Phase  h k l Name    h k l  Constant Deviation Square Dev. Obs Freedom
 1   SIO2   1 0 2  SI     3 1 1   0.5014   0.7536

Meas Analyte Line Standard Line    RIR  Standard Root Mean Num Deg of
 #   Phase  h k l Name    h k l  Constant Deviation Square Dev. Obs Freedom
 1   SIO2   2 0 0  SI     3 1 1   0.5014   0.7206

         Wt Mean Over the standard lines used:
Meas Analyte Line Standard Line    RIR  Standard Root Mean Num Deg of
 #   Phase  h k l Name    h k l  Constant Deviation Square Dev. Obs Freedom
     SIO2   1 1 0  SI     all     0.5014   0.4723   0.0000    2   0
     SIO2   1 0 2  SI     all     0.5014   0.5122   0.0000    2  -1
     SIO2   2 0 0  SI     all     0.5014   0.4878   0.0000    2  -1

         Overall Wt Mean over all lines for each phase:
Meas Analyte Line Standard Line    RIR  Standard Root Mean Num Deg of
 #   Phase  h k l Name    h k l  Constant Deviation Square Dev. Obs Freedom
     SIO2   all    SI     all     0.5014   0.2829   0.0000    6 - 2
```

Title: AL2O3 WITH .50000 WT FRACTION SI AS INT STD SNYDER RUN File: SIAL2O3
File: SIAL2O3 collected on: 25-MAY-88 18:41:01 # Repetitions=1 # of Peaks = 5

Data collected on unit 1-A with a CU Target with a wavelength of 1.54056 A
Divergence Slit of type 1   Dead Time = 0.00

Negative numbers in the following table imply quantities which are meaningless for the current run, or to be calculated

Ln #	fl #	Two Theta Bkglo	Theta Peak	Region Bkghi	Delt Ang	Phase Name	h	k	l	Rel Int	Sig Rel	Intstd Conc	Ref Conc	Reference Intensity	Sig of Refint
1	1	27.46	28.42	29.52	.010	SI	1	1	1	-1.	-1.0	0.500	-1.000	-1.0000	-1.0000
2	1	34.30	35.12	35.91	.010	ALOX	1	0	4	-1.	-1.0	-1.000	1.000	-1.0000	-1.0000
3	1	37.00	37.75	38.30	.010	ALOX	1	1	0	-1.	-1.0	-1.000	1.000	-1.0000	-1.0000
4	1	42.77	43.32	43.88	.010	ALOX	1	1	3	-1.	-1.0	-1.000	1.000	-1.0000	-1.0000
5	1	46.43	47.27	48.32	.010	SI	3	1	1	-1.	-1.0	0.500	-1.000	-1.0000	-1.0000

cnt Typ	Peak Int.	Sig PkI	Intg Area	Sig Area	Req Err	Obs Err	Bkg. Time	Bkg Pts	Peak Time	Pk. Pts	Max Tim
Int	5336	544	72257	186	0.5	0.3	12.98	10	1.82	5	10.
Int	5790	847	136161	300	0.5	0.2	2.39	10	0.61	5	10.
Int	2831	372	73042	244	0.5	0.3	7.16	10	2.04	5	10.
Int	5995	944	226141	481	0.5	0.2	0.95	10	0.44	5	10.
Int	2211	295	43175	147	0.5	0.3	14.12	10	1.88	5	10.

Relative Intensity Analysis:
The following line(s) will be used in scaling
Ln Phase h k l   Method
 4 ALOX   1 1 3 INTEGRT
 1 SI   1 1 1 INTEGRT

```
Meas Analyte Line Irel=100 Line  Relative  Standard Root Mean Number of
 #   Phase  h k l Name    h k l Intensity Deviation Square Dev. Measurements
 1   ALOX   1 0 4 ALOX    1 1 3   60.211    23.656

Meas Analyte Line Irel=100 Line  Relative  Standard Root Mean Number of
 #   Phase  h k l Name    h k l Intensity Deviation Square Dev. Measurements
 1   ALOX   1 1 0 ALOX    1 1 3   32.300    14.851

Meas Analyte Line Irel=100 Line  Relative  Standard Root Mean Number of
 #   Phase  h k l Name    h k l Intensity Deviation Square Dev. Measurements
```

Figure 5. Runfile Creation Output (continued).

```
 1  ALOX  1 1 3   ALOX  1 1 3   100.000   29.400
```

```
Meas Analyte Line  Irel=100 Line   Relative  Standard  Root Mean  Number of
 #   Phase h k l   Name   h k l  Intensity  Deviation  Square Dev. Measurements
 1    SI   1 1 1    SI    1 1 1   100.000    75.853
```

```
Meas Analyte Line  Irel=100 Line   Relative  Standard  Root Mean  Number of
 #   Phase h k l   Name   h k l  Intensity  Deviation  Square Dev. Measurements
 1    SI   3 1 1    SI    1 1 1   59.752    51.723
```

Summary of relative intensities
```
      Phase  h k l   Method       Intensity   e.s.d.   RMS Dev  #
      ALOX   1 0 4   INTEGRT       60.211    23.656    0.000    1
      ALOX   1 1 0   INTEGRT       32.300    14.851    0.000    1
      ALOX   1 1 3   INTEGRT      100.000    29.400    0.000    1

      Phase  h k l   Method       Intensity   e.s.d.   RMS Dev  #
      SI     1 1 1   INTEGRT      100.000    75.853    0.000    1
      SI     3 1 1   INTEGRT       59.752    51.723    0.000    1
```

RIR Calibration for quantitative analysis by the internal standard method
***** Line by line RIR Analysis *****
```
Meas Analyte Line  Standard Line      RIR    Standard  Root Mean  Num Deg of
 #   Phase h k l   Name   h k l    Constant Deviation  Square Dev. Obs Freedom
 1   ALOX  1 0 4    SI    1 1 1     3.1297   3.3248
```

```
Meas Analyte Line  Standard Line      RIR    Standard  Root Mean  Num Deg of
 #   Phase h k l   Name   h k l    Constant Deviation  Square Dev. Obs Freedom
 1   ALOX  1 1 0    SI    1 1 1     3.1297   3.4888
```

```
Meas Analyte Line  Standard Line      RIR    Standard  Root Mean  Num Deg of
 #   Phase h k l   Name   h k l    Constant Deviation  Square Dev. Obs Freedom
 1   ALOX  1 1 3    SI    1 1 1     3.1297   3.1183
```

```
Meas Analyte Line  Standard Line      RIR    Standard  Root Mean  Num Deg of
 #   Phase h k l   Name   h k l    Constant Deviation  Square Dev. Obs Freedom
 1   ALOX  1 0 4    SI    3 1 1     3.1297   3.8029
```

```
Meas Analyte Line  Standard Line      RIR    Standard  Root Mean  Num Deg of
 #   Phase h k l   Name   h k l    Constant Deviation  Square Dev. Obs Freedom
 1   ALOX  1 1 0    SI    3 1 1     3.1297   3.9471
```

```
Meas Analyte Line  Standard Line      RIR    Standard  Root Mean  Num Deg of
 #   Phase h k l   Name   h k l    Constant Deviation  Square Dev. Obs Freedom
 1   ALOX  1 1 3    SI    3 1 1     3.1297   3.6237
```

              Wt Mean Over the standard lines used:
```
Meas Analyte Line  Standard Line      RIR    Standard  Root Mean  Num Deg of
 #   Phase h k l   Name   h k l    Constant Deviation  Square Dev. Obs Freedom
     ALOX  1 0 4    SI     all      3.1297   2.5031    0.0000     2  -1
     ALOX  1 1 0    SI     all      3.1297   2.6141    0.0000     2  -1
     ALOX  1 1 3    SI     all      3.1297   2.3636    0.0000     2   0
```

        Overall Wt Mean over all lines for each phase:
```
Meas Analyte Line  Standard Line      RIR    Standard  Root Mean  Num Deg of
 #   Phase h k l   Name   h k l    Constant Deviation  Square Dev. Obs Freedom
     ALOX   all    SI     all      3.1297   1.4360    0.0000     6  -2
```

Listing of the information output to RUN file: INTSTD
RUN FILE FOR ANALYSIS OF AL2O3 AND SIO2 WITH SI AS INT STD   SNYDER

```
26-MAY-88 08:52:55
Peak  Phase h k l    Irel     Sigma     RIR    Sigma   Cnt Err  Max Time  Overlap
28.420 SI    1 1 1  100.000   52.713   0.000   0.000   0.500    10.000     0
35.120 ALOX  1 0 4   60.211   23.656   3.130   1.436   0.500    10.000     0
36.510 SIO2  1 1 0  100.000   74.090   0.501   0.283   0.500    10.000     0
37.750 ALOX  1 1 0   32.300   14.851   3.130   1.436   0.500    10.000     0
39.430 SIO2  1 0 2   71.774   60.239   0.501   0.283   0.500    10.000     0
42.420 SIO2  2 0 0   74.426   58.025   0.501   0.283   0.500    10.000     0
43.320 ALOX  1 1 3  100.000   29.400   3.130   1.436   0.500    10.000     0
47.270 SI    3 1 1   59.010   35.458   0.000   0.000   0.500    10.000     0
```

        8 Lines output to the runfile with   2 unoverlapped standard lines

Figure 5: Runfile Creation Output (continued).

run is shown in Figure 6. The Al_2O_3 concentration is averaged to be 48.7% with a root-mean-square deviation of 1.5%. It should be noted that the estimate of the standard deviation from counting statistics only is very large (i.e., 24%) using this small number of measurements. However, the SiO_2 analysis illustrates even more interesting effects. The SiO_2 concentration is seen to vary from 54 to 72%! This variation alone indicates serious systematic errors. The average measured concentration of 60% is 10% higher than actual. Since the Al_2O_3 results are correct to within one standard deviation, it is clear that the problem is not with the internal standard but lies completely with the SiO_2 measurement. We include this "honest" example to illustrate the typical difficulties one encounters on setting up for analyzing a particular type of unknown. To isolate the source of this error one must repeat the measurements varying the grinding technique, the sample preparation, etc. In general this wide variation is due to preferred orientation.

EXAMPLE USE OF THE REFERENCE INTENSITY RATIO OR CHUNG METHOD

The RIR or Chung method is simply a version of the internal-standard method where all phases are analyzed and it is assumed that the sum of all of the crystalline weight fractions equals 1.0. In the AUTO/QUANT84 system, a RUN file must first be prepared containing the RIR and I^{rel} values for each of the phases in the unknown, in addition to the line measurement information. The information required is exactly the information that an internal-standard calibration procedure puts onto a RUN file. Thus, the procedures described above to measure the I^{rel} and RIR values for alumina and silica would be followed here. But in the RIR method, one of the phases to be analyzed is selected arbitrarily to act as an internal standard. There is no internal standard added in this method, however, the formalism of the RIR method is easiest to understand if we designate one of the phases in the unknown as if it were an internal standard.

Suppose that our 1:1:1 specimen used in the internal standard example is an unknown containing only crystalline Si, SiO_2 and Al_2O_3. In the Chung method we need to get the RIR and I^{rel} values for each of the lines to be analyzed for each phase. Rather than take the relatively inaccurate I/I_c and I^{rel} values from the literature we decide to measure our own. Then the samples of SiO_2 – Si and Al_2O_3 – Si are calibrated exactly as for the internal-standard method.

1. Choose one of the three phases to be a reference in the determination of the RIRs. We will choose Si.

2. Select an internal standard which has resolved lines not overlapping the resolved lines of Al_2O_3 and SiO_2.

3. Select diffraction lines to be measured for each phase.

4. Calibrate the Al_2O_3–Si and SiO_2–Si standards.

5. Process the Al_2O_3–Si and SiO_2–Si RAW data files to create a RUN file.

Quantitative Analysis by the Internal Standard or Intensity Ratio Methods
Written by R. L. Snyder and C. R. Hubbard
run on 26-MAY-88   at 11:20:03

Title: ANALYSIS OF A 50:50 SIO2/AL2O3 SAMPLE WITH 33% SI AS STD   SNYDER
File: INTSTDUNK collected on: 25-MAY-88 19:27:36 # Repetitions=1 # of Peaks = 8
Internal Standard Analysis
Resolved lines only

Line by line composition analysis:

Meas #	Analyte Phase	Line h k l	Standard Name	Line h k l	Weight Fraction	Standard Deviation	Root mean Square Dev.	Number of Measurements
1	ALOX	1 0 4	SI	1 1 1	0.4744	0.0000		
1	SIO2	1 1 0	SI	1 1 1	0.5422	0.0000		
1	ALOX	1 1 0	SI	1 1 1	0.4524	0.0000		
1	SIO2	1 0 2	SI	1 1 1	0.6287	0.0000		
1	SIO2	2 0 0	SI	1 1 1	0.5417	0.0000		
1	ALOX	1 1 3	SI	1 1 1	0.4621	0.0000		
1	ALOX	1 0 4	SI	3 1 1	0.5428	0.7026		
1	SIO2	1 1 0	SI	3 1 1	0.6204	0.9661		
1	ALOX	1 1 0	SI	3 1 1	0.5176	0.7013		
1	SIO2	1 0 2	SI	3 1 1	0.7193	1.1827		
1	SIO2	2 0 0	SI	3 1 1	0.6198	0.9990		
1	ALOX	1 1 3	SI	3 1 1	0.5287	0.6493		

Wt mean over the standard lines used:

Meas #	Analyte Phase	Line h k l	Standard Name	Line h k l	Weight Fraction	Standard Deviation	Root mean Square Dev.	Number of Measurements
	ALOX	1 0 4	Wt mean	all	0.4989	0.4206	0.0328	2
	SIO2	1 1 0	Wt mean	all	0.5722	0.5983	0.0380	2
	ALOX	1 1 0	Wt mean	all	0.4763	0.4241	0.0314	2
	SIO2	1 0 2	Wt mean	all	0.6640	0.7376	0.0442	2
	SIO2	2 0 0	Wt mean	all	0.5719	0.6215	0.0380	2
	ALOX	1 1 3	Wt mean	all	0.4853	0.3835	0.0318	2

Overall Wt mean over all lines for each phase:

Meas #	Analyte Phase	Line h k l	Standard Name	Line h k l	Weight Fraction	Standard Deviation	Root mean Square Dev.	Number of Measurements
	ALOX	all	Wt mean	all	0.4868	0.2356	0.0149	6
	SIO2	all	Wt mean	all	0.5955	0.3721	0.0252	6

Figure 6. Internal-Standard Analysis controlled by a RUN file.

6. Use the RUN file template to control collection of data on an unknown sample containing *only* these three phases.

Since the calibration procedure is exactly the same as illustrated for the internal-standard method we used the same RUN file to do a Chung analysis on the 1:1:1 sample. The results are shown in Figure 7. Again only a single sample was measured to conserve space, so the resulting precision is low.

The Chung method using direct input of RIR and I^{rel} values

Another approach to the Chung method is to take the literature values for the I/I_c and I^{rel} values (as given on the PDF cards) and input them directly to a RUN file. This allows direct one-step analysis of an unknown with no calibration, or what is usually called a "standardless" quantitative analysis. The difficulty here is that the published I/I_c and I^{rel} values are notoriously inaccurate. This method at best should be called semi-quantitative and at worst simply wrong. However, if the RIR and I^{rel} values have been previously determined in your laboratory, the method can have all of the accuracy of the internal-standard method.

ABSORPTION-DIFFRACTION METHOD

In the internal-standard method, Equation 1 was written twice, once for the unknown and again for the internal standard. The ratio of these gave Equation 3 in which the absorption coefficient of the unknown sample canceled out. In the absorption-diffraction method we again write Equation 1 twice: once for line i of phase α in the unknown and again for line i of a pure sample of phase α. The ratio here gives:

$$\frac{I_{i\alpha}}{I_{i\alpha}^\circ} = \frac{(\frac{\mu}{\rho})_\alpha}{(\frac{\mu}{\rho})_m} X_\alpha, \tag{18}$$

where $I_{i\alpha}^\circ$ is the intensity of line i in pure phase α. Equation 18 is the basis of the absorption-diffraction method for quantitative analysis. There are several methods for implementing Equation 18. Five cases will be discussed in some detail. Each of these cases is dependent only on the validity of Equation 18, and since we have made no assumptions in the derivation, this equation is rigorous.

Analysis of a mixture of known chemical composition

It is more common than one might imagine that quantitative X-ray diffraction phase analysis is carried out on samples whose chemical composition is known. When this is the case the mass absorption coefficients of the pure phase to be analyzed and of the mixture can be computed from

$$(\frac{\mu}{\rho})_m = \sum_{j=1}^{\#of elements} X_j (\frac{\mu}{\rho})_j, \tag{19}$$

Program NBS*QUANT88 - QUANT

Quantitative Analysis by the Internal Standard or Intensity Ratio Methods

Written by R. L. Snyder and C. R. Hubbard

run on 26-MAY-88   at 11:47:48

Title: 50:50 AL2O3 & SIO2 WITH 33"% SI   SNYDER   RUN File: INTSTD
File: CHUNGUNK collected on: 26-MAY-88 10:29:15 # Repetitions=1 # of Peaks = 8

Chung Version of Reference Intensity Ratio Method
Resolved lines only

Line by line composition analysis:

Meas	Analyte	Line	Weight	Standard	Root Mean	Number of
#	Phase	h k l	Fraction	Deviation	Square Dev.	Measurements
1	SI	1 1 1	24.7860	25.7238		

Meas	Analyte	Line	Weight	Standard	Root Mean	Number of
#	Phase	h k l	Fraction	Deviation	Square Dev.	Measurements
1	ALOX	1 0 4	34.8576	28.3545		

Meas	Analyte	Line	Weight	Standard	Root Mean	Number of
#	Phase	h k l	Fraction	Deviation	Square Dev.	Measurements
1	SIO2	1 1 0	38.4828	46.0008		

Meas	Analyte	Line	Weight	Standard	Root Mean	Number of
#	Phase	h k l	Fraction	Deviation	Square Dev.	Measurements
1	ALOX	1 1 0	32.9919	29.9663		

Meas	Analyte	Line	Weight	Standard	Root Mean	Number of
#	Phase	h k l	Fraction	Deviation	Square Dev.	Measurements
1	SIO2	1 0 2	42.4949	56.2840		

Meas	Analyte	Line	Weight	Standard	Root Mean	Number of
#	Phase	h k l	Fraction	Deviation	Square Dev.	Measurements
1	SIO2	2 0 0	34.5347	45.0450		

Meas	Analyte	Line	Weight	Standard	Root Mean	Number of
#	Phase	h k l	Fraction	Deviation	Square Dev.	Measurements
1	ALOX	1 1 3	33.4472	23.5448		

Meas	Analyte	Line	Weight	Standard	Root Mean	Number of
#	Phase	h k l	Fraction	Deviation	Square Dev.	Measurements
1	SI	3 1 1	36.8385	39.8576		

Overall Wt mean over all lines for each phase:

Meas	Analyte	Line	Weight	Standard	Root Mean	Number of
#	Phase	h k l	Fraction	Deviation	Square Dev.	Measurements
	SI	all	28.3300	21.6134	5.4913	2
	ALOX	all	33.7469	15.5019	0.5292	3
	SIO2	all	37.9525	27.9390	2.2002	3

Figure 7. Chung analysis using RIR and I^{rel} values measured in a RUN file calibration.

where X_j is the weight fraction of element j whose mass absorption coefficient is $\left(\frac{\mu}{\rho}\right)_j$ which may be obtained from tables. Thus, if $I_{i\alpha}$ and $I^o_{i\alpha}$ are measured from the unknown and a pure sample of α respectively, the weight fraction of the analyte (i.e., X_α) may be computed from Equation 18.

Analysis of a mixture of unknown chemical composition by measuring $\left(\frac{\mu}{\rho}\right)_j$

Leroux et al. (1953) were the first to propose that quantitative analysis may be performed on unknowns by measuring the density and determining the linear absorption coefficients of the unknowns through absorption experiments on specimens of different thickness using the mass absorption law:

$$\ln\left(\frac{I}{I_o}\right) = e^{-\mu x}. \qquad (20)$$

The difficulty with this approach is that the measurement of μ is extremely error prone. The low precision of this measurement dramatically limits the accuracy of a quantitative analysis.

Analysis of a sample whose mass absorption coefficient is the same as that of the pure phase

In this case Equation 18 reduces to

$$\frac{I_{i\alpha}}{I^o_{i\alpha}} = X_\alpha, \qquad (21)$$

since $\left(\frac{\mu}{\rho}\right)_\alpha$ is exactly the same as $\left(\frac{\mu}{\rho}\right)_m$. This case should be kept in mind because the analysis of cristobalite in a quartz and amorphous SiO_2 matrix or the analysis of the cubic, tetragonal, and monoclinic forms of ZrO_2 in pure ZrO_2 bodies are examples of common analyses. In cases where we have mixtures of polymorphs, the absorption-diffraction method of analysis becomes very attractive.

Analysis of thin films or lightly loaded filters

In this case the sample approximates a monolayer of sample and presents a special case of Equation 18. As long as all the particles lie alongside each other and do not shade each other from the X-ray beam, then all matrix absorption effects are eliminated. There can of course be no preferential absorption. Each crystallite of each phase diffracts as if it were a pure phase. Thus the absorption coefficient for each phase is the same as for the pure phase. This causes Equation 21 in case 3 to become operable. A plot of $\frac{I_{i\alpha}}{I^o_{i\alpha}}$ vs. X_α will be linear and immediately allows for quantitative analysis. This procedure is valid up to the limit of concentration of particles where crowding invalidates the assumption that $\left(\frac{\mu}{\rho}\right)_\alpha = \left(\frac{\mu}{\rho}\right)_m$. This method is commonly used, for example, in the analysis of respirable silica in air-filtered samples collected on silver membranes.

Analysis of a binary mixture of two known phases

In this case Equation 1 becomes:

$$I_{i\alpha} = \frac{K_{i\alpha} X_\alpha}{\rho_\alpha[X_\alpha(\frac{\mu}{\rho})_\alpha - (\frac{\mu}{\rho})_\beta] + (\frac{\mu}{\rho})_\beta}.$$

Equation 1 for the pure phase α may be written as

$$I_{i\alpha}^\circ = \frac{K_{i\alpha}}{\rho_\alpha(\frac{\mu}{\rho})_\alpha}.$$

Dividing these two equations gives

$$\frac{I_{i\alpha}}{I_{i\alpha}^\circ} = \frac{X_\alpha(\frac{\mu}{\rho})_\alpha}{X_\alpha(\frac{\mu}{\rho})_\alpha + X_\beta(\frac{\mu}{\rho})_\beta}. \tag{22}$$

Equation 22 can easily be used to compute the standard curve of $\frac{I_{i\alpha}}{I_{i\alpha}^\circ}$ vs X_α for all possible compositions of mixtures of α and β.

METHOD OF STANDARD ADDITIONS OR SPIKING METHOD

Lennox (1957) was the first to develop the spiking method which has been widely adopted in X-ray fluorescence spectroscopy as the Method of Standard Additions. The method, like the internal-standard method, is perfectly general applying to any phase α in a mixture. The only requirement is that the mixture also contain, among other phases, a phase β which has a diffraction line unoverlapped by any line from α, which may be used as a reference. Phase β is not analyzed and does not even need to be identified.

In a sample containing α and β, the ratio of the intensities from a line from each phase can be obtained from Equation 1,

$$\frac{I_{i\alpha}}{I_{j\beta}} = \frac{K_{i\alpha}\rho_\beta X_\alpha}{K_{j\beta}\rho_\alpha X_\beta}. \tag{23}$$

Using the method of standard additions, some of the pure phase α is added to the mixture containing the unknown concentration of α. After adding Y_α grams of the α phase per gram of the unknown, the ratio $\frac{I_{i\alpha}}{I_{j\beta}}$ becomes

$$\frac{I_{i\alpha}}{I_{j\beta}} = \frac{K_{i\alpha}\rho_\beta(X_\alpha + Y_\alpha)}{K_{j\beta}\rho_\alpha X_\beta}, \tag{24}$$

where,

- X_α = The initial weight fraction of phase α,
- X_β = The initial weight fraction of phase β,
- Y_α = The number of grams of pure phase α added per gram of the original sample.

In general, the intensity ratio is given by

$$\frac{I_{i\alpha}}{I_{j\beta}} = K(X_\alpha + Y_\alpha), \tag{25}$$

where K is the slope of a plot of $\frac{I_{i\alpha}}{I_{j\beta}}$ versus Y_α, with Y_α in units of grams of α per gram of sample. Multiple additions are made to prepare a plot in which the negative x intercept is X_α, the desired concentration of the phase α in the original sample.

Amorphous phase analysis

The area under the amorphous scattering hump in the approximate range between 15° and 30° 2θ is related to the concentration of the amorphous phase present in exactly the same manner as the intensity of a diffraction peak is related to concentration. If an unobscured region of the amorphous scattering region is integrated, the intensity may be treated according to any of the above methods. The only consideration which must be emphasized is the points at which background is measured. Usually only a small error will result if a low-angle background in the vicinity of 10° and a high-angle background in the range of 40° are chosen. This technique was used by Hubbard to estimate the amount of amorphous silica in an NBS Standard Reference sample of quartz.

ERROR PROPAGATION IN QUANTITATIVE ANALYSIS

Error propagation in quantitative analysis by X-ray powder diffraction is one of the areas most prone to mistakes. The exact propagation of error due to the counting statistics is the subject we will address here. The systematic errors which so commonly creep into quantitative analysis must be assessed by looking at the variation of results from one repetition to another (i.e., the root-mean-square deviation). It is recommend that in all computations that variances (σ^2) be propagated and that the square root only be taken to obtain the standard deviation after all other computations are complete. The reason for this is that variances propagate linearly in the common equations of quantitative analysis. We consider it important to include the exact formulae for the error terms most commonly encountered in quantitative analysis as derived by Snyder and Hubbard (1984) in this chapter.

The equations required to calculate the error due to counting statistics are:

1. Count Rate R'_i at Point i, for C^i = the number of counts observed at point i, and t_i = the count time at point i, and ϵ = the instrument stability in percent.

$$R'_i = C_i/t_i, \tag{26}$$

$$\sigma^2_{R'_i} = \frac{C_i + C_i(\epsilon^2 \cdot 10^{-4})}{t_i^2}. \tag{27}$$

2. Dead-Time Correction, for τ = the detector dead time in microseconds.

$$R_i = \frac{R_i'}{1 - R_i' \cdot \tau \cdot 10^{-6}}, \qquad (28)$$

assuming a negligible error in the dead time:

$$\sigma_{R_i}^2 = \sigma_{R_i'}^2 / (1 - R_i' \cdot \tau \cdot 10^{-6})^2. \qquad (29)$$

3. Net Peak Intensity.

$$I_P = \frac{\sum_{i=1}^{N_{pk}} R_i}{N_{pk}} - \frac{\sum_{j=1}^{N_{bkg}} R_j}{N_{bkg}}, \qquad (30)$$

$$\sigma_{I_P}^2 = \frac{\sum_{i=1}^{N_{pk}} \sigma_{R_i}^2}{N_{pk}^2} + \frac{\sum_{j=1}^{N_{bkg}} \sigma_{R_j}^2}{N_{bkg}^2}. \qquad (31)$$

4. Net Integrated Intensity, where $\Delta 2\theta$ is the step width between the N_{pk} points collected over the profile.

$$I_I = \Delta 2\theta \left\{ \sum_{i=1}^{N_{pk}} R_i - \frac{N_{pk} \sum_{j=1}^{N_{bkg}} R_j}{N_{bkg}} \right\}, \qquad (32)$$

$$\sigma_{I_I}^2 = \Delta 2\theta^2 \left\{ \sum_{i=1}^{N_{pk}} \sigma_{R_i}^2 + \left(\frac{N_{pk}}{N_{bkg}}\right)^2 \cdot \sum_{j=1}^{N_{bkg}} \sigma_{R_j}^2 \right\}. \qquad (33)$$

5. Theta-Compensating Slit Correction.
If RIR values from published sources or different instruments are to be used, then the effects of theta-compensating slits on the intensities must be removed. Intensities can be corrected to those collected with a constant volume of sample irradiated (i.e., the fixed slit case) as follows.

$$I_o = I \frac{\sin(30)}{\sin 2\theta_p}, \qquad (34)$$

$$\sigma_{I_o}^2 = \sigma_I^2 \left[\frac{\sin(30)}{\sin 2\theta_p}\right]^2. \qquad (35)$$

where $2\theta_p$ is the peak location and I is either the net peak or net integrated intensity. The $\sin(30°)$ term puts the intensities *approximately* on the same scale that would be observed with a one-degree incident slit.

6. Relative Intensity, where $I_{i,\alpha}$ is the intensity of line i of phase α and $I_{x,\alpha}$ is the intensity of the strongest line.

$$I_{i,\alpha}^{rel} = \frac{I_{i,\alpha}}{I_{x,\alpha}}, \qquad (36)$$

$$\sigma^2_{I^{rel}_{i,\alpha}} = \left(\frac{I_{i,\alpha}}{I_{x,\alpha}}\right)^2 \left[\frac{\sigma^2_{I_{i,\alpha}}}{I^2_{i,\alpha}} + \frac{\sigma^2_{I_{x,\alpha}}}{I^2_{x,\alpha}}\right]. \tag{37}$$

7. Reference Intensity Ratios.

α represents the phase being calibrated and s, the reference phase. These equations assume zero uncertainty in the weight fractions (X) of the two phases.

$$RIR_{\alpha,s} = \left(\frac{I_{i,\alpha}}{I_{j,s}}\right)\left(\frac{X_\alpha}{X_s}\right)\left(\frac{I^{rel}_{j,s}}{I^{rel}_{i,\alpha}}\right), \tag{38}$$

$$\sigma^2_{RIR_{\alpha,s}} = RIR^2_{\alpha,s}\left\{\frac{\sigma^2_{I^{rel}_{i,\alpha}}}{(I^{rel}_{i,\alpha})^2} + \frac{\sigma^2_{I^{rel}_{j,s}}}{(I^{rel}_{j,s})^2} + \frac{\sigma^2_{I_{i,\alpha}}}{(I_{i,\alpha})^2} + \frac{\sigma^2_{I_{j,s}}}{(I_{j,s})^2}\right\}. \tag{39}$$

8. Weighted Average Values.

for any quantity x

$$<x> = \frac{\sum x_i/\sigma^2_{x_i}}{\sum 1/\sigma^2_{x_i}}, \tag{40}$$

$$\sigma^2_{<x>} = \frac{1}{\sum 1/\sigma^2_{x_i}}. \tag{41}$$

9. Weighted Root-Mean-Square Deviations for Weighted Average Values.

for any quantity x

$$\sigma^2_{<x>RMS} = \frac{\sum_{i=1}^n \left[\frac{(x_i - <x>)^2}{\sigma^2_i}\right]^{1/2}}{(n-1)\sum_{i=1}^n 1/\sigma^2_i}. \tag{42}$$

FULL-PATTERN FITTING

With the availability of automated X-ray powder diffractometers, digital diffraction data are now routinely available on computers and can be analyzed by a variety of numerical techniques. Complete digital diffraction patterns provide the opportunity to perform quantitative phase analysis using all data in a given pattern rather than considering only a few of the strongest reflections. As the name implies, full-pattern methods involve fitting the entire diffraction pattern, often including the background, with a synthetic diffraction pattern. This synthetic diffraction pattern can either be calculated from crystal structure data (Bish and Howard, 1986, 1988; Hill and Howard, 1987; O'Connor and Raven, 1988) or can be produced from a combination of observed standard diffraction patterns (Smith et al., 1987).

The use of complete diffraction patterns in quantitative phase analysis has several significant advantages over conventional methods that often use only a few isolated reflections or groups of reflections. In addition to eliminating the troublesome requirement of extracting reproducible intensities from complex patterns, the traditional problems of line overlap, primary extinction, and preferred orientation are minimized by using complete diffraction patterns.

The two different full-pattern fitting methods have evolved separately and each method has its own advantages and disadvantages. Both methods require the use of a computer and digital diffraction data, and data analysis is more involved than conventional quantitative analysis methods. However, the full-pattern fitting methods have the potential for providing considerable information on samples. In ideal cases, the methods can provide quantitative phase results approaching the precision and accuracy of X-ray fluorescence elemental analyses. Because the two methods are significantly different in approach, they will be treated separately here.

Quantitative phase analysis using the Rietveld method

Quantitative phase analysis using calculated patterns is a natural outgrowth of the Rietveld method (Rietveld, 1969) originally conceived as a method of refining crystal structures using neutron powder diffraction data. Refinement is conducted by minimizing the sum of the weighted, squared differences between observed and calculated intensities at every step in a digital powder pattern. The Rietveld method requires a knowledge of the approximate crystal structure of all phases of interest (not necessarily all phases present) in a mixture, a trivial requirement for most phases today. The input data to a refinement are similar to those required to calculate a diffraction pattern (e.g., POWD10, (Smith et al., 1982) i.e., space group symmetry, atomic positions, site occupancies, and lattice parameters. In a typical refinement, individual scale factors (related to the weight percents of each phase) and profile, background, and lattice parameters are varied. In favorable cases, the atomic positions and site occupancies can also be successfully varied.

Theory. The methodology involved in quantitative analysis using the Rietveld method is analogous to conventional Rietveld refinement (see Post and Bish, Von Dreele, this volume), and the quantitative analysis theory is identical to that implemented in most conventional quantitative analyses (e.g., Klug and Alexander 1974; Cullity 1978). The method consists of fitting the complete experimental diffraction pattern with calculated profiles and backgrounds, and obtaining quantitative phase information from the scale factors for each phase in a mixture.

The $K_{i\alpha}$ of Equation 2 can be divided into two terms. The first is the k shown in Equation 43, which depends only on the experimental conditions and is independent of angle and sample effects.

$$k = \left\{ \left(\frac{I_o \lambda^3}{32\pi r} \right) \left(\frac{e^4}{2m_e^2 c^4} \right) \right\}. \qquad (43)$$

The second part $R_{hk\ell}$ given in Equation 44 depends on the crystal structure and specific reflection in question.

$$R_{hk\ell} = \frac{M}{V^2} \mid F \mid^2 \left(\frac{1 + cos^2 2\theta cos^2 2\theta_m}{sin^2\theta \cos\theta} e^{-2m} \right)_{hk\ell}, \quad (44)$$

where $2\theta_m$ is the diffraction angle of the monochromator. Equation 1 can now be written in terms of 43 and 44 as Equation 45.

$$I_{hk\ell} = k(1/2\mu)R_{hk\ell}. \quad (45)$$

In a mixture, the intensity of the $hk\ell$ reflection from the phase is given as

$$I_{\alpha,hk\ell} = C_\alpha k(1/2\mu_m)R_{\alpha,hk\ell}, \quad (46)$$

where C_α is the volume fraction of the α phase and μ_m is the linear absorption coefficient of the mixture. In terms of weight fraction, Equation 46 can be rewritten as

$$I_{\alpha,hk\ell} = \frac{W_\alpha}{\rho_\alpha} k \frac{\rho_m}{2\mu_m} R_{\alpha,hk\ell}. \quad (47)$$

In Rietveld refinement, the quantity minimized is

$$R = \sum w_i \left| y_i(o) - y_i(c) \right|^2, \quad (48)$$

where $y_i(o)$ and $y_i(c)$ are the intensity observed and calculated, respectively, at the ith step in the data and w_i is the weight. Therefore, it is perhaps more appropriate for Rietveld quantitative analysis to consider the intensity at a given 2θ step rather than for a given reflection. The intensity at a given step is determined by summing the contributions from background and all neighboring Bragg reflections as

$$y_i(c) = S \sum_K p_K L_K \left| F_K \right|^2 G(\Delta\theta_{iK}) P_K + y_{ib}(c), \quad (49)$$

where S is the scale factor, L_k is the Lorentz and polarization factors for the kth Bragg reflection, F_k is the structure factor, p_k is the multiplicity factor, P_k is the preferred orientation function, θ_k is the Bragg angle for the kth reflection, $G(\Delta\theta_{ik})$ is the reflection profile function, and $y_{ib}(c)$ is the background (Wiles and Young, 1981). The Rietveld scale factor, S, includes all of the constant terms in Equation 1 and for X-rays can be written

$$S = K/V^2\mu, \quad (50)$$

where V is the unit cell volume and μ is the linear absorption coefficient for the sample. For a multi-phase mixture, Equation 50 can be rewritten summing over the p phases in a mixture (e.g., Hill and Howard, 1987) as

$$y_{ic} = y_{ib} + \sum_p S_p \sum_k J_{kp} L_{kp} \left| F_{kp} \right|^2 G_{ikp}. \quad (51)$$

The scale factor for each phase can now be written

$$S_\alpha = C_\alpha K/V_\alpha^2 \mu_m, \tag{52}$$

where C_α is the volume fraction of the α phase and μ_m is the linear absorption coefficient of the mixture. Recasting Equation 52 in terms of weight fractions and the mass absorption coefficient of the mixture we obtain

$$S_\alpha = W_\alpha K/(\rho_\alpha V_\alpha^2 \mu^*), \tag{53}$$

where μ^* is the mass absorption coefficient of the sample, W_α is the weight fraction of phase α, and ρ_α and V_α are the density and unit cell volume, respectively, of phase α. Alternatively, the unit cell volume can be incorporated into the variable, phase-specific parameters as outlined by Bish and Howard (1988). Therefore, in a Rietveld analysis of a multicomponent mixture, the scale factors contain the desired weight fraction information. However, the value of K and the sample mass absorption coefficient cannot easily be determined so that an analysis of an unknown sample is usually performed by constraining the sum of the weight fractions of the phases considered to unity. Thus, for a two-phase mixture,

$$W_\alpha = W_\alpha/(W_\alpha + W_\beta). \tag{54}$$

Equation 54 can be solved for the weight fractions of the α and β phases to yield an expression for the weight fraction of phase α in terms of the scale factor information determined in the Rietveld analysis.

$$W_\alpha = S_\alpha \rho_\alpha V_\alpha^2 / (S_\alpha \rho_\alpha V_\alpha^2 + S_\beta \rho_\beta V_\beta^2). \tag{55}$$

In general, the weight fraction for the ith component in a mixture of n phases can be obtained from:

$$W_i = S_i \rho_i V_i^2 / \sum_j S_j \rho_j V_j^2. \tag{56}$$

This method is exactly analogous to the adiabatic principle of Chung (1974a,b,1975) in which Reference Intensity Ratios are measured prior to analysis. Instead of measuring Reference Intensity Ratios to put all intensities on an absolute scale, the Rietveld method calculates absolute intensities.

A second method of Rietveld quantitative analysis requires that a known weight fraction of a crystalline internal standard be added to the unknown mixture. The internal standard can be any well-crystallized material that is readily available in pure form. Bish and Howard (1988) used Si, but corundum appears to be preferable (Linde A corundum). If W_α is known, then an additional parameter (C) can be evaluated from the internal standard:

$$C = S_\alpha \rho_\alpha V_\alpha^2 / W_\alpha = K/\mu^*. \tag{57}$$

This parameter can then be used to determine the weight fractions for other phases in the sample. For example, the weight fraction for the s phase is determined by

$$W_\beta = S_\beta \rho_\beta V_\beta^2 / C. \tag{58}$$

S_β is a refined parameter, ρ_β can be calculated from the composition and cell parameters of phase β, and C is determined using an internal standard and Equation 57. Therefore, the weight fraction of phase β (W_β) can be easily determined. This second method does not constrain the sum of the weight fractions, as does the first method, and it is analogous to internal-standard methods commonly used to perform quantitative analysis such as the matrix–flushing method of Chung (1974b). The total weight fraction of any amorphous components can also be determined with this method if the amorphous profile can be fit with the Rietveld background polynomial. The difference between the sum of the weight fractions of the crystalline components and 1.0 is the total weight fraction of the amorphous components.

O'Connor and Raven (1969) used this method in slightly modified form, instead choosing to determine the constant parameter, K, from a single sample, to use the refined cell parameters and cell contents to evaluate density and volume, and to calculate mass absorption coefficients using the known compositions of the two phases in their single sample. In light of their results on a 50:50 quartz:corundum mixture, in which they concluded that their quartz contained 18% amorphous component, it appears that some pitfalls may exist with this approach.

Applications. The Rietveld method of analysis provides numerous advantages over conventional quantitative analysis methods. Since the method uses a pattern-fitting algorithm, all lines for each phase are explicitly considered, and even severely overlapped lines are not problematic. The use of all lines in a pattern minimizes the uncertainty in the derived weight fractions and the effects of nonlinear detection systems. The effects of primary extinction are also minimized, as all reflections from each phase are used in the analysis rather than just the strongest ones. Because each phase in a mixture must be explicitly included in the analysis, failure to consider a phase will yield obvious differences between the observed and calculated diffraction patterns. The Rietveld quantitative analysis method can also be used in the traditional manner for refining the structural parameters of each phase in the mixture. Thus, information obtained by this method can include atom positions, site occupancies, and precise lattice parameters. This feature allows one to calculate tailor-made standards to match the phases in the mixture to be analyzed.

Preferred orientation of crystallites, a problem inherent in conventional quantitative analyses, is much less of a problem with full-pattern methods. The Rietveld method uses all classes of reflections and orientation effects tend to cancel. In addition, the method presents the opportunity to correct for preferred orientation using the March function (Dollase, 1986). It is important to note, however, that older preferred orientation corrections do not perform properly with multi-phase refinements. To our knowledge, it has not been verified experimentally that the March function can be used to improve quantitative analyses.

Few applications of Rietveld quantitative analyses have been published, particularly for geological samples, although the power and versatility of the method suggest that it will be widely applied to geologic samples. A number of researchers (Hill and Howard, 1987; Bish and Chipera, 1988; Bish and Howard, 1988; Bish and Post, 1988) have presented examples of quantitative analysis of multicomponent mixtures, and the method has been successfully applied to mixtures containing up to 10 components (S. Howard, personal communication).

Bish and Post (1988) presented the most thorough results on geological samples to-date illustrating the power of the Rietveld quantitative analysis method. Because their results were in abstract form, the complete results are presented here as examples. They applied the method to a variety of samples, including synthetic mixtures of clinoptilolite and corundum, quartz and corundum, hematite and corundum, and biotite and corundum. They also analyzed natural samples containing two volcanic feldspars, a mixture of aragonite and two distinct calcites, and the standard G-1 granite. Their data were obtained using Cu Kα X-rays on a conventional powder diffractometer. Results for the binary mixtures were excellent with the exception of hematite-corundum (see below). Final results for the 50:50 quartz-corundum mixture gave a composition of 49.8% quartz and 50.2% corundum and cell parameters of a = 4.9110(1) and c = 5.4021(4) Å for quartz and a = 4.7565(1) and c = 12.984(1) Å for corundum. These cell parameters are of high precision and compare well with literature values for quartz and corundum cell parameters. The plot of the observed and calculated patterns (R_{wp}=18.8%) for this mixture is shown in Figure 8.

Analysis of the 50:50 clinoptilolite:corundum mixture illustrates one of the most significant advantages of the method in analyzing complex mixtures, the ability to accommodate explicitly severely overlapping reflections. Figure 9 is the plot of the observed and calculated diffraction patterns (R_{wp} = 14.9%) for this mixture, illustrating the very large number of reflections (>300) used in the analysis. This analysis yielded 50.1% clinoptilolite and 49.9% corundum, excellent results considering the severely overlapped nature of the clinoptilolite pattern. Cell parameters for clinoptilolite were a = 17.6377(1), b = 17.962(1), c = 7.4002(1) Å, and β = 116.221(1)$°$, and those for corundum were a = 4.7584(1) and c = 12.990(1) Å. Figure 10 is an expansion of the region between 21.0 and 25.0$°$ 2θ illustrating the significant peak overlap in this region. The most intense doublet in this region is composed of six independent reflections, a situation that would make indexing this pattern difficult even when applying techniques such as profile refinement (Chipera and Bish, 1988; Chapter 8, this volume). This overlap would preclude highly precise and accurate unit-cell refinement by conventional methods, but the Rietveld method produced very precise cell parameters.

The analysis of the 50:50 biotite:corundum mixture is an example of the effects of preferred orientation on Rietveld quantitative analyses. Figure 11 is the

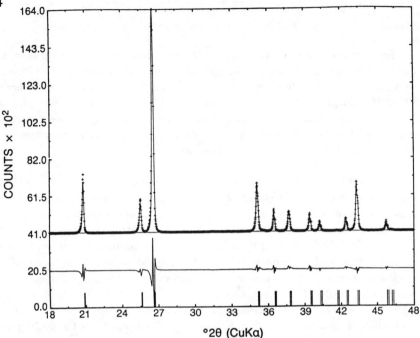

Figure 8. Observed (as pluses) and calculated (line) patterns for a 50:50 quartz corundum mixture. The lower curve shows the difference between the observed and calculated patterns. Vertical marks at the bottom indicate the positions of allowed $K\alpha 1$ and $K\alpha 2$ lines. Intensity has been displaced for clarity.

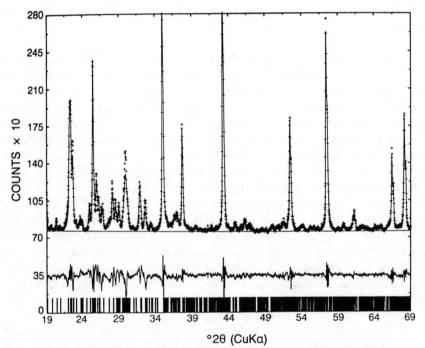

Figure 9. Observed and calculated patterns for a 50:50 clinoptilolite:corundum mixture using the conventions stated in Figure 8.

135

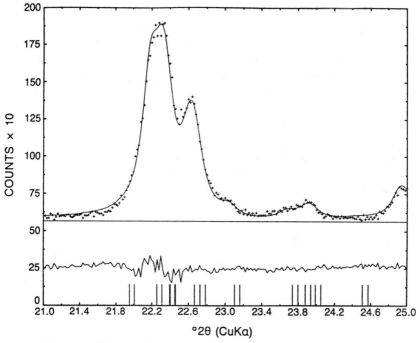

Figure 10. Close up of the 21.0 to 25.0° 2θ range for the 50:50 clinoptilolite:corundum mixture using the conventions stated in Figure 8.

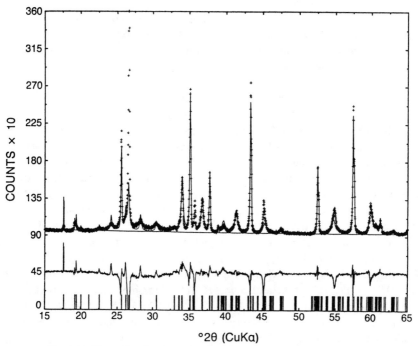

Figure 11. Observed and calculated patterns for a 50:50 biotite:corundum mixture using the conventions stated in Figure 8.

plot of the observed and calculated data, illustrating the very poor fit obtained in this analysis due to preferred orientation ($R_{wp} = 32.2\%$). In spite of the poor fit and the preferred orientation, quantitative results were good (47.5% biotite and 52.5% corundum). Cell parameters for biotite were a = 5.349(2), b = 9.248(3), c = 10.208(1) Å, β = 100.25°, and those for corundum were a = 4.7594(1) and c = 12.995(1) Å. This analysis highlights one of the strengths of the Rietveld method and points out a major weakness of traditional methods that rely upon ratios of only a few reflections.

The final synthetic binary mixture, 50:50 hematite:corundum, illustrates a potential problem with the Rietveld method when analyzing mixtures with one or more phases having a linear absorption coefficient significantly greater than the average for the sample. Figure 12 shows the good agreement between observed and calculated patterns for this mixture (R_{wp}=21.4%). In spite of this good agreement, the quantitative results, 44% hematite and 56% corundum, were relatively poor. The poor results for this sample are due to microabsorption, yielding low relative weight percents for the high linear absorption coefficient material, hematite. This problem can be minimized by grinding samples to very fine particle sizes or choosing a different radiation. Initial refinements with data obtained on <400 mesh material yielded even worse results. This mixture is a good example of a case in which traditional methods will implicitly correct for the problem during standardization, therefore potentially yielding results superior to those obtained with the Rietveld method. In spite of the microabsorption problem, precise cell parameters were still obtained (hematite: a = 5.0352(1), c = 13.7469(2) Å; corundum: a = 4.7589(1), c = 12.9911(3) Å). Because the Rietveld method provides an opportunity to calculate the linear absorption coefficient for every phase in a mixture, it may be possible to correct for microabsorption problems assuming an average crystallite shape and size (see Klug and Alexander, 1974, p. 541-542).

Natural geologic samples are more difficult to use in demonstrating the accuracy of the Rietveld method of quantitative analysis due to difficulties in determining the true quantities of phases in mixtures, but such samples do illustrate the versatility of the method. For the samples examined by Bish and Post (1988) optical point counts were used as an estimate of the true phase contents. The analysis of a mixture of two volcanic feldspars separated from a tuff again illustrated the power of the method in dealing with severely overlapped patterns. Fig. 13 shows the observed and calculated patterns for the mixture of sanidine and high albite; the fit is good (R_{wp}=26.6%) considering the complexity of the pattern and the large number of reflections. Rietveld analysis yielded 93% sanidine and 7% high albite, agreeing (perhaps fortuitously) well with the point count (100 grains) results of 93% sanidine and 7% albite. Cell parameters for sanidine were a = 8.4499(4), b = 13.008(1), c = 17.1755(4) Å, and β = 116.07°; parameters for high albite were a = 8.211(3), b = 12.926(6), c = 7.128(2)Å, α = 93.35(5), β = 116.37(3), and γ = 90.15(4)°. The cell parameters determined from such an analysis can be used, for example, with the familiar b-c plots to verify composition and state of ordering of the feldspars.

137

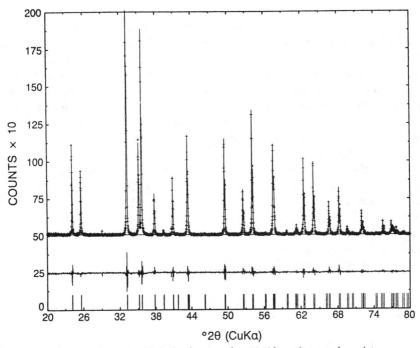

Figure 12. Observed and calculated patterns for a 50:50 hematite:corundum mixture using the conventions stated in Figure 8.

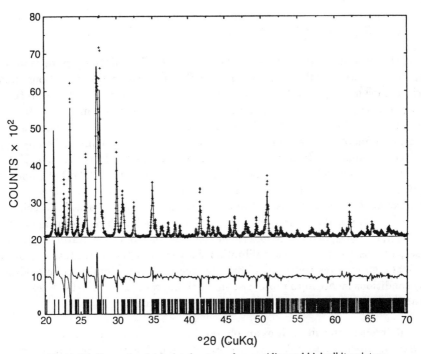

Figure 13. Observed and calculated patterns for a sanidine and high albite mixture using the conventions stated in Figure 8.

The analysis of the G-1 standard granite provided a more rigorous test of the method and involved analysis for four complex phases. The results of the Rietveld analysis for quartz, albite, microcline, and biotite (Figure 14) agreed with the modes of Chayes (1951) within one standard deviation and also provided precise cell parameters for all of the major phases in the rock. A final test involved evaluating the ability to distinguish closely overlapping reflections from similar phases. Bish and Post (1988) analyzed a synthetic mixture of 50% coralline algae and 50% echinoid calcite. The coralline algae contained both a Mg-calcite and aragonite and the echinoid contained a Mg-calcite of crystallite size significantly different from the calcite in the coralline algae. Results of the Rietveld analysis (Figure 15) gave 49.1% echinoid calcite, 41.0% coralline calcite, and 9.9% aragonite. Application of the cell parameters obtained for the two calcites to the determinative curve of Goldsmith et al. (1961) gave a Mg content of 20 mol% for the coralline calcite and 10 mol% for the echinoid calcite. Because of peak-width variations between the two distinct calcites and severe peak overlap, analysis of this sample by conventional methods would be extremely difficult.

Quantitative phase analysis using observed patterns

Quantitative phase analysis using observed patterns differs from the Rietveld quantitative analysis method in that it involves fitting a synthetic diffraction pattern constructed from appropriate observed standard diffraction patterns to an observed diffraction pattern. The concepts employed in this method have been described in detail by Smith et al. (1987) and are included here in brief form.

The procedure first involves collection of standard data on pure materials using fixed instrumental conditions, preferably identical to the conditions under which unknown samples will be analyzed. These standard patterns are processed to remove background and any artifacts, and the data may be smoothed. In some cases, reflections from unwanted phases may be removed from the standard pattern. For standard materials that are unavailable in appropriate form, diffraction patterns can either be simulated from Powder Diffraction File data and the program SIMUL (see chapter 7 this volume) or calculated using the program POWD10 (Smith et al., 1982). The next step in standardization is analogous to procedures used in conventional Reference Intensity Ratio quantitative analyses. Data are collected on each of the standard samples that has been mixed with a known amount of corundum, α-Al_2O_3, in order to determine the Reference Intensity Ratio (see above). These patterns yield the basic calibration data used in the quantitative analysis. The final step involves collection of data for the unknown samples to be analyzed, using conditions identical to those used in obtaining the standard data. Background and artifacts are also removed from unknown-sample data.

Reference Intensity Ratios are obtained in the same manner that quantitative analyses of unknown samples are performed, assigning the standard material a

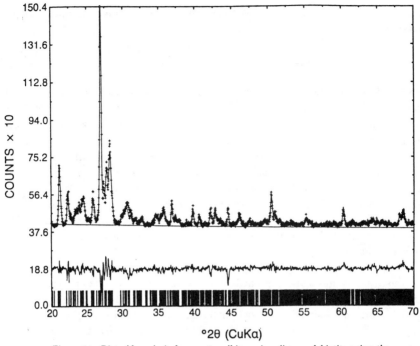

Figure 14. Rietveld analysis for quartz, albite, microcline, and biotite using the conventions stated in Figure 8.

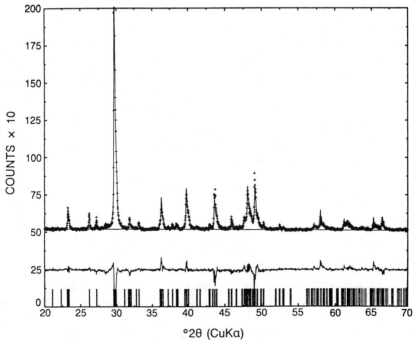

Figure 15. Rietveld analysis for echinoid calcite, coralline calcite, and aragonite using the conventions stated in Figure 8.

Reference Intensity Ratio of 1.0. The least-squares method involves minimizing the expression

$$\delta(2\theta) = I_{unk}(2\theta) - \sum_p W_p C_p I_p(2\theta), \qquad (59)$$

where $I_{unk}(2\theta)$ and $I_p(2\theta)$ are the diffraction intensities at each 2θ interval for the unknown and each of the standard phases, p, respectively. C_p is the Reference Intensity Ratio for phase p and W_p is the weight fraction. For analysis of Reference Intensity Ratios, the standard phase is assigned a ratio of 1.0 and the correct Reference Intensity Ratio is determined from the resultant weight fractions. For example (Smith et al., 1987), a standard mixture of 50% ZnO and 50% corundum yielded results of 57.8% ZnO and 42.2% corundum. The Reference Intensity Ratio for ZnO is therefore $57.8/42.2 = 1.37$. It is important to note that the Reference Intensity Ratios derived using this procedure are equivalent to peak-height RIRs rather than integrated-intensity RIRs because the minimization is conducted on a step-by-step basis.

Just as with most other quantitative analysis methods, the analyst must first determine what phases are present in the mixture before performing quantitative analysis. Using the selected standard patterns, the weighted sum of these patterns is least-squares fit to the pattern of the unknown. Weight fractions are obtained from the Reference Intensity Ratios, and results are typically reported on the basis of a material that is 100% crystalline, i.e., the sum of all phases is constrained to total to 100%. This analysis is analogous to the adiabatic method of Chung (1974b). One of the components of the sample may be given a fixed weight fraction if an internal standard has been added. This procedure, analogous to the matrix-flushing method of Chung (1974a), does not constrain the total weight fraction and can thus indicate the presence of amorphous components or unidentified phases. Unlike the Rietveld method of quantitative analysis, this full-pattern method does not explicitly vary the cell parameters of the phases in the simulated pattern. A routine has been incorporated that allows movement of the unknown pattern along the 2θ scale to give better fits, but Smith et al. (1987) commented that quantitative results are little affected. Reference patterns may be broadened by convolution with a crystallite size function prior to fitting.

Results reported by Smith et al. (1987,1989) for several mixtures are encouraging. As their most rigorous test, they analyzed a six-component mixture containing 16.7% each of fluorite (CaF_2), calcite ($CaCO_3$), corundum ($\alpha\text{-}Al_2O_3$,), $Ca(OH)_2$, halite (NaCl), and quartz ($\alpha\text{-}SiO_2$). Absolute standard deviations were below 2% for all phases, with relative errors less than 12%. Smith et al. (1988) analyzed a four-component mixture of 30% fluorite, 10% calcite, 20% corundum and 40% ZnO. They tested the method using: 1) experimental reference patterns; 2) calculated reference patterns (Smith et al., 1982); 3) simulated PDF reference patterns (SIMUL). Results for all three sets of standard patterns were excellent, with maximum absolute errors of <3%. The ability to use calculated or simulated reference patterns is a significant advantage of this method. Smith et al. (1989) included chemical constraints with this method of full-pattern quantitative analysis, resulting in improved accuracy and some information on the chemical composition of individual phases. They analyzed six mixtures containing between five and eight common minerals and obtained very good results. The use of total chemistry with the X-ray results yielded lower errors for all mixtures, with average total absolute errors of about 6%. Experience with this method has shown that it is relatively insensitive to preferred orientation effects because of the effects of averaging over numerous reflections, just as is the Rietveld method. In addition, preferred orientation and microabsorption effects are further minimized by using standards that are comparable in these respects to the unknowns. The method should also be insensitive to several other factors such as extinction and peak overlap because the method uses the full diffraction pattern. Perhaps the biggest advantage of this method is that is does not rely on calculating the three-dimensional diffraction pattern of each component of a mixture. Thus this full-pattern fitting method can theoretically treat materials of unknown structure, clay minerals, and even amorphous solids. Clay minerals and related layer-structure materials are unique in that they often exhibit two-dimensional diffraction effects (see chapter 6, this volume) that cannot be approximated using three-dimensional diffraction patterns. Smith et al. (1987) described the application of this method to both qualitative and quantitative analysis of clay minerals. Another advantage of this method is that is does not require a thorough understanding of the theory behind calculated diffraction patterns nor does it require a knowledge of the crystal structures of each of the phases present in a mixture.

In summary, full-pattern methods of quantitative analysis have several significant advantages over more conventional methods employing at most a few of the strongest reflections from each phase in a mixture. Because these methods use the full observed diffraction patterns, problems in obtaining reproducible intensities and in dealing with severely overlapped profiles are eliminated. Ambiguities in determining background position during intensity measurement are also largely eliminated. In addition, some of the more troublesome effects in conventional quantitative analysis, such as preferred orientation and extinction, are minimized. The Rietveld method provides the added benefit of yielding high-precision cell parameters for each phase in a mixture and has the potential for providing information on the structures of individual crystalline components. We believe that these methods of quantitative analysis may spur a Renaissance in analysis of crystalline components of mixtures such as rocks and will allow data previously unobtainable to be within the reach of any researcher with powder diffraction capabilities.

REFERENCES

Alexander, L. and Klug, H.P. (1948) X-ray diffraction analysis of crystalline dusts. Anal. Chem. 20, 886-894.
Bish, D.L. and Chipera, S.J. (1988) Problems and solutions in quantitative analysis of complex mixtures by X-ray powder diffraction. Adv. in X-ray Anal. 31, 295-308.
_____ and Howard, S.A. (1986) Quantitative analysis via the Rietveld method. Workshop on Quantitative X-ray Diffraction Analysis, Nat'l Bur. Stds, June 23-24.
_____ and Howard, S.A. (1988) Quantitative phase analysis using the Rietveld method. J. Appl. Cryst. 21, 86-91.
_____ and Post, J.E. (1988) Quantitative analysis of geological materials using X-ray powder diffraction data and the Rietveld refinement method. Geol. Soc. Am. Abstr. Programs 20, A223.
Boski, T. and Herbillon, A.J. (1988) Quantitative determination of hematite and goethite in lateritic bauxites by thermodifferential X-ray powder diffraction. Clays & Clay Minerals 36, 176-180.
Braun, G.E. (1986) Quantitative analysis of mineral mixtures using linear programming. Clays & Clay Minerals 34, 330-337.
Brindley, G.W. (1980) Quantitative X-ray mineral analysis of clays. In G.W. Brindley and G. Brown, eds., Crystal Structures of Clay Minerals and Their X-ray Identification. Mineralogical Society, London, p. 411-438.
Carter, J.R., Hatcher, M.T., and Di Carlo, L. (1987) Quantitative analysis of quartz and cristobalite in bentonite clay based products by X-ray diffraction. Anal. Chem. 59, 513-519.
Chayes, F. (1951) Modal analyses of the granite and diabase test rocks. U.S. Geol. Surv. Bull. 980, 59-68.
Chipera, S.J. and Bish, D.L. (1988) Pitfalls in profile refinement of clay mineral diffraction patterns. Clay Minerals Soc. 25th Annual Meeting Abstr. 39.
Chung, F.H. (1974a) Quantitative interpretation of X-ray diffraction patterns. I. Matrix-flushing method of quantitative multicomponent analysis. J. Appl. Cryst. 7, 519-525.
_____ (1974b) Quantitative interpretation of X-ray diffraction patterns. II. Adiabatic principle of X-ray diffraction analysis of mixtures. J. Appl. Cryst. 7, 526-531.
_____ (1975) Quantitative interpretation of X-ray diffraction patterns. III. Simultaneous determination of a set of reference intensities. J. Appl. Cryst. 8, 17-19.
Clark, G.L. and Reynolds, D.H. (1936) Quantitative analysis of mine dusts. Ind. Eng. Chem., Anal. Ed. 8, 36-42.
Cline, J.P. and Snyder, R.L. (1983) The dramatic effect of crystallite size on X-ray intensities. Adv. in X-ray Anal. 26, 111-118.
_____ and Snyder, R.L. (1985) Powder characteristics affecting quantitative analysis. In R.L. Snyder, R.A. Condrate and P.F. Johnson, eds., Advances in Material Characterization II. Plenum Press, New York, p. 131-144.
_____ and Snyder, R.L. (1987) The effects of extinction on X-ray powder diffraction intensities. Adv. in X-ray Anal. 30, 447-456.
Copeland, L.E. and Bragg, R.H. (1958) Quantitative X-ray diffraction analysis. Anal. Chem. 30, 196-206.
Cullity, B.D. (1978) Elements of X-ray Diffraction. Addison-Wesley, Reading, Massachusetts.
Davis, B.L. and Walawender, M.J. (1982) Quantitative mineralogical analysis of granitoid rocks: a comparison of X-ray and optical techniques. Am. Mineral. 67, 1135-1143.
Dollase, W.A. (1986) Correction of intensities for preferred orientation in powder diffractometry: application of the March model. J. Appl. Cryst. 19, 267-272.
Garbauskas, M.F. and Goehner, R.P. (1982) Complete quantitative analysis using both X-ray fluorescence and X-ray diffraction. Adv. in X-ray Anal. 25, 283-288.
Goldsmith, J.R., Graf, D.L., and Heard, H.C. (1961) Lattice constants of the calcium-magnesium carbonates. Am. Mineral. 46, 453-457.
Gehringer, R.C., McCarthy, G.J., Garvey, R.G., and Smith, D.K. (1983) X-ray diffraction intensity of oxide solid solutions: application to qualitative and quantitative phase analysis. Adv. in X-ray Anal. 26, 119-128.
Hill, R.J. and Howard, C.J. (1987) Quantitative phase analysis from neutron powder diffraction data using the Rietveld method. J. Appl. Cryst. 20, 467-474.

Hodgson, M. and Dudeney, A.W.L. (1984) Estimation of clay proportions in mixtures by X-ray diffraction and computerized chemical mass balance. Clays & Clay Minerals 32, 19-28.

Hussey, G.A., Jr. (1972) Use of a simultaneous linear equations program for quantitative clay analysis and the study of mineral alterations during weathering. Unpublished Ph.D. thesis, The Pennsylvania State University, University Park, Pennsylvania.

Hubbard, C.R., Evans, E.H., and Smith, D.K. (1976) The Reference Intensity Ratio, I/I_c for computer simulated patterns. J. Appl. Cryst. 9, 169-174.

_____, Robbins, C.R., and Snyder, R.L. (1983) XRD quantitative analysis using the NBS*QUANT82 system. Adv. in X-ray Anal. 26, 149-157.

_____ and Snyder, R.L. (1988) Reference Intensity Ratio—measurement and use in quantitative XRD. Powder Diffraction 3, 74-78.

Johnson, L.J., Chu, C.H., and Hussey, G.A. (1985) Quantitative clay mineral analysis using simultaneous linear equations. Clays & Clay Minerals 33, 107-117.

Klug, H.P. and Alexander, L.E. (1974) X-ray Diffraction Procedures, 2nd edition. John Wiley and Sons, New York, (a) p. 549-553, (b) p. 365-368.

Lennox, D.H. (1957) Monochromatic diffraction-absorption technique for direct quantitative X-ray analysis. Anal. Chem. 29, 767-772.

Leroux, J., Lennox, D.H., and Kay, K. (1953) Applications of X-ray diffraction analysis in the environmental field. Anal. Chem. 25, 740-748.

Maniar, P.D. and Cooke, G.A. (1987) Modal analyses of granitoids by quantitative X-ray diffraction. Am. Mineral. 72, 433-437.

Navias, A.L. (1925) Quantitative determination of the development of mullite in fired clays by an X-ray method. J. Am. Ceram. Soc. 8, 296-302.

O'Connor, B.H. and Raven, M.D. (1988) Application of the Rietveld refinement procedure in assaying powdered mixtures. Powder Diffraction 3, 2-6.

Parker, R.J. (1978) Quantitative determination of analcime in pumice samples by X-ray diffraction. Mineral. Mag. 42, 103-106.

Pawloski, G.A. (1985) Quantitative determination of mineral content of geological samples by X-ray diffraction. Am. Mineral. 70, 663-667.

Rietveld, H.M. (1969) A profile refinement method for nuclear and magnetic structures. J. Appl. Cryst. 2, 65-71.

Renault, J. (1987) Quantitative phase analysis by linear regression of chemistry on X-ray diffraction intensity. Powder Diffraction 2, 96-98.

Smith, D.K., Nichols, M.C. and Zolensky, M.E. (1982) POWD10. A FORTRAN IV program for calculating X-ray powder diffraction patterns–version 10. The Pennsylvania State University, University Park, Pennsylvania.

_____, Johnson, G.G., Jr., and Ruud, C.O. (1986) Clay mineral analysis by automated powder diffraction analysis using the whole diffraction pattern. Adv. in X-ray Anal. 29, 217-224.

_____, _____, Scheible, A., Wims, A.M., Johnson, J.L., and Ullmann, G. (1987) Quantitative X-ray powder diffraction method using the full diffraction pattern. Powder Diffraction 2, 73-77.

_____, _____, and Wims, A.M. (1988) Use of full diffraction spectra, both experimental and calculated, in quantitative powder diffraction analysis. Aust. J. Phys. 41, 311-321.

_____, _____, Kelton, M.J., and Anderson, C.A. (1989) Chemical constraints in quantitative X-ray powder diffraction for mineral analysis of the sand silt fractions of sedimentary rocks. Adv. in X-ray Anal. 32, 489-496.

Smith, S.T., Snyder, R.L., and Brownell, W.E. (1979a) Minimization of preferred orientation in powders by spray drying. Adv. in X-ray Anal. 22, 77-88.

_____, _____, _____ (1979b) Quantitative phase analysis of Devonian shales by computer controlled X-ray diffraction of spray dried samples. Adv. in X-ray Anal. 22, 181-191.

Snyder, R.L. and Carr, W.L. (1974) A method of determining the preferred orientation of crystallites normal to a surface. In V.D. Frechette, ed., Surfaces and Interfaces of Glass and Ceramics, p. 85-99.

_____ and Hubbard, C.R. (1984) NBS*QUANT84: A system for quantitative analysis by automated X-ray powder diffraction. Nat'l Bur. Stds Special Publication.

_____, _____, and Panagiotopoulos, N.C. (1981) AUTO: A Real Time Diffractometer Control System. Nat'l Bur. Stds IR 81-2229.

_____, _____, _____ (1982) A second generation automated powder diffraction control system. Adv. in X-ray Anal. 25, 245-260.

Visser, J.W. and deWolff, P.M. (1964) Absolute Intensities. Report 641.109, Technisch Physische Dienst, Delft, Netherlands.

Wiles, D.B. and Young, R.A. (1981) A new computer program for Rietveld analysis of X-ray powder diffraction patterns. J. Appl. Cryst. 14, 149-151.

6. DIFFRACTION BY SMALL AND DISORDERED CRYSTALS
R.C. Reynolds, Jr.

INTRODUCTION

Chapter 1 described the fundamentals of three-dimensional X-ray diffraction from randomly oriented crystalline powders composed of crystallites or crystal fragments of finite size. In this chapter, the equations that arise from these principles are modified so that diffraction phenomena can be treated for one- two- and three-dimensional examples of crystals that contain disorder of different kinds and whose dimensions differ markedly from the shapes of equi-dimensional parallelepipeds. Consideration is given here to (1) two-dimensional diffraction from crystals that are highly disordered in one direction, (2) one-dimensional diffraction from structures with different kinds of unit cells in various stacking patterns, (3) the modification of "normal" diffraction patterns by defects that consist of unit cell displacements whose magnitudes and frequencies are random, and (4) the degradation of long-range structural order that leads to a glass. These crystal structure characteristics are partially or wholly manifested by diffraction line shape, or in the change in line shape with respect to the diffraction angle 2θ. The term "shape" includes peak breadth at half maximum height (β), the symmetry or lack thereof about the intensity maximum, and the mathematical form of the diffraction profile which may be Lorentzian or Gaussian or a form intermediate between these two. The need to extract pure line shape from the experimental profile has prompted a brief discussion of procedures for removing instrumental effects. Accurate diffraction intensity data are vital to any structural interpretation, so the effects are discussed of preferred orientation on the relative intensities of the basal diffraction series for the clay minerals, which are the most troublesome in this respect.

The approach used here is not easily extended to the most practical methods for calculating diffraction patterns of disordered structures. Such calculations are most efficiently done by the matrix techniques developed by Kakinoki and Komura (1952), which have been rigorously applied by Plançon and Tchoubar (1977), Plançon (1981), and more recently by Plançon et al. (1988). Their work is not described here because it is difficult for the non-specialist. The writer stays with the elementary (but just as valid) approach of James (1965) and Brindley (1980), with the disadvantage that numerical calculations are likely to be more time-consuming. All diffraction patterns shown in this chapter were calculated for $CuK\alpha$ radiation.

DIFFRACTION BY SMALL CRYSTALS

Equi-dimensional parallelepiped crystals

Equation 1 below and its components describe three-dimensional diffraction from a random powder aggregate of finite crystals. The calculated diffraction patterns are shown

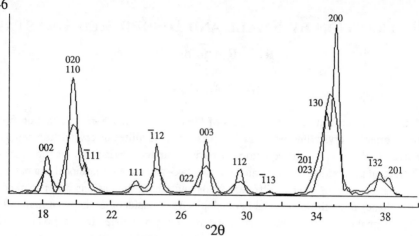

Figure 1. Calculated three-dimensional diffraction patterns for dehydrated Na-montmorillonite. Sharp trace is for $N_1 = 40$, N_2 and $N_3 = 20$; broad trace is for $N_1 = 20$, N_2 and $N_3 = 10$.

by Figure 1. The summations are made in reciprocal space as shown by Figure 5, Chapter 1, and Figure 2 below. The latter has been drawn for the special case of two-dimensional diffraction, but the required definitions are relevant to it. h, k, and l are continuous variables, a, b, and c are the unit cell dimensions and β is the monoclinic angle. The distances on Figure 2 of $1/a\sin\beta$, $1/b$, and $1/c\sin\beta$ are the reciprocal lattice lengths a^*, b^* and c^*. F is the amplitude of scattering from the unit cell in the direction hkl, and N_1, N_2, and N_3 refer to the number of unit cells that make up the crystal in, respectively, the X, Y, and Z directions. The summations are taken over the ranges of γ_1 and γ_2 required by the unit cell symmetry. For this case, the space group $C2/m$ allows the elimination of those regions of reciprocal space for which k and l are negative numbers, that is, $\gamma_1 > 90°$ and $\gamma_2 > 180°$. The reader should refer to Chapter 1 for a more detailed description of these variables and the nature of the problem.

$$I = \frac{1+\cos^2 2\theta}{N_1 N_2 N_3 \sin\theta} \sum_{\gamma_1=0}^{\gamma_1=90} 2\pi R_1 \sum_{\gamma_2=0}^{\gamma_2=180} |F(hkl)|^2 \, \Phi_1 \Phi_2 \Phi_3 \, \Delta\gamma_1 \Delta\gamma_1 \quad . \quad (1)$$

The terms Φ_1, Φ_2, and Φ_3 refer to the interference function in, respectively, the h, k, and l directions. These are

$$\Phi_1 = \frac{\sin^2(\pi h\, N_1)}{\sin^2(\pi h)} \quad , \quad \Phi_2 = \frac{\sin^2(\pi k\, N_2)}{\sin^2(\pi k)} \quad , \quad \Phi_3 = \frac{\sin^2(\pi l\, N_3)}{\sin^2(\pi l)} \quad . \quad (2)$$

Other relations required for the unit transformations among d, θ, γ_1, γ_2, h, k, and l are given below.

$$d = \lambda / 2\sin\theta$$

$$R_1 = (1/d) \cos\gamma_1$$

$$\beta^* = 180 - \beta$$

$$h = a\sin\beta \, (R_1 \cos\gamma_2 - \sin\gamma_1 / d\tan\beta^*)$$

$$k = R_1 b \sin\gamma_2$$

$$l = (c/d) \sin\gamma_1$$

The Lorentz-polarization factor is accounted for by the quantity $(1 + \cos^2 2\theta)/\sin\theta$ because the form of the summation contains, implicitly, the point-by-point single crystal factor which is $1/\sin\theta$. For a randomly oriented powder, then, the Lorentz factor for this method of calculation is reduced to the powder ring distribution factor, $1/\sin\theta$. The calculation is done by selecting a value for θ which gives $1/d$. The summations are made over γ_1 and γ_2, the sum recorded, another value for θ is selected etc.

Figure 1 shows the diffraction pattern for a dehydrated sodium-montmorillonite for which $1M$ stacking is assumed with $a = 5.21$ Å, $b = 9.00$ Å, $c = 9.90$ Å, and $\beta = 101.6°$. The "sharp" pattern was calculated for $N_1 = 40$ and $N_2 = N_3 = 20$ unit cells, an approximately parallelepiped crystal. Idealized hexagonal symmetry was assumed for the atomic positions in each of the atomic sheets, that is, there is no tetrahedral tilt or rotation. False peaks in the diffraction pattern represent ripples in the interference function and would be eliminated by summing the interference function over a range of N_1, N_2, and N_3. The diffraction pattern for crystals that are one-half as large in each of the three dimensions is superimposed on this. As seen, the peaks are approximately twice as broad at half-height, in accordance with the Scherrer equation (Chapter 1), and division by $N_1 N_2 N_3$ in Equation 1 has resulted, as it should, in equal areas for each of the corresponding reflections hkl. If the crystal dimensions are increased, at some point the laborious calculation of Equation 1 becomes needless, because the intrinsic peak breadths (due to crystal size) become sharp, but their breadths are dominated by the ever-present diffractometer broadening function. For such crystals, the diffraction peaks are completely described by the Lorentz-polarization factors times $|F|^2$.

Two-dimensional diffraction from turbostratic structures

Interesting and instructive experiments can be performed by calculating the patterns for crystals that are markedly restricted in one or two dimensions with respect to the other(s). An example of practical importance is the case of montmorillonite with turbostratic stacking along Z which gives rise to two-dimensional diffraction bands. For such a structure, the 2:1 layers are rotated about Z or displaced in the X-Y plane, by random increments, to produce a stacking similar to that which would result from applying a rotary and shearing motion to a stack of playing cards. Coherence is maintained along Z, but no fixed phase relations remain between layers in the X and Y and Z directions. The model for this is one in which the crystals are a single 2:1 layer thick, with respect to diffraction angles for which h and/or $k > 0$, but consist of N layers in a coherent array along Z. To calculate the diffraction effects for such an arrangement, we need to set $N_3 = 1$ in Equation 1, but the result will not include the necessary 00l contributions, which can be added later.

Figure 2 shows the reciprocal space construction for the interference function (Φ) for turbostratic montmorillonite. In this example the crystals scatter as single X-Y layers, so the reciprocal spots are drawn out in the c^* direction to form rods. These rods extend indefinitely upward, but have been arbitrarily truncated on Figure 2 for clarity. The diameters of the rods in the a^* and b^* directions are equal to, respectively, $1/N_1$ and $1/N_2$, where N_1 and N_2 are the numbers of unit cells in the X and Y crystal directions that make up the crystal. The diffuse rods in Figure 2 depict diagrammatically the contribution of the interference function to the distribution of diffraction intensity in reciprocal space, that is, the intensity distribution that would result if the unit cell scattered with unit intensity at all values of h, k, and l. The diffuse rods of Figures 2 and 3 are highly idealized, but they are useful for visualizing the concept of two-dimensional diffraction discussed below. The complete formulation for intensity requires that each point of the reciprocal space Φ-distribution be multiplied by the square of the magnitude of the scattering from the unit cell at the appropriate reciprocal space coordinate hkl.

The radius R_1 produces a circle, the summation along which passes through the centers of the 11 a rod at the positions $l = 1$, and perhaps catches the outer edge of the 02 rod. Summation over all values of γ_1 and γ_2 at this fixed value of $1/d$ produces the combined intensities of the interference function for the 11 and 02 bands at that specific value of 2θ. The total diffraction intensity is obtained at the corresponding increment of 2θ if each step of the summation is multiplied by the appropriate value of $F(hkl)^2$, and the final result is corrected for Lorentz-polarization effects. This procedure is identical to that proposed by Brindley (1980) except that his analytical procedure is performed by summing along circles whose radii lie in a plane normal to X-Y (see Fig. 3).

The Brindley (1980) method is described by Figure 3, which shows a reciprocal space construction for the interference function. Reciprocal lattice rods for the $-h$ positions have been omitted for clarity. This approach is described because the development of the band shapes is clear. It is shown, however, that the choice of summation directions selected for Equation 1 and Figure 2 leads to a more efficient algorithm for calculating diffraction patterns for crystals with disorder or irregularities in the stacking along Z. Summation along the circle with the radius $1/d_1$ produces a small but significant intensity value at the band-head, or the low 2θ limit of the band. The integration circle just barely intercepts the reciprocal lattice rod, and the length of interception is very short. The next angular increment of $1/d_2$ involves a long summation through the core of the 11 rod with a resulting large increase in intensity that corresponds roughly to the band-head. Further increases of $1/d$ cause these larger radii summation circles to cut smaller and smaller cross-sections from the rod resulting in intensity that diminishes continuously with increasing $1/d$ and thus increasing 2θ. This procedure gives the intensity profile of the rod with respect to 2θ along a specific direction that is fixed by γ_2. The entire calculation requires summing such intensity distributions over different increments of γ_2 (Fig. 3).

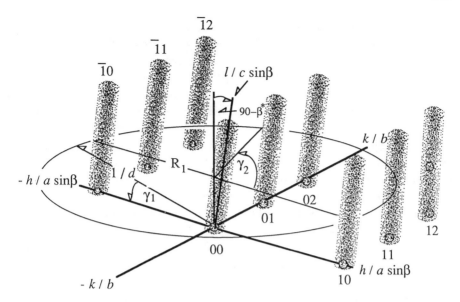

Figure 2. Reciprocal space construction for Eq. 1. The diffuse rods schematically represent the intensity distribution of the interference function for turbostratic montmorillonite. They are not spots because the crystal is only one unit cell thick in the c^* direction.

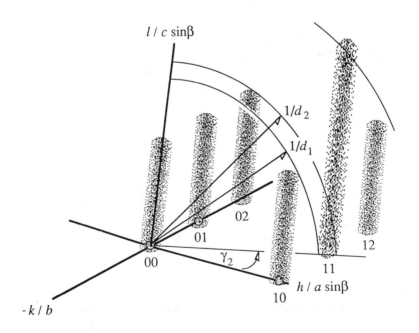

Figure 3. Brindley's (1980) method for calculating the diffraction profiles of two-dimensional bands.

Figure 4 shows the two-dimensional band characteristics of montmorillonite with $N_1 = 20$, $N_2 = 10$, and $N_3 = 1$. This should be compared with the three-dimensional diffraction pattern of Figure 1. The bands in Figure 4 have their characteristic shapes: a rapid rise in intensity toward increasing 2θ up to the band head, followed by a exponential intensity decline beyond. If F were constant with respect to l along the reciprocal lattice rods (Fig. 2), then all such bands would display the "normal" shape described. The shapes are often not this simple because F varies along c^* and because, for most mica-like structures, each band is a composite of two bands whose heads have different values of d. The near-constant F with respect to c^* makes the 20;13 the better choice for studies of three-dimensional structural organization or lack thereof in montmorillonite. The decrease in F with respect to c^* causes the 02;11 to be quite symmetrical even for complete turbostratic stacking, hence its shape is relatively insensitive to turbostratic disorder. The high-background region on the high-angle side of the 02;11 can be useful for interpretative purposes, but it is often obscured by a strong $00l$ reflection which is not included in the pattern of Figure 4.

Figure 5 represents a realistic two dimensional diffraction pattern for randomly oriented dehydrated turbostratic montmorillonite. The $00l$ pattern has been included by adding the results of a separate calculation. This procedure is valid because Equation 1 deals with intensity, not amplitude, and the intensities from the different regions of reciprocal space are additive. The basal pattern was computed according to the defect-broadening model of Ergun (1970), which is described below, though equally valid results could have been obtained by setting, for example, $N_3 = 10$. The basal diffraction series is calculated by summing Equation 1 only over the region of reciprocal space for which $|h|$ and k are < 0.5, which may be achieved by setting the intensity to zero for the reciprocal space region that is out of bounds. This procedure can be visualized by inserting a cylinder of elliptical (radii h = ±0.5 and k = ±0.5) cross section about the 000 rod of Figure 2 and confining the summations to the region inside of this cylinder. A point-by-point addition of this pattern to the profile of Figure 4 produced the results of Figure 5.

Figures 1 and 5 depict, respectively, the diffraction patterns for three and two-dimensional crystals. A reasonable question, at this point, is, what is the nature of diffraction from something in between, that is, a crystalline powder in which there are single layers, some two-layer crystallites, and perhaps a smaller number of thicker ones? To answer this question, we need to consider the interference function as a Fourier series, and the application of Ergun's model of defect broadening (Ergun, 1970).

<u>Two-dimensional diffraction with some third-dimensional influence</u>

The three sine-squared quotients in Equation 1 define the interference function in each of the three directions h, k, and l. Any one or all of these can be replaced by their Fourier series equivalents, but here we choose only the expression for l and designate it

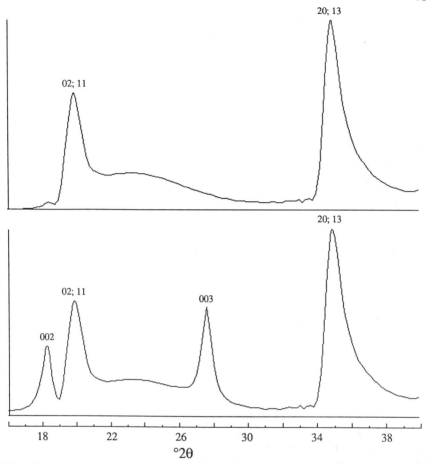

Figure 4 (upper pattern). Band structures for dehydrated turbostratic Na-montmorillonite. $N_3 = 20$, $N_3 = 10$ and $N_3 = 1$.

Figure 5 (lower pattern). Calculated diffraction pattern for turbostratic Na-montmorillonite. $00l$ series is based on the defect-broadening parameters $\delta = 6$, $N = 30$.

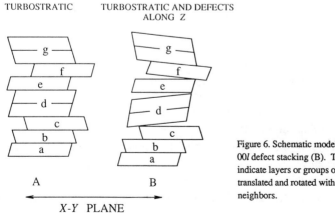

Figure 6. Schematic models for turbostratic (A) and $00l$ defect stacking (B). The designations a, b, c, etc. indicate layers or groups of layers that are randomly translated and rotated with respect to their adjacent neighbors.

Φ_3. This can be written (James 1965)

$$\Phi_3 = N_3 + 2 \sum_{n=1}^{n=N_3-1} (N_3 - n) \cos(2\pi nl) , \qquad (3)$$

where N_3 in this case is the number of mica-like silicate layers in a coherent scattering array, and n is the number of interlayer spacings separating any two of them. The summation is taken over all layer pairs. This expression is the Debye equation solved for a crystal that contains identical layers (identical F for all) that are centrosymmetric on projection to Z (James, 1965). It is useful for many crystallographic calculations. The solution requires more steps than the sine-squared quotient, but it makes layer-by-layer phase interactions mathematically accessible, and these are essential for treatments of stacking disorder along Z and the mixed-layering of two (or more) different kinds of unit layers along Z. Ergun (1970) has considered the effect of randomly distributed stacking defects along Z that break a large crystal up into a statistical assemblage of smaller coherent scattering domains. If the separation of the stacking defects is random, then the probability of a defect-free stacking, δ, is given by exp (-n / δ). The exponential term modifies the weighting term (N - n), so that if N is large compared to δ, the hypothetical crystal consists of a special particle-size distribution of smaller diffracting domains. The overall result is

$$\Phi_3 = N_3 + 2 \sum_{n=1}^{n=N_3-1} (N_3 - n) e^{-n/\delta} \cos(2\pi nl) . \qquad (4)$$

The distribution exp (-n / δ) fixes the probability of occurrence at unity for single layers (n = 0), whose contribution to the diffraction pattern is included by the constant N_3 in Equation 4. Larger and larger layer separations become exponentially less and less probable so that, for practical purposes, N need not exceed 7δ. Ergun has shown that a one-dimensional peak profile has a Lorentzian or Cauchy shape for pure defect broadening (N $\geq$ 7δ). The line shape approaches a Gaussian form as the diffraction domains become more and more particle-size limited (N < 7δ). The 00l portion of the diffraction pattern of Figure 5 was calculated by substituting Equation 4 into Equation 1, for Φ_3, and setting δ = 6 and N_3 = 42.

The defect theory can be applied to two-dimensional as well as one-dimensional diffraction calculations. Figure 6 shows a simplified schematic representation of two kinds of stacking disorder in montmorillonite. Figure 6A represents random displacements of unit layers in the X-Y plane and random rotation angles, but the sketch is a simplification made necessary by drafting constraints. What is intended is the description of turbostratic stacking in which random translations occur in any direction in the X-Y plane, and rotations of any angle are allowed. These random displacements are labelled a, b, c, d - - - etc. Correctly displayed, this structural model contains individual layers, two-layer 1M-type crystallites, and perhaps thicker 1M arrays. The two-dimensional diffraction bands from such a structure show some modulation of the bands into definite maxima as the thicker arrays dominate any hypothetical domain-size distribution, but the 00l peaks are relatively sharp because they represent the diffraction from the entire stack of layers. Figure 6B represents the introduction of another kind of stacking fault. These superimpose on the turbostratic stacking, and they separate the crystal into smaller coherent domains along Z, resulting in broader 00l diffraction maxima. These defects increase the turbostratic stacking disorder, but for montmorillonite, the type B defects are much more widely spaced that the type A displacements, thereby allowing the effects of the B defects to be ignored for the calculation of two-dimensional diffraction bands. The calculations for the bands have excluded the contributions from the 00l series.

Figure 7 shows the change in the 20;13 band structure as a function of δ. In the context of the discussion here (Eqs. 1 and 4), δ is the average number of interfaces along Z connecting ordered neighboring layers. If $\delta = 0$, the structure contains only individual layers. If δ is very large compared to N_3 (e.g. 1000), the crystal has an ordered 1M three-dimensional structure. The profiles of Figure 7 have been calculated for $N_1 = 40$, $N_2 = 20$, and $N_3 = 20$. It is interesting that only limited order is necessary to develop noticeable three-dimensional modulation of the 20;13 band. These results suggest that the common, well-defined turbostratic band structure of montmorillonite means that there are very few layers organized even into pairs. For $\delta = 0.5$, 87 % of the layers are separate (n = 0), that is, most layers are randomly displaced with respect to their neighbors, and for $\delta = 1.0$, 64 % are single and 23 % are involved in pairs.

The changes in the diffraction patterns of Figure 7 suggest that decreasing δ has caused asymmetrical increases in broadening for both the 200 and 130 reflections, and that the intensity of the 200 has decreased with respect to the intensity of the 130. The final 20;13 band lies at almost the same d as the three-dimensional 130 reflection. Brindley (1980) explained that an ideal two-dimensional band is displaced to larger d than $d(hk0)$. He noted that this behavior requires that the magnitude of F be nearly constant with respect to c^*. The patterns shown in Figure 7 do not seem to follow this rule. But we are dealing here with the behavior of a composite of two bands (or four). Figure 8 shows the composite parts of the 20;13. The results were obtained by calculating the diffraction patterns separately for the 20 and the 13 bands. We see that the relative intensities of the 20 and 13 have remained nearly constant, but the 20 has been displaced to larger d whereas the 13 has not. The nearly symmetrical shape of the 13 indicates that $|F(13l)|^2$ decreases along c^*, and this decrease partially compensates for the intrinsically asymmetrical nature of the interference function. The 13 two-dimensional band is actually very similar to the 130 reflection because $|F(13l)|^2$ and thus intensity diminishes as l is increased. Consequently, the d values are similar for the 13 and 130 diffraction maxima, and reducing the crystal thickness along Z has little effect of the d-spacing for the 130 peak. A simple explanation of such effects is complicated by the unit cell symmetry because $d(20l) \neq d(\overline{2}0l)$ and $d(13l) \neq d(\overline{1}3l)$ (except when $l = 0$), so the sum of the diffraction patterns of four bands are involved, the 20l, $\overline{2}0l$, 13l and $\overline{1}3l$. This example demonstrates that care must be applied when interpretations of structure are based on the

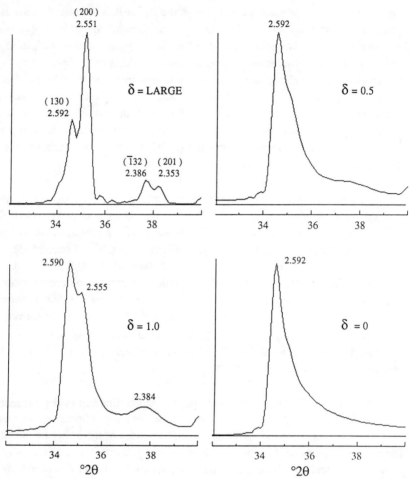

Figure 7. Changes in the 20;13 band for different degrees of third-dimensional dependence. $N_1 = 40$ and $N_2 = 20$. δ refers to the mean defect-free distance along l where δ is expressed in the units "number of defect-free interfaces per coherent scattering domain".

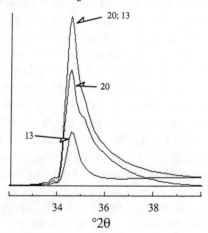

Figure 8 (left). The $20l$ and $13l$ bands for dehydrated turbostratic Na-montmorillonite.

casual study of two-dimensional diffraction phenomena.

The models represented by Figure 6A and 6B are *not* a description of what, for the micas, is called the 1*Md* polytype. The 1*Md* structure, which Méring has called "semi-ordered", involves randomly distributed adjacent layer rotations of n60° where n is integral. Alternatively, the 1*Md* structure can be developed by randomly distributed translations of the upper tetrahedral sheet of the silicate layer by the distance $a/3$ in the six hexagonal directions $X_1, X_2, X_3, -X_1, -X_2$, and $-X_3$ (Bailey, 1988). The rotations for the turbostratic model have values where n is any number and translations are not restricted to $a/3$. This model as portrayed by Figure 6A may be a good one for interstratified illite / smectite which possibly has 1*M* stacking within the zones of coherent illite layers, or the fundamental particles as Nadeau and his coworkers call them (Nadeau et al, 1984), but has turbostratic displacements at the "smectite" inter-fundamental particle junctions.

ONE-DIMENSIONAL DIFFRACTION FROM SIMPLE CLAY MINERALS

Mathematical formulation

The most common type of X-ray diffraction analysis of clay minerals involves the study of the 00l or basal diffraction series. Oriented powder aggregates are used in which the mean orientation of the X-Y planes of the crystals are parallel to the sample surface. The importance of the basal series arises from the structures of the simple clay minerals (excluding the amphibole-like clay minerals, e.g. palygorskite, sepiolite) which are very similar in the X and Y crystallographic directions, but quite different in the Z direction, consequently, the basal diffraction pattern is the most diagnostic. Determination of polytypes and measurements of the 060 or 06l reflection to establish octahedral type, of course, require three-dimensional studies.

For the one-dimensional case, the form of the equations for F and Φ are modified so that they are expressed in the units of 2θ instead of h, k, and l. For this application, the Bragg Law is written

$$l\lambda = 2d \sin\theta, \text{ and } l = 2d \sin\theta / \lambda , \qquad (5)$$

where l is the order of the reflection and $d = d(001)$. Substitutions are made as follows:

$$F(00l) = \sum_j n_j f_j \cos(2\pi z_j l / D) \; ; \quad F(\theta) = \sum_j n_j f_j \cos(4\pi z_j \sin\theta/\lambda) \; , \qquad (6)$$

and

$$\Phi(00l) = \frac{1}{N}\left[N + 2 \sum_{n=1}^{n=N-1} (N-n) \cos(2\pi n l) \right] ; \qquad (7a)$$

$$\Phi(\theta) = \frac{1}{N}\left[N + 2\sum_{n=1}^{n=N-1} (N-n)\cos(4\pi n D \sin\theta / \lambda) \right] . \qquad (7b)$$

f_j is the scattering factor of an atom that lies z_j Å from the origin of the calculation which is a center of symmetry on projection to Z, n_j is the number of such atoms at that location, N is the number of layers stacked in coherent array along Z, and $D = d(001)$. Note that here, the atomic coordinates (z_j) are expressed as distances from the origin of the unit cell along the direction normal to 001 and not as fractions of the unit cell dimension c or $c\sin\beta$. The sine series for F can be omitted if, as is usually the case, the mica-like layers are centrosymmetric on projection to Z. Under these conditions, we need not worry about the square of the magnitude of F but simply equate intensity to F^2. These two equations are combined with the random powder form (or other) of the Lorentz-polarization factor to give the overall one-dimensional diffraction equation for the intensity $I(\theta)$.

$$I(\theta) = Lp(\theta)\, F(\theta)^2\, \Phi(\theta) . \qquad (8)$$

Figure 9 shows the the assignments for Equation 7b of the spacings, $R = nD$ where $D = d(001)$, and the multiplier $(N - n)$. For a simple clay mineral or mica, all of the F values are identical and hence the intensity factor due to the unit cells is F^2 which can be factored out to give the form of Equation 8. Mixed-layered minerals do not allow this simplification, and for them, calculations require that the F terms be moved inside of the summation of Equation 7b.

Equation 7b describes the interference function for an infinitely thick aggregate of crystallites each of which is N layers thick. This is a physically unrealistic condition, and a more meaningful solution requires summing the diffraction from an infinitely thick aggregate of crystallites that are distributed over a range of thickness N. A distribution of thickness can be handled by the use of the summation given in Equation 9b, and this replaces the quantity $(N - n)$ for each iteration of n in Equation 7b (MacEwan, 1958). $q(N)$ is the fraction of the aggregate composed of crystallites N layers thick, and this distribution is normalized so that the sum of all $q(N) = 1$ (Eq. 9a). $\langle N \rangle$ is the mean N for the aggregate, and N_1 and N_2 are redefined for the one-dimensional case as, respectively, the thinnest (smallest N) and the thickest (largest N) crystallites. The summation of Equation 7b is taken to the upper limit of $N - 1$ and the constant (N) and the ratio $1/N$ are replaced by $\langle N \rangle$ and $1/\langle N \rangle$.

$$\sum_{N=N_1}^{N=N_2} q(N) = 1 ; \qquad \langle N \rangle = \sum_{N} q(N)\, N \qquad (9a)$$

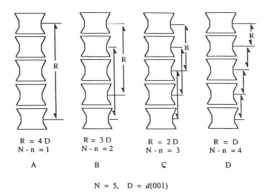

Figure 9. The spacings, R, in a crystallite for which N = 5.

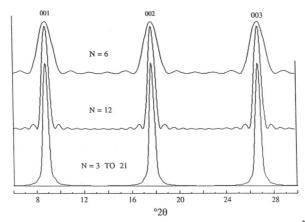

Figure 10. The effects of particle size on the one-dimensional interference function (Φ) for a 10 Å structure.

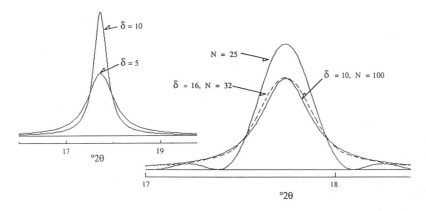

Figure 11 (left). The 002 reflections of the interference function for a 10 Å structure with two different mean defect-free distances, δ. N = 7δ.

Figure 12 (right). The 002 reflections of the interference function for 10 Å structures. Peak shapes are due to particle size (N = 25), defect broadening (δ = 10), and the effects of both (δ = 16, N = 32).

$$\sum_{N=N_1}^{N=N_2} q(N)(N-n) \qquad n \leq (N-1) \ . \qquad (9b)$$

Peak shapes and widths, corrected for instrumental distortion, are controlled by (1) particle size, (2) defect density, (3) strain and (4) mixed-layering. Each is discussed below.

The 00l series and particle size effects

Figure 10 shows the interference function (Φ) for a 10 Å 2:1 phyllosilicate mineral such as muscovite or illite. The top trace represents the single crystallite thickness $N = 6$ unit cells, and the center profile depicts $N = 12$. The integrated areas of all peaks are identical and hence intensity is independent of N. Peak heights are proportional to N and peak breadths β, which are measured as the peak widths at half height, are proportional to $1/(Nd(001)\cos\theta)$ in accordance with the Scherrer equation. A reminder is necessary here. This is what the diffraction pattern of illite would resemble if the Lorentz-polarization factor were unity, and if the individual illite unit cells scattered with an amplitude of +1 at all diffraction angles. The calculated profiles incorporate only Φ (see Eq. 7b).

The top two traces of Figure 10 are modulated by ripples between the peaks. There are $N - 2$ of them, and they result from the single values of N upon which the calculations are based. The lower profile was calculated for a range of N that varies from 3 to 21 ($q(N) = 1$). For the different values of N, the ripples occur at different values of 2θ with the result that they tend to merge into a continuous background between the 00l reflections. Integrated intensities are equal for the different orders and are identical to those calculated for single values of N if $\langle N \rangle$ is properly substituted into Equation 7b.

The 00l series and defect broadening

The defect-broadening treatment of Ergun (1970) was introduced above, and its implementation here requires that $N_1 = N_2 = N$ (Eqs. 9a and 9b) and that the term $(N - n)$ in Equation 7b be multiplied by $\exp(-n/\delta)$ where δ is the mean defect-free distance in units of "number of coherent interfaces". Profiles for the mica 002 reflection (Φ only) are shown by Figure 11 for two different values of δ ($N = 7\delta$). Pure defect broadening is adequately described by $N = 7$ or 8 times δ, for then, the exponential weighting quantity is so small that extending the series further has no detectable effect on the diffraction pattern. The defect model produces a perfectly Lorentzian or Cauchy shape (Ergun, 1970), as evidenced by its "peaked" maximum and the wide tails at the base. The peak height is proportional to δ and β is proportional to $1/(\delta d(001) \cos\theta)$; the integrated peak area is independent of δ and θ.

Figure 12 compares the peak shapes of the Φ function for the mica 002 reflection calculated for fixed N, pure defect-broadening ($\delta = 10$, $N = 100$), and an intermediate case that incorporates defect-broadening in conjunction with finite crystal-size effects. Parameters were selected to give all three cases the same β. The profile for $N = 25$ is almost perfectly Gaussian down to intensities that are about one-tenth of the peak height, and the pure defect-broadened peak fits a Cauchy function. The profile for the intermediate case is described by a Voigt function. Strictly speaking, a Voigt function is the convolution of Gaussian and Cauchy functions, but as Henke et al. (1978) have shown, the Voigt function can be very accurately portrayed by the simple point-by-point sum of the two functions if their widths are similar and if each is normalized to give an intensity of unity at its maximum. This result is called a pseudovoigt function and it is very useful in diffraction applications. Equation 10 is a pseudovoigt function with the breadth constants adjusted to give the same β for each profile type. The overall profile is centered on $\varepsilon = 0$, which is expressed in 2θ units on each side of the peak. To express this profile in 2θ space, simply add $2\theta_0$ to ε where $2\theta_0$ is the Bragg position for a given reflection. G is the fraction of Gaussian component and $(1 - G)$ is the Cauchy fraction. In the writer's experience, clay mineral (non-interstratified) 00l reflections are accurately portrayed by $G = 0.2$ to 0.4, which result from model calculations in which N is close to 5δ.

$$I(\varepsilon) = G\, e^{-2.77\, \varepsilon^2 / \beta^2} + \frac{1 - G}{1 + 4\, \varepsilon^2 / \beta^2} \quad . \tag{10}$$

The 00l series and strain broadening

Diffraction peak breadth is increased by lattice strains which result from displacements of the unit cells about their nominal positions (Ergun, 1970). For this model of strain-broadening (Warren and Averbach, 1950), the integrity of the unit cell structure is maintained, and the distortions are propagated from one cell to the next according to a mean-square displacement parameter in Å which will be called α. α is the standard deviation of a Gaussian displacement function that affects adjacent unit cells, and α^2 is the variance, which if multiplied by n (Eq. 7b) describes the relative displacements between any two unit cells separated by the distance nd(001). Returning to Equation 7b, consider the term for a pair of *adjacent* layers, that is, n = 1:

$$\cos(4\pi (D + \varepsilon) \sin\theta / \lambda) \quad .$$

The variable ε is an incremental distance that modifies the nominal separation $D = d(001)$. To visualize this concept, refer to Figure 9D, and imagine that one of the separations has been lengthened or shortened by ε. The cosine expression containing $D + \varepsilon$ can be separated into the product of two cosine terms as follows:

$$\cos(4\pi D \sin\theta / \lambda) \cos(4\pi\varepsilon \sin\theta / \lambda) \quad .$$

The second term is integrated over all values of ε, from plus to minus infinity, and each increment of the integration is weighted by a normalized Gaussian distribution. The integral can be solved analytically because the equal distribution of $+\varepsilon$ and $-\varepsilon$ eliminates the imaginary component.

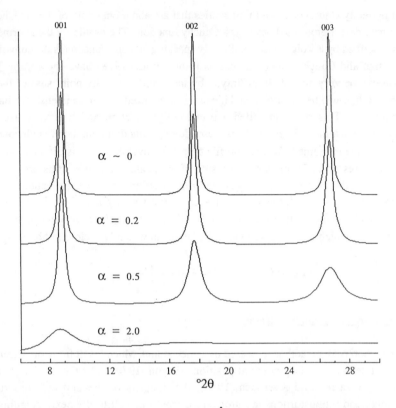

Figure 13. The one-dimensional interference functions for 10 Å structures with different mean displacements in Å (α) between adjacent layers.

$$\frac{1}{\alpha\sqrt{2\pi}} \int_{-\infty}^{\infty} e^{-\varepsilon^2/2\sigma^2} \cos(4\pi\varepsilon\sin\theta/\lambda)\, d\varepsilon = e^{-8\pi^2\alpha^2 \sin^2\theta/\lambda^2} \quad . \tag{11}$$

The integration has been made for n = 1, but because the displacements are propagated as $n\alpha^2$, we need to insert n into the exponent of the result to make Equation 11 applicable to any n. If this is not done, then the displacements are independent of n and the exponential term becomes the well known Debye-Waller temperature correction for the thermal vibration of scattering centers (for example, atoms) in which the thermal vibration constant $B = 8\pi^2\alpha^2$. Thermal effects cause no line broadening but diminish intensity with respect to 2θ. On the other hand, strain effects broaden the higher-angle peaks so much that they merge with the baseline. The peaks become broader in linear proportion to tanθ (Cullity, 1978).

The effects of strain-broadening require the weighting term $\exp(-8\pi^2 n\alpha^2 \sin^2\theta/\lambda^2)$ in Equation 12. Figure 13 shows the results of calculations for strain-broadening on the interference function for the illite 00*l*. Note that a mean adjacent unit cell displacement of only 0.2 Å has produced noticeable line-broadening at even these low diffraction angles. Because broadening increases rapidly with respect to 2θ, high-angle diffraction patterns of clay minerals that show no broadening in addition to the 1/cosθ effect due to particle size, represent structures with very, very uniform unit cell separations. The bottom trace of Figure 13 depicts the almost total degradation of the diffraction pattern due to large unit cell displacements along Z. The pattern approaches that of a glass because the random displacements of the unit cells have destroyed the long-range order. In the true glass (or liquid) structure, even the short-range order is diminished because random displacements exist *between atoms*, so that the unit cell itself is only statistically definable.

This analysis of strain broadening is applicable to other types of disorder that involve random displacements between scattering entities. In addition, it forms a suitable frame of reference from which to discuss the concept of coherence or partial coherence. To anticipate the next section a bit, let us multiply Φ by the square of the scattering magnitude for the unit cell and ignore the Lorentz-polarization factor. Then, from Equation 7b and the strain term,

$$I(\theta) = \frac{F(\theta)^2}{N}\left[N + 2\sum_{n=1}^{n=N-1}(N-n)\,e^{-8\pi^2 n\,\alpha^2\sin^2\theta/\lambda^2}\cos(4\pi nD\sin\theta/\lambda)\right]. \quad (12)$$

Inspection of Equation 12 shows that if α is large, the summation equals zero and the intensity is equal to $F(\theta)^2$, that is, the N unit cells scatter with no phase interactions and their scattering relationship is said to be incoherent. Alternatively, if α is near zero, the strain term approaches unity for all values of n and the N unit cells form a coherent array. If α is between these two extremes, the unit cells form a scattering array that is *partially coherent*. The entity represented by F is, in this case, the unit cell. But the analysis is identical if F is defined for atoms. If the separations of atoms are modulated by a strain term, the resulting diffraction profile is that of a glass (of liquid). If the strain term modulates unit cell separations, then the result depicts strain broadening or any other type of disorder that is expressed by random unit cell displacements. In the latter connection, a possible application is the 00*l* diffraction series for a (hypothetical) smectite whose solvation complex produces a distribution of thicknesses.

Examination of high-resolution transmission electron microscope lattice fringe images may show, for example, two parallel mica crystals tens of Å apart, or perhaps a mica crystal next to a quartz crystal which by chance has an *hkl* plane oriented parallel to the mica 00*l*. How do such arrangements diffract? The crystals in the isolated view do indeed diffract coherently to produce a diffraction pattern different from that of either of the pure phases, but the X-ray diffraction experiment averages over billions of similar but nonidentical arrangements with the result that phase effects between adjacent crystals cancel out, leaving only the $F^2 N$ term of Equation 12. Consequently, intensities add, but not amplitudes, and the final diffractogram is the simple sum of the intensities from the pure

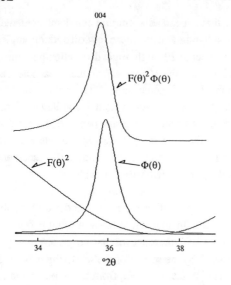

Figure 14 (left). The illite 004 reflection and its components $F(\theta)^2$, and $\Phi(\theta)$. Vertical scales are arbitrary.

Figure 15 (below). Two different locations in the mica structure for the origin of $F(\theta)$.

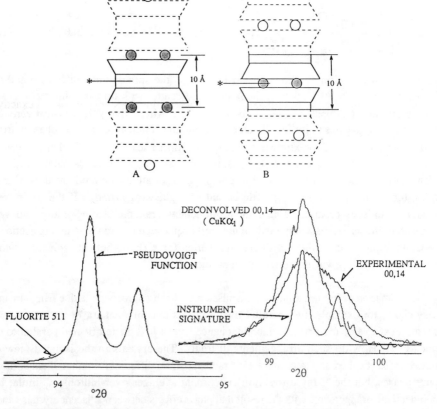

Figure 16 (left). Fit of an empirical pseudovoigt function to the 511 reflection of fluorite.

Figure 17 (right). The experimental profile of a chlorite 00,14 peak, the instrument signature, and the deconvolved pure diffraction profile. Instrument signature is not to scale.

phases present. The difference between additive amplitudes (coherence) and additive intensities (incoherence) can be seen from a simple mathematical example. Let A and B be the amplitudes of scattering of two points. If the amplitudes add, intensity $I = (A + B)^2 = A^2 + 2AB + B^2$, and if the intensities add, the result is $I = A^2 + B^2$. The difference lies in the cross products 2AB. Suppose we introduce a weighting factor, p, to give

$$I = A^2 + 2pAB + B^2 \ .$$

Phase relations are conserved and the scattering adds coherently if $p = 1$, and the scattering adds incoherently if $p = 0$. Values of p between 0 and 1 define the region of *partial coherence*. The analogy between this simple case and Equation 12 is as follows. A^2 and B^2 are related to the term $F^2 N$ and 2AB is the counterpart of the summation in Equation 12; p is represented by the exponential term of Equation 12 which also has the limits of 0 and 1. In summary, if integration is made over all of the small plus and minus relative displacements of the crystallites that comprise the irradiated volume, the cosine series sums to zero and the residual intensity is due to the sum of the intensities from each of the crystallites. They scatter incoherently with respect to each other.

The effect of the continuous function $F(\theta)^2$ on peak shape and displacement

The peaks of the interference function accurately describe diffraction line shape if F^2 is constant over the angular range of the peak. If F^2 changes markedly over this interval, the peaks become asymmetrical and are shifted in the direction of increasing F^2. Such peak shifts may make the pattern irrational, that is, peak position does not correspond exactly to that predicted by the Bragg Law, and the pattern may be mistakenly taken as evidence of interstratification. The effect has been discussed by Reynolds (1968), who points out that rapid increases in the Lorentz factor at low diffraction angles can augment the effects of small crystallite size on peak displacements. The amounts of asymmetry and peak shift become greater as peak breadth increases, regardless of the cause of broadening, and the effects are enhanced as the slope of the F^2 function increases. Figure 14 shows the interference function for the illite 004 reflection, the F^2 function, and the corresponding 004 intensity profile demonstrating graphically the behavior described (refer to Eq. 8). The line shape and the intensity level of the background on opposite sides of the peak contain information on the magnitude, the direction of slope, and the magnitude of the slope of the F^2 function, information that cannot be obtained from thick, defect-free crystals because their sharp diffraction peaks sample F^2 at essentially single points, namely, where l is integral.

The 004 illite reflection shown on Figure 14 was generated by locating the origin of the calculation for F at the projected center of symmetry that lies in the middle of the octahedral sheet (Fig. 15A). An alternative is the potassium plane at the other projected center of symmetry in the interlayer region (Fig. 15B). For a thick crystal, the selection of either of these (or any other point if the imaginary part of F is included) makes no difference in the calculated intensity. But the selection of origins is meaningful for thin crystals. The calculation of F^2 based on the Fig 15B produces an F^2 function that slopes in the opposite direction and goes through zero on the opposite side of the peak than does the function shown on Figure 14, consequently, the regions of high and low background are reversed, as is the peak symmetry and direction of shift in 2θ. These effects are real because the two crystallites of Figure 15 are physically different. One is a crystallite that terminates on an interlayer region (Fig 15A) and other terminates at the center of the octahedral sheet (Fig 15B). Experimental diffraction patterns of illite show relatively high background levels at the low 2θ and low background at the high 2θ sides of the illite 004 reflection. In addition, the experimental 004 peak is shifted to lower 2θ values as shown on Figure 14. All of this gives credence to the not too surprising conclusion that illite crystallites are terminated at the interlayer plane. This is what one would expect for cleavage fragments, but it is not the only possible form for crystal growth habits. These effects have been described by Tettenhorst and Reynolds (1971), and they suggest that, under favorable circumstances, studies of peak shape can yield information of the nature of crystal surfaces. Figure 14 also suggests that measurable intensity at diffraction angles well-removed from the Bragg position is not entirely instrumental background such as air scatter, fluorescence, or Compton scatter, but is part of the diffraction pattern.

The pure diffraction line profile

The analysis of line shape and breadth requires diffraction data that are undistorted by instrumental aberrations. The instrumental signature can be referred to the diffraction profiles of peaks from a standard material that contains no defects, strain, or particle size broadening. Theoretically, such a crystalline powder produces diffraction lines that are delta functions, that is, the lines have no width, and their observed profiles are characterized by instrumental distortions and can be treated as signatures of the instrument function for a specific value of 2θ. The effect of the instrumental signature on the pure diffraction line from a clay mineral aggregate is described mathematically by the convolution of one profile on the other. Very approximately, the breadth of the convolved line is equal to the square root of the sums of the squares of the breadths of the undistorted line and the instrumental profile. Low to intermediate 2θ peaks for clay minerals are of the order of 0.3 to $0.6°2\theta$ wide at half height, and a single line of the instrumental profile (e.g. $CuK\alpha_1$) is approximately 0.07° wide if a well-aligned instrument is set up in high resolution mode such as two Soller slits, aperture slit 0.05° or less, and divergence slit 1.0° or less. At low angles, the effect of the instrument is small enough to be neglected, particularly if the clay mineral peaks are broad. But instrumental corrections must be made if the experimental peak breadths are 0.3° or less, and/or if high-angle reflections are used because $CuK\alpha_1$-$CuK\alpha_2$ broadening quickly dominates line breadth at high diffraction angles. The older literature describes different semi-empirical methods for making such corrections, but in these days of easily available computer power there is no excuse for avoiding the rigorously correct solution.

Figure 16 shows the instrumental signature for a Siemens D-500 based on powdered fluorite that was ground under acetone to minimize strain. The profile was fit to a pseudovoigt function (Eq. 10) by computing two peaks with identical parameters, located at the $K\alpha_1$ and $K\alpha_2$ positions, with the $K\alpha_1$ twice as intense as the $K\alpha_2$. The fit leaves little to be desired. The mathematical equation for the profile is used to calculate the correct instrumental signature at other values of 2θ, because a given clay mineral will not have reflections located at the same diffraction angles as those from the fluorite standard. The individual $K\alpha_1$ and $K\alpha_2$ signatures change slightly with 2θ, so good experimental procedure requires that you map the instrumental signature for your diffractometer, with a specific set of slits, over the 2θ range anticipated for a clay mineral study.

The instrumental signature is extracted from the experimental profile by deconvolution to yield the pure diffraction profile. The traditional method employed was the Stokes Fourier Series, but it is difficult to set up and requires extensive "number-crunching". The Ergun method of iterative folding (Ergun 1968) is now almost always the technique of choice (Klug and Alexander, 1974). It is described clearly and in detail by Klug and Alexander (1974) and will not be discussed further here. Figure 17 shows the experimental profile for the 00,14 reflection from a clay-sized chlorite, the instrumental signature (not to vertical scale), and the final, deconvolved chlorite reflection. Convolution is a smoothing process and deconvolution is a noise-generating process, therefore random noise in the experimental profile is exacerbated by the deconvolution to such an extent that very good experimental data are essential here -- data collected for long count-times at small 2θ increments. The experimental parameters required depend on the peak-to-background ratio, but for the clay minerals, two-theta increments of $0.01°$ are appropriate together with count-times that give at least 10000 counts for the integrated peak area.

The deconvolved profile is too noisy to allow accurate measurements of peak shape, breadth, and the accurate calculation of d(which is now computed from the $CuK\alpha_1$ wavelength). A simple statistical least-squares fit to a pseudovoigt function will provide these parameters by using all of the data, not just a few points. Figure 18 shows such a fit to the deconvolved profile of Figure 17. The writer has obtained results for β (the peak-width at half maximum) that are accurate to $\pm 0.02°$ for 16 basal reflections from a chlorite, based on the standard deviation of results from those predicted by the Scherrer equation. Data processing time for deconvolution and curve-fitting of one peak with, for example, a Macintosh II, is approximately two minutes.

The Lorentz factor for oriented clay mineral aggregates

The discussion so far has dealt with the structure factor (F) and the interference function (Φ). The Lorentz factor is the final quantity required for a complete model of diffraction. As we have seen in Chapter 1, it is the product of two expressions, the single crystal factor which is $1/\sin 2\theta$ for integrated peak intensities (James, 1965), and $1/\sin\theta$ for point-by-point corrections on a 2θ continuous diffraction pattern (Cullity, 1978). The second portion is the powder ring distribution factor, Ψ, which is proportional to the

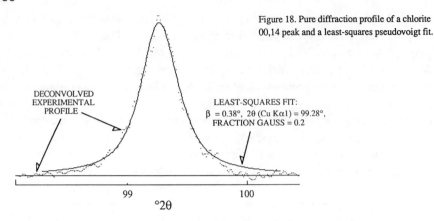

Figure 18. Pure diffraction profile of a chlorite 00,14 peak and a least-squares pseudovoigt fit.

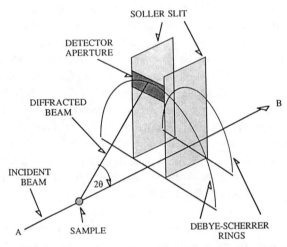

Figure 19. Soller slit geometry for a flat-specimen diffractometer in the reflection mode.

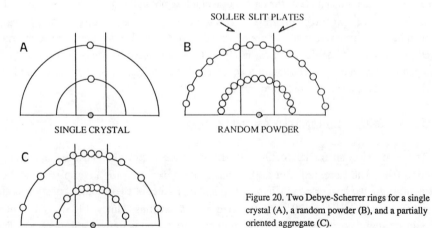

Figure 20. Two Debye-Scherrer rings for a single crystal (A), a random powder (B), and a partially oriented aggregate (C).

number of crystallites oriented in such a way as to contribute to the intensity of a Debye-Scherrer diffraction ring (James, 1965). We assume here the 2θ-continuous form and neglect any constants that are independent of 2θ.

Figure 19 reviews the Debye-Scherrer geometry for a flat specimen diffractometer operating in the reflection mode. The line A-B defines the path of the incident and undeviated X-ray beams. The detector aperture is the receiving slit which moves along a circle defined by the goniometer radius in the plane parallel to and between the Soller slit plates. The sample rotates at half the rate of the receiving slit motion and intercepts the different Debye-Scherrer rings. Only two Soller slit plates are shown, but that is sufficient for this analysis because the other plate pairs duplicate the diffraction geometry. The incident beam is shown as a single ray, but in fact it has divergence in the planes parallel to and normal to the diffracted-beam Soller slit. The divergence in the Soller slit plane is unimportant to this discussion, but the divergence normal to this is controlled by the primary beam Soller slit or lack thereof and must be taken into account in any analysis of Ψ. It is neglected here for simplicity, as are complications that arise from the use of a diffracted-beam monochromater which causes additional Soller slit (and detector aperture slit) collimation.

Figure 20 shows two-dimensional representations of two Debye-Scherrer rings for three cases. This figure is meant to portray the view one would get by looking along the line A-B on Figure 19. Suppose that the sample consists of a suitably oriented single crystal of mica (Fig. 20A), and that the large ring represents the 002 and the smaller the 001 reflections. We see that all of the intensity from each reflection (and any other $00l$ peak) lies within the Soller slit spacing, so there is no angle-dependent intensity factor, in other words, $\Psi = 1$ and the Lorentz factor equals $1/\sin\theta$.

Figure 20B shows rings for the same two reflections from a random mica powder. Any crystallite that is oriented so that it produces one $00l$ reflection is oriented so that it will produce all others. Consequently, all $00l$ diffraction rings contain the same number of diffraction spots, and for the random case, these are evenly distributed along the powder rings. But the spots are spread out more as the diffraction angle increases (Fig. 20B) because the radii of the rings and their circumferences are proportional to θ. The Soller slit intercepts the same linear distance regardless of θ, so the fixed plate separation admits more of the ring, and hence larger numbers of spots, as the diffraction angle is decreased. It is easy to show that the fraction of the ring admitted is proportional to $1/\sin\theta$ (James, 1965) which is Ψ, the powder ring distribution factor for random orientations. The Lorentz factor, then, is

$$L = \Psi / \sin\theta = 1/\sin^2\theta \ . \tag{13}$$

A rough, intuitive appreciation of the Lorentz factor for oriented aggregates can be gained by the following:

$$L = (1/\sin\theta)(1/\sin^x\theta) \ . \tag{14}$$

If x = 1, Equation 14 applies to the random powder condition, and if x = 0, the result is the single-crystal form. For partial orientation, x is some number between 0 and 1. Unfortunately, x decreases monotonically but nonlinearly with respect to the degree of orientation. Nevertheless, deviations from the random powder form are small unless the orientation is extreme.

The orientation of powder aggregates of micas or clay minerals follows a Gaussian distribution (Reynolds 1986). σ^* is defined as the standard deviation of the tilt angles of the crystallites' *X-Y* planes from their mean position which parallels the sample surface. Calculation of the Lorentz factor for oriented aggregates requires that σ^* be measured, which is easy to do with a two-circle diffractometer (Reynolds, 1986). The crystallite orientation function (defined by σ^*) is integrated for all of the crystallites that are so oriented as to diffract into the Soller slit aperture at each required value of θ. Such an integration includes five spots on the large ring and seven spots on the small ring shown by the example of Figure 20C. Suppose that the orientation is more extreme than that shown by Figure 20C, that is, the spots are more concentrated at the apices of the powder rings. Then perhaps you can see that there will be some low diffraction angle below which all of the spots lie within the Soller slit aperature. Below that angle, the single-crystal form of the Lorentz factor applies, and with increasing angles above it, the Lorentz factor becomes more and more like the random powder form. Calculations of the Lorentz factor for oriented aggregates are beyond the scope of this chapter. The interested reader should consult Reynolds (1986).

Sample-to-sample Lorentz factor variations usually pose no serious problems for quantitative analysis or one-dimensional structure determinations using oriented micaceous aggregates, so long as a few strategies are employed. The *relative* intensities of the lowest angle peaks are the most affected by preferred orientation in a plane parallel to the sample surface (for reflection diffractometry), so these are best avoided in quantitative analysis. For structure determinations, highly oriented aggregates will be needed to provide good data on the weak high-angle diffraction orders. Use a highly oriented aggregate for only the high-angle peaks and apply the random powder correction. Prepare a poorly oriented aggregate for recording the low angle peaks, and apply the random powder form because it is very nearly correct unless the orientation is extreme. Include some overlap in the two diffraction series and normalize them to the base of a common reflection. Alternatively, σ^* can be measured and the appropriate Lorentz factor corrections made for the entire diffraction pattern.

Figure 21 shows calculated diffraction patterns for a trioctahedral chlorite that contains no Fe. All patterns were normalized to give the same intensities for their 004 reflections. The absolute intensities are very different because intensity is proportional to $(1/\sigma^*)^2$

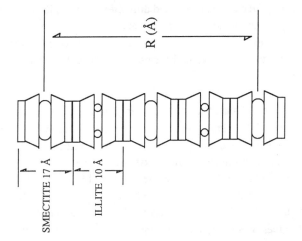

Figure 21 (left). The effect of preferred orientation (σ^*) on the relative intensities of the 00l series of chlorite. Vertical scales are normalized to a constant intensity for the 004 reflections.

$R = (17/2) + 10 + 17 + 10 + (17/2) = 54$ Å

Figure 22 (above). The spacing, R, for one possible array in a mixed-layered illite/smectite crystallite.

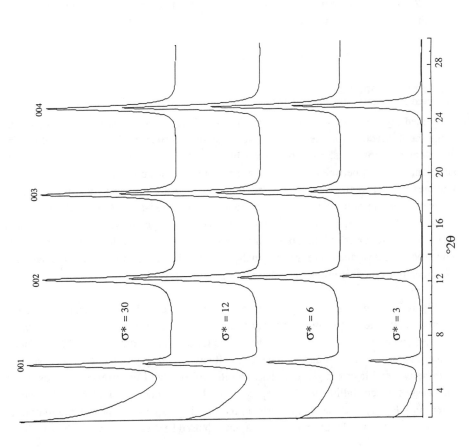

(Reynolds, 1986). Thus, the absolute intensities of the bottom trace are 100 times as great as those of the top trace. The calculated diffraction patterns include the low-angle intensity loss caused by a short sample (3.4 cm in conjunction with a one-degree divergence slit for a goniometer radius of 20 cm). The parameters were selected to simulate commonly used laboratory conditions and have no bearing on the Lorentz factor because the same effects are present in all.

Figure 21 shows that little relative change in intensity occurs between orientations described by $\sigma^* = 30°$ (which gives results very close to those from a random powder) and $\sigma^* = 12°$ (a common value for oriented clay mineral aggregates). The relative intensities of the low-angle peaks diminish rapidly as σ^* is decreased below 12°, but note that even for such extreme orientations, the relative intensities of the 003 and 004 (and higher orders) are not much affected. Experience with more than one hundred different clay mineral samples has shown that only a few percent have sufficiently good platy crystal morphology to provide σ^* values of 3 to 4°, and the samples must be prepared by a centrifuge method to achieve these orientations. Almost all clay mineral aggregates will orient with σ^* in the range of 7 to, perhaps 20, if other methods of oriented sample preparation are used, and as Figure 21 shows, relative intensities of the higher diffraction orders are not greatly affected by such variations in orientation.

MIXED-LAYERED MICAS AND CLAY MINERALS

Mathematical formulation

Mixed-layered clay minerals consist of two different kinds of silicate layers stacked along the Z direction. In the general case, these have different thicknesses and different chemical compositions, both of which produce two different values of F(00l). In addition, spacings between two layers are not simple multiples of $d(001)$ because two thicknesses of layers make up the spacing, and they are probabilistically distributed depending on the proportions of the two types in the mixed-layered mineral, and on the order or lack thereof in the stacking sequence. Therefore Equation 7b must be modified to make it suitable for calculating basal diffraction patterns of mixed-layered clay minerals. The procedure described here was originally developed by MacEwan (1958) and has been used in various modifications since that pioneer work (Reynolds, 1980; 1985).

Figure 22 shows one stacking sequence out of many that collectively make up a mixed-layered crystallite. The layer types for this example are illite (I) and ethylene glycol-solvated smectite (S), and the respective values for $d(001)$ are indicated. This particular sequence consists of two I and 3 S layers, and it has S layers on each end. The separation of these two end layers is given by R_{jk}, and its probability of occurrence depends of the thickness of the crystallite (N), the proportions of illite and smectite in the phase (P_I and P_S where $P_I + P_S = 1$), and the type of order in the stacking sequence. The spacing multiplier (N - n; Eq. 7b) is a little different for the approach used here and is equal to (N + 1 - S - I) where S and I refer to the numbers of illite and smectite layers in the sequence. The overall equation for calculating the basal diffraction patterns of mixed-layered clays is represented by Equation 15 (neglecting the Lorentz-polariztion factor), in which general designations (j and k) are used for two types of layers A and B.

$$I = \left(NF_A^2 P_A + NF_B^2 P_B\right) + 2\sum_{A=0}^{A=N}\sum_{B=0}^{B=N}(N+1-A-B)\sum_{jk} F_j F_k \sigma_{jk} \cos(4\pi R_{jk} \sin\theta/\lambda) \quad (15)$$

The summations are taken for all values of A and B for which $(A + B) > 1$ and $(A + B) < (N + 1)$. The structure factors can no longer be factored out of the summation because there are two of them. In Equation 15, F_j can be either F_A or F_B and F_k can be F_A or F_B. Both F_j and $F_k = F_S$ (smectite) for the single array shown by Figure 22. Many of the terms in the double summation have three possible layer terminations, i.e., A-A, B-B, and A-B which are coded here as jk. These different terminations produce a maximum of three values of R_{jk} and σ_{jk} for each term AB in the double sum, namely R_{AA}, R_{BB}, R_{AB} and σ_{AA}, σ_{BB} and σ_{AB}. There obviously can be no spacing that has A layers on each end (R_{AA}) for a term such as A = 1, B = 5.

The results calculated from Equation 15 are somewhat unrealistic because they are based on crystallites that terminate at the center of the octahedral sheet (see Fig. 15). To set it up otherwise complicates the discussion so much that the improvement in results is not worth the loss of clarity. In general terms, however, the problem can be solved by "grafting" on one half of a silicate layer to the terminal unit cells for each crystallite. Alternatively, the layers and interlayer sheets can be treated as separate entities so that problem becomes one of three components: the silicate layers, the ethylene glycol solvation complex, and the potassium plane (see Reynolds, 1980). Note that Equation 15 is valid only for layer types that are centrosymmetric on projection to Z. The more complicated treatment required by serpentine and kaolinite is discussed by Reynolds (1985).

The values of R_{jk} are easy to calculate, but the σ_{jk} coefficients pose the greatest difficulties in using Equation 15. These are best explained by example. A two-component mixed-layered mineral can be described, in terms of composition, by the proportions of A and B, and $P_A + P_B = 1$. We follow MacEwan (1958) and introduce the concept of the transition probability $P_{A.B}$ which reads as *the probability of a B given an A*. The quantity P_{AB} denotes the probability of an AB pair occurring, and is equal to $P_A P_{A.B}$. We write

$$P_{A.B} + P_{A.A} = 1 \quad \text{and} \quad P_{B.A} + P_{B.B} = 1 \quad , \quad (16)$$

because something must follow an A or B. In addition, $P_{AB} = P_{BA}$ because inversion does not affect the probability of occurrence of a pair. So

$$P_{AB} = P_{BA} = P_A P_{A.B} = P_B P_{B.A} \quad \text{and} \quad P_{A.B} = P_B P_{B.A}/P_A \quad . \quad (17)$$

Specifying P_A or P_B and one of the four transition probabilities allows the calculation of the other three. Note that if, for example, $P_A = P_{B.A} = P_{A.A}$, the interstratification is random because the occurrence of an A does not depend on the preceding layer type. This is designated Reichweite = 0 or R0 (Jadgozinski, 1948). A composition of >0.5 A is ordered, R1, if $P_{B.A} = 1$ for then all B layers are followed by A types. But note that not all A types are followed by B because for <0.5 B, there are not enough B layers to go around. At $P_A = 0.5$, R1 ordering is described by $P_{A.B} = P_{B.A} = 1$. Structures are known for which R > 1 and these are described by longer range probability values such as $P_{AAB.A}$, which reads *the probability of an A following the sequence AAB*. The applications are given Reynolds (1980).

Suppose that P_I and one transitional probability are specified, and we wish to calculate the frequency of occurrence of the layer array SISIS shown by Figure 22. Then

$$\sigma_{SS} = P_S\, P_{S.I}\, P_{I.S}\, P_{S.I}\, P_{I.S}\ .$$

To follow this example to its conclusion, note that for S = 3 and I = 2 there are nine other arrays that occur. These are SSIIS, SIISS, ISSSI, SSISI, SSSII, SISSI, ISSIS, IISSS, and ISISS. The spacing values are $R_{SS} = 54$ Å, $R_{II} = 61$ Å, and $R_{IS} = 57.5$ Å. The individual ten values for σ depend on the proportion of I and S and on ordering. For a given composition, all of the ten σ values are identical because frequency of occurrence depends only on P_I. For R1 ordering, $P_I \geq 0.5$, all arrays are forbidden that contain an SS pair and for $P_I \leq 0.5$ II pairs are illegal. If $P_I = 0.5$ and $P_{I.S} = 1$, the only array that remains is SISISI (Fig. 22). The frequency coefficients are best calculated by a recursive method described by Bethke and Reynolds (1986). Coefficients can be inserted into Equation 15 to incorporate the effects of a range of N and to simulate the effects of defect density or strain along Z.

<u>Interpretation of basal series diffraction patterns of mixed-layered clay minerals</u>

Figure 23 shows the diffraction pattern for randomly interstratified biotite-chlorite that contains equal proportions of each component (biotite(0.5)/chlorite). This pattern and the ones that follow for mixed-layered minerals were calculated by NEWMOD (Reynolds, 1985). The mineral shown probably does not exist in nature, but it is a good choice for exemplifying the effects of interstratification on diffraction patterns. The interpretations are based on those of Méring (1949) which remain the basic principles for this subject. The first noteworthy point is the irrationality of the pattern, that is, the positions of the reflections fit no sensible periodic pattern that can be described by the indices 00*l*, and the Bragg Law does not apply. This is a sensitive test for interstratification, and the mineral is said to be interstratified (Bailey, 1982) if the standard deviation of the peak positions about their nominal Bragg positions exceeds 0.75%. Figure 23 is marked with the nominal positions of the 00*l* series for pure biotite and chlorite, and one can see that the mixed-

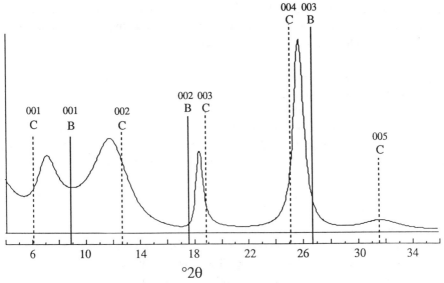

Figure 23. Calculated basal diffraction pattern for randomly interstratified biotite(0.5)/chlorite. Vertical lines labelled C refer to the chlorite 00l positions and lines B refer to biotite.

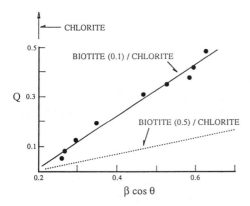

Figure 24. The relation Q vs. line breadth (β) for randomly interstratified biotite/chlorite.

Table 1. Calculation of the broadening descriptor, Q. The quantity 0.704 is the ratio of $d(001)$ values for biotite and chlorite.

Order	Order x 0.704	Q
1	0.704	0.296
2	1.408	0.408
3	2.113	0.113
4	2.817	0.183
5	3.521	0.479
6	4.225	0.225
7	4.930	0.070
8	5.634	0.366
9	6.338	0.338
10	7.042	0.042

layered reflections lie between the limits imposed by these end-member peak locations. If composition is altered, peaks migrate between these limits so, to a good approximation, peak positions can be used to estimate the proportions of the two components.

Additional evidence of interstratification is provided by peak breadth. A pure mineral produces 00l diffraction peaks whose breadths increase by the factor 1/cos θ (for crystallite-size broadening) in accordance with the Scherrer equation. This effect is present with interstratification, but is usually overwhelmed by mixed-layer broadening (Fig. 23). Note that, like peak migration, peak broadening is controlled by the limits defined by the end-member reflections. The greater their separation, the greater the broadening effect. If by chance, the 00l from one end member lies at the same 2θ as an 00l from the other, the resulting peak contains no mixed-layered broadening, and its breadth is due only to particle-size or defect-density effects. If a reflection from one end member lies midway between two reflections of the other, the mixed-layered reflection has maximum broadening, and the amount is inversely proportional to the fraction of the component whose nominal 00l position is so flanked. The principle can be quantified. The subject has been given brief mention in previous work (Reynolds, 1988), but studies by the writer since that time suggest that line broadening is a sensitive and quantitative descriptor of interstratification and is applicable to the study of very small amounts of one component in the other. It should be possible to detect only a few percent of "foreign" layers in a host mineral.

Table 1 demonstrates a method for detecting and quantifying minor amounts of biotite in a randomly interstratified biotite/chlorite mineral. Divide d(001) for biotite by d(001) for chlorite to give 10/14.2 = 0.704. Then multiply this fraction by the integers l that refer to the indices of the chlorite 00l reflections. Finally, compute the quantity Q that is a measure of the extent of departure of the previous results from integral values. The maximum Q is thus 0.5 and the minimum 0.0.

This analysis shows that the broadest peaks are the chlorite 002 and 005 reflections and the sharpest are the 003 and 007 and 00,10 (cf. Fig. 23). Figure 24 is a plot of Q vs. calculated peak breadth β for a randomly interstratified biotite(0.1)/chlorite. Breadths have been multiplied by cosθ to remove the effects of particle size broadening and a linear regression line has been fit to the data. The figure includes a solid line for the relation for biotite(0.5)/chlorite, and the results for a pure chlorite that are represented by a line coincident with the Y axis. The slopes of these three relations are inversely proportional to the fraction of biotite in mixed-layered biotite/chlorite, and they demonstrate the extreme sensitivity of line breadth to mixed-layering for this mineral system. The analysis also allows extrapolation to Q = 0 which gives the line breadth due to only particle size effects, for if Q = 0, there is no mixed-layered line broadening.

In a strange way the Q relation is "self-focusing". Consider a mixed-layered talc/biotite that contains two percent talc. The locations of the 00l biotite reflections will be indistinguishable from the nominal Bragg positions for biotite, so the question arises -- how can a measure be obtained of $d(001)$ for talc, if indeed it is talc and not, say, pyrophyllite or wonesite? The procedure is to compute a Q vs. β plot for the three alternatives, and even others for which $d(001)$ is varied over small increments. One of these will produce a good relation ($r^2 > 0.8$) and the other incorrect (or poorer) models should produce very poor correlations between Q and β.

Examples of mixed-layered clay minerals

Calculated diffraction patterns for ethylene glycol-solvated illite(0.5)/smectite are plotted on Figure 25. Trace A depicts the ordered ($P_I = 0.5$, $P_{I.S} = 1$) case which defines a superstructure whose $d(001)$ is equal to the sums of d(00l) for illite and smectite. Natural minerals of this type are named K-rectorite. No disorder is present, so the diffraction peaks are sharp and their positions follow the Bragg relation for an 00l pattern based on $d(001) = 27$ Å. Trace B is "half-ordered", that is, $P_{I.S}$ is less than unity ($P_{I.S} = 0.75$) but greater than it would be for the random case that is described by $P_{I.S} = P_S = 0.5$. Trace C was calculated for random interstratification. Note that the position of the reflection near 16° is essentially the same for all of these. For illite/smectite, this peak position is most sensitive to composition, and its location is used for estimating P_I. The low-angle region is the most sensitive to ordering type for most mixed-layered minerals, and it is here that the differences between the patterns of Figure 25 are manifested. The partially ordered structure contains some layer sequences of SS, SSS, etc. which have diffraction harmonics with maxima at 17 Å; these have moved the 002 superstructure reflection toward 5°2θ (17 Å). Otherwise, patterns A and B are very similar except for peak breadth. Trace C shows a well-developed reflection corresponding to 17 Å because sequences such as SS and SSS are just as probable as the superstructure sequences IS, ISI or any others. The increased disorder of this structure (trace C) has produced the broadest peaks of the three examples.

Figure 26 shows calculated diffraction patterns for other possible illite/smectite structures in which the disorder is of a different type. Trace A is a reiteration of Trace A on Figure 25; it has been plotted here to provide reference. Diagnostic superstructure reflections are noted by asterisks. Trace B is fully ordered ($P_{S.I} = 1$) but is based on 0.65 illite. Such a structure contains no SS sequences, but it is disordered to the extent that II and III sequences are present. It can be thought of a random interstratification of illite with K-rectorite, that is, all smectite layers are separated by *one or more* illite layers. Evidence of this effect is seen by the location of the second-order superstructure peak which has been displaced toward the illite 001 position at 8.8°2θ. If R1 ordering is retained and the proportion of illite is increased, the average number of illite layers between smectite layers is increased.

Trace C on Figure 26 is a calculated pattern for illite(0.9)/smectite that is ordered on the R3 scheme. All smectite layers are separated by at least three illite layers, or $P_{SII.I} = 1$. Disorder is present because $P_I > 0.25$. Again, it is the low-angle reflections that convey information on ordering type. The two starred peaks contain components of 00l reflec-

176

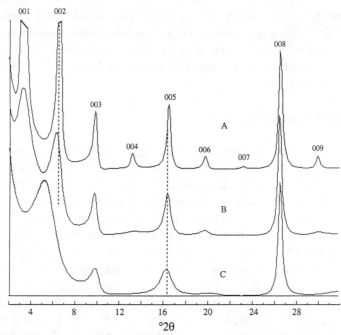

Figure 25. Calculated one-dimensional diffraction patterns for ethylene-glycol solvated illite(0.5)/smectite with different amounts of disorder. Trace A is R1 ordered, trace B is "half-ordered", and trace C is random (R0).

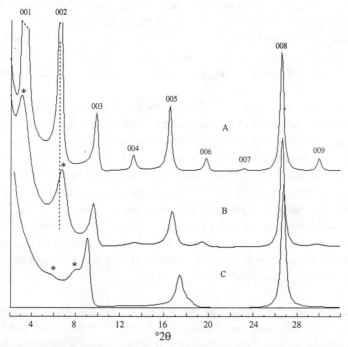

Figure 26. Calculated one-dimensional diffraction patterns for ethylene-glycol solvated illite/smectite. Trace A represents R1 ordering, PI = 0.5, trace B is R1, $P_I = 0.65$, and trace C is R3, $P_I = 0.9$.

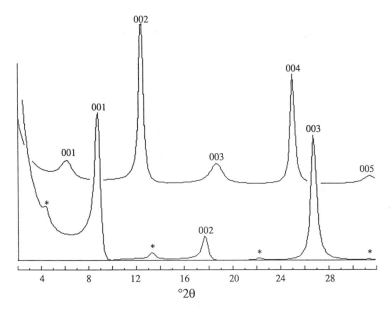

Figure 27. Two unique types of interstratification. Top pattern depicts randomly interstratified serpentine(0.2)/chlorite, and the bottom trace is an example of a compositional superstructure, in this case, the ordered 1:1 interstratification of high- and low-Fe illite.

tions from a 47-Å superstructure because of the abundance of sequences such as SIII. Note that the reflection near 17°2θ displays the anticipated migration toward the illite 002 position as the proportion of illite is increased in these model structures.

Figure 27 shows two examples of anecdotal but interesting mixed-layered structures. The top trace represents a randomly interstratified serpentine(0.2)/chlorite for which the ratio of $d(001)$ values is 0.5. All even-order chlorite reflections are superimposed on the serpentine 00l series, whereas the odd-order chlorite peaks are equidistant from adjacent serpentine positions. This produces a Q distribution of 0.5, 1, 0.5, 1, 0.5, ... etc., with respect to increasing l for chlorite, causing the line broadening pattern of broad-sharp-broad-sharp-, etc.

The bottom trace of Figure 27 is based on the ordered interstratification (R1) of a 50/50 composition of high-Fe (1.0 atoms per Si_4O_{10}) and low-Fe (0 Fe) illite, that is, the structure consists of the regular alternation of high and low-Fe unit cells. This arrangement defines what may be called a compositional superstructure, as distinct from the previous examples in which the superstructure is based mostly on different layer thicknesses. We can think of the diffraction pattern as a normal Bragg series, based on a new 20-Å unit cell, in which the starred peaks are, respectively, the odd-order reflections. Alternatively, the pattern can be described as a normal Bragg sequence, based on $d(001) = 10$ Å, that contains forbidden reflections for which l equals 0.5, 1.5, 2.5, ... etc. The latter frame of reference is a good one to explain the intensities of the forbidden reflections. Recall that the addition of amplitudes from two different but adjacent centrosymmetric unit cells is given by $F_1 + F_2 \cos(2\pi l)$. If l is integral, the intensity is proportional to $(F_1 + F_2)^2$, but if l equals 0.5, 1.5, ... etc., the result is zero if $F_1 = F_2$. In the case under consideration, however, F_1 differs from F_2 by the scattering due to the Fe (f_{Fe}) minus the scattering due to the substituted Al (f_{Al}). The Fe is located at the center of the octahedral sheet and that lies on a center of symmetry on projection to Z, causing the full magnitude of the Fe scattering to express itself in F. In other words, when l is equal to 0.5, 1.5, ... etc., the sum of the unit-cell amplitudes is equal to ($f_{Fe} - f_{Al}$) and the intensity of the starred peaks is proportional to ($f_{Fe} - f_{Al}$)^2.

DIFFRACTION FROM GLASS

<u>Mathematical formulation</u>

Diffraction from glasses or liquids is modified from that of the crystalline state by random atomic displacements. The atoms have mean positions that define a unit cell and, by translation, a crystal. But between adjacent atoms there exists a statistical range of positions in three-dimensional space, which is conveniently treated as a Gaussian distribution whose standard deviation in Å is α (see Eq. 12). If α is the standard deviation between neighbors, then $n\alpha^2$ is the variance in positions between two atoms separated by $(n - 1)$ others. The situation is identical to the case for strain broadening treated above except that, in that case, the displacements in positions occur between unit cells, the structural integrity of which is maintained. Here, the displacements are between atoms so that the unit cell itself is only statistically defined. Calculations of three-dimensional diffraction patterns for glasses are complicated indeed, for any but the simplest unit cells, because the diffraction intensities must be added for every atom with respect to every other atom instead of for every unit cell with respect to every other unit cell. The calculation is based on the Debye equation (James, 1965; Klug and Alexander, 1974).

The simplest possible case is selected here to illustrate diffraction from glass, which is developed in the discussion by the progressive degradation of order in a crystalline starting material. We assume a primitive cubic unit cell ($a = 2.8$ Å) containing one kind of atom (K) for the structure of the crystalline form. This is not a very practical example for mineralogists, namely, potassium glass with a primitive instead of body-centered structure -- perhaps someday on another planet. The atomic scattering factors are factored out of the summation because they are all identical, and each atom is treated as a single unit cell. A stack of such cells along $X, Y,$ and Z produces a primitive unit cell with a equal to the atomic diameter of potassium. The problem does not involve the orientation of unit cells in reciprocal space because individual atoms do have not definable orientations. The intensity distribution in reciprocal space (Eq. 1) can be integrated analytically under these circumstances to account for all possible orientations of diffracting domains, and the result expresses diffraction intensity as a function of only the diffraction angle, θ (James, 1965). Equation 18 is one way to do the calculations.

$$I(\theta) = P\, f^2 \left[N^3 + \sum_{N_1=0}^{N_1=N-1} \sum_{N_2=0}^{N_2=N-1} \sum_{N_3=0}^{N_3=N-1} M(N-N_1)(N-N_2)(N-N_3)\, \sigma(R)\, \frac{\sin(4\pi R \sin\theta/\lambda)}{4\pi R \sin\theta/\lambda} \right] \quad (18)$$

where $\sigma(R) = e^{\left(-8\pi^2 \alpha^2 (R/a) \sin^2\theta/\lambda^2\right)}$.

The quantities N_1, N_2, and N_3 refer to the numbers of atoms in the directions X, Y, and Z of a cube-shaped crystal that contains N^3 atoms. The summations are taken over all values of N_1, N_2, and N_3 for which $(N_1 + N_2 + N_3) > 0$ and $(N_1 + N_2 + N_3) < N$. The spacing R is given by

$$R = \frac{a}{\sqrt{N_1^2 + N_2^2 + N_3^2}} . \quad (19)$$

The coefficient M is a multiplicity factor, and it is equal to 8 if all three N_j quantities are greater than zero, 4 if one of them equals zero, and 2 if two of them equal zero. It can be eliminated if the three summations are taken from lower limits of $-N_j$, but then, the calculation is eight times as lengthy. α is the standard deviation in Å of the distribution of distances that separate adjacent atoms. P is the polarization factor and f^2 is the square of the atomic scattering factor. The random-powder Lorentz factor is implicit in the method of calculation and should not be included in Equation 18.

Examples of diffraction patterns from glass

Figure 28 shows the solution of Equation 18 for the structures described below. The trace labelled $\alpha = 0$ applies to a cube-shaped perfect crystal that contains 40 atoms on each side with no atomic displacements. For $\alpha = 0.2$ Å, broadening of the different reflections is obvious and the breadths increase with θ. Increasing α to 0.5 Å produces a scattering profile that is characteristic of glass. Note that the maxima in the glass profile do not necessarily coincide with peak positions for the crystalline state. Figure 29 carries the analysis farther. It demonstrates the effects of large increases in α. If α is very large, the triple summation of Equation 18 is equal to zero and we are left with the scattering pattern of N^3 atoms that scatter independently of each other, as is the case for a gas at low pressures. This discussion shows that there are degrees of "glassiness", and the statement that a solid is "amorphous" describes some poorly defined crystallinity parameter that is a segment of a continuum.

The diffraction patterns of Figures 28 and 29 are unrealistic in one sense. In a close-packed structure, it is easy to see how atoms can be displaced away from each other, but it is not obvious how they can be displaced towards each other beyond the limits imposed by their electron-defined diameters. This observation suggests that the atomic separations must be larger in glasses than in chemically identical crystals, and that is consistent with the fact that glasses are usually less dense than their crystalline counterparts. The diffraction patterns shown here do not reflect this requirement, though it could have been included.

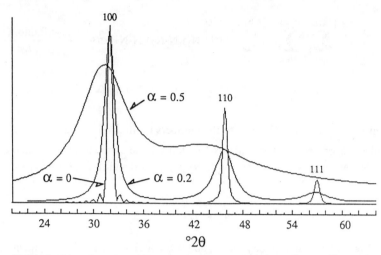

Figure 28. Calculated three-dimensional diffraction patterns of a hypothetical crystalline metallic potassium phase and its transition to glass with increasing values of α, the mean atomic displacement in Å.

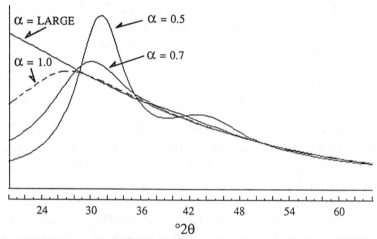

Figure 29. Calculated three-dimensional diffraction patterns of a hypothetical metallic potassium glass and its transition to a gas-like phase with increasing values of α, the mean atomic displacement in Å.

Diffraction studies of glasses are rarely conducted by modeling the diffraction phenomena. Invariably, a procedure is used that inverts that mathematics and proceeds by a calculation that is conceptually similar to that of a Patterson transformation. Clay mineralogists will recognize this procedure in the MacEwan Direct Fourier transform (MacEwan, 1956). The analysis is called a radial distribution analysis and the results provide information on the atomic density difference between the glass and a random distribution of atoms, as a function of the magnitude of the distances that radiate outward from a given atom. From these data, the number of atoms segregated at each distance provides the coordination scheme about a given atom, and the crystal structure can be deduced. The subject of radial distribution analysis encompasses a large and highly specialized literature, and the required experimental techniques are demanding. The subject is beyond the scope of this chapter. The interested reader will profit by the excellent and detailed summary given by Klug and Alexander (1974).

REFERENCES

Bailey, S. W. (1982) Nomenclature for regular interstratifications. Amer. Mineral. 67, 394-398.
____ (1988) X-ray diffraction identification of the polytypes of mica, serpentine, and chlorite. Clays & Clay Minerals 36, 193-213.
Bethke, C. M. and Reynolds, R. C. (1986) Recursive method for determining frequency factors in interstratified clay diffraction calculations. Clays & Clay Minerals 34, 224-226.
Brindley, G. W. (1980) Order-Disorder in clay mineral structures. Ch. 2 In: Crystal Structures of Clay Minerals and Their X-Ray Identification. Brindley, G. W. and G. Brown, eds. Mineralogical Society, London, 495 pp.
Cullity, B. D. (1978) Elements of X-Ray Diffraction. Addison Wesley Publishing Co. Inc. 555 pp.
Ergun, S. (1968) Direct method for unfolding convolution products--its application to X-ray scattering intensities. J. Appl. Crystallogr. 1, 19-23.
____ (1970) X-ray scattering by very defective lattices. Phys. Rev. B, 131, 3371-3380.
Henke, B. L., Perera, R. C. C., Gullikson, E. M. and Schattenburg, M. L. (1978) High-efficiency low-energy X-ray spectroscopy in the 100-500-ev region. J. Appl. Physics 49(2), 480-494.
Jadgozinski, H. (1949) Eindimensionale Fehlordnung in Kristallen und ihr Einfluss auf die Röntgeninterferenzen. I. Berechnung des Fehlordnungsgrades aus der Röntgenintensitäten. Acta Cryst. 2, 201-207.
James, R. W. (1965) The Optical Principles of the Diffraction of X-Rays. Cornell Univ. Press, Ithaca, New York, 664 pp.
Kakinoki, J. and Komura, Y. (1952) Intensity of X-ray diffraction by a one-dimensionally disordered crystal. J. Geol. Soc. Japan 7, 30-35.
Klug, H. P. and Alexander, L. E. (1974) X-Ray Diffraction Procedures. J. Wiley and Sons Inc., New York, 996 pp.
MacEwan, D. M. C. (1956) Fourier transform methods for studying scattering from lamellar systems. I A direct method for analyzing interstratified mixtures. Kolloidzeitschrift 149, 96-108.
____ (1958) Fourier transform methods for studying scattering from lamellar systems. II. The calculation of diffraction effects for various types of interstratification. Kolloidzeitschrift 156, 61-67.
Méring, J. (1949) L'Intérference des Rayons X dans les systems à stratification désordonnée. Acta Cryst. 2, 371-377.
Nadeau, P. H., Tait, J. M., McHardy, W. J. and Wilson, M. J. (1984) Interstratified X-ray diffraction characteristics of physical mixtures of elementary clay particles. Clay Minerals 19, 67-76.
Plançon, A. (1981) Diffraction by layer structures containing different kinds of layers and stacking faults. J. Appl. Cryst. 14, 300-304.
____ and Tchoubar, C. (1977) Determination of structural defects in phyllosilicates by X-ray powder diffraction. I Principle of calculation of the diffraction phenomenon. Clays & Clay Minerals 25, 430-435.
____, Giese, R. F. and Snyder, R. (1988) The Hinckley index for kaolinites. Clay Minerals 23, 249-260.
Reynolds, R. C. (1968) The effect of particle size on apparent lattice spacings. Acta Cryst. A24, 319-320.
____ (1980) Interstratified clay minerals. Ch. 4. In: Crystal Structures of Clay Minerals and Their X-Ray Identification; Brindley, G.W. and G. Brown, eds. Mineralogical Society, London, 495 p.
____ (1985) NEWMOD© a Computer Program for the Calculation of One-Dimensional Diffraction Patterns of Mixed-Layered Clays. R.C. Reynolds, 8 Brook Rd., Hanover, NH 03755, U.S.A.
____ (1986) The Lorentz factor and preferred orientation in oriented clay aggregates. Clays & Clay Minerals 34, 359-367.
____ (1988) Mixed layer chlorite minerals. Ch. 15. In: Hydrous Phyllosilicates (Exclusive of Micas). Bailey, S.W. ed. Reviews in Mineralogy, 19. Mineralogical Society of America, 725 p.
Tettenhorst, R. and Reynolds, R. C. (1971) Choice of origin and its effect on calculated X-ray spacings for thin montmorillonite crystals. Amer. Mineral. 56, 1477-1480.
Warren, B. E. and Averbach, B. L. (1950) The effect of cold-work distortion on X-ray patterns. J. Appl. Physics 21, 595-599.

7. COMPUTER ANALYSIS OF DIFFRACTION DATA

D.K. Smith

INTRODUCTION

Much of the theory of geometrical crystallography was developed in the 1800s, and much of the theory of diffraction was developed in the early 1900s shortly after von Laue and his associates proved, in 1912, that crystals diffract X-rays. As the computer impacted the field in the 1940s, experiments caught up with the theory, and the field advanced rapidly and is still advancing as our ability to process numbers improves.

In early diffraction experiments, computers were used for data interpretation only. Elaborate experiments were cumbersome, time consuming, and required direct supervision by the operator. Today the computer also runs the experiment, and in some cases it can process and interpret the data without any input from the operator. It is not easy to say when the computer had its biggest influence on our use of diffraction experiments, as it has played such a vital role in all aspects of the advancement of the science.

This presentation will be concerned with only the data analysis aspects of computer applications. The modern automated powder diffractometer (APD) is controlled by a computer that not only directs the instrument to collect the desired data, it usually tests the instrument to assure the operator that the system is working properly. These systems have data analysis programs as part of the software that will do most of the steps usually desired by the user. These programs are so integrated with the APD system that they are usually not separated from the package and distributed independently and are not available to the user who has not purchased the APD. However, many of the APD analysis programs have been derived from stand-alone routines written earlier for specific applications, and there are many other programs that have not been incorporated that are also available to the researcher who does not desire to be dependent on a specific APD package. These latter routines are the subject of this presentation.

References to programs will be handled in two ways. Where there is a literature reference describing the program, there will be a year given in parentheses with the authors. Where there is only a program manual or no general documentation, the authors will be listed with no date. The list of program sources at the end of this Chapter should then be searched for the individual or company to contact for further information. If the program name is marked with (*) or (**), the program is known to require some payment for a copy that may range from supplying the medium to a commercial price. Even when requesting freeware, a requestor should supply the medium and ask if a small remittance would be appropriate.

This presentation will follow the sequence of analysis that would usually be followed in the interpretation of diffraction data. Initially, the discussion will consider the many sources of crystallographic and diffraction information that are available in computer-readable form. Then it will consider programs to be used when nothing is known about the material under study followed by programs for specific purposes when some information is known. Finally, it will consider the routines used for preparing presentation material for publications. In each category, there will be no attempt to compare all the routines that are available. Only the general aspects of the particular analysis will be considered, pointing out differences in concepts rather than actual practice. Some sections will be deferred to other chapters in this volume where much more detail will be presented. The routines will be listed here only for completeness.

This list of programs is not complete. Those included are based in part on the author's experience with the many aspects of diffraction analysis; but many are the result of a request to over 100 mineralogists and diffractionists throughout the world. Without their help some programs would have been overlooked. There is still the probability that some significant programs have been missed, and an apology is due to the authors. This list has been supplied to the Commission on Powder Diffraction of the International Union of Crystallography to initiate a program information center. Without the exchange of programs, this field would advance much more slowly than it currently does with the ready and willing distribution of routines that we enjoy today.

During this discussion, keep in mind that there are three types of information in a diffraction pattern: the position, the intensity, and the shape of the diffraction peaks. The positions of the diffraction peaks are determined by the geometry of the crystal lattice, i.e., the size and shape of the unit cell. The intensities of the peaks in a pattern are related to the specific atoms in a crystal and their arrangement in the unit cell of the crystal structure. The diffraction profile is related to the size and perfection of the crystallites under study and to various instrumental parameters. The peak positions are fairly easy to determine accurately by diffraction methods; whereas the intensities and the profiles are considerably more difficult to measure accurately. Consequently, most analytical methods emphasize the position variables (the d-spacings) and rely less on the other two parameters unless the intensities (e.g., quantitative analysis or structure refinement) or the profile parameters (e.g., crystallite size and strain) are specifically required.

TYPES OF NUMERICAL DIFFRACTION DATA

The raw diffraction data consist of a record of intensity of diffraction versus diffraction angle. In the early days, this information was recorded on film or on a chart record from which the user measured the desired information. This information was usually the peak position that was converted to a d-value and the peak height that was considered the intensity. Peak positions of complex profiles were estimated by methods that ranged from "guesstimation" to graphical curve fitting. Intensities were derived from charts by planimetry, counting squares or cutting out the peak and weighing the cutout. Intensities from films were estimated visually or obtained by using calibrated film strips for matching. Only very elaborate experiments provided more precise d-values and integrated intensities. These data were adequate for the identification of phases and refinement of lattice geometry. Structure determination and the determination of structure-related properties require the resolution of all peaks and the determination of integrated intensities and details of the diffraction profile. Extensive use of this information waited until APD's or densitometers were available to collect digitized diffraction data that could be processed by computer methods.

As early as 1839 with Dana's *System of Mineralogy,* 1922 with the Barker index of crystals, and *Structurbericht* (Niggli et al., 1931), crystallographers recognized the need to collect and categorize crystallographic data. Collecting data directly applicable to powder diffraction began in the 1930s, and these early efforts have led to several currently available databases. The data banks will be described in terms of the type of information that is incorporated. Most diffraction pattern databases use discrete d-I information. New ones are being developed that will employ the digitized trace information. Some of the databases use derived crystallographic data.

The data analysis portion of this presentation will be divided into two sections. The first will deal with those programs that use only d-spacings to determine information on the geometrical lattice or those that use the d's with intensities for phase comparison. These programs are useful for any d-I data including those derived by APD systems. The second will examine those routines that use the digitized trace as the starting point in the analysis. These routines may yield d-I information for use with the preceding routines, or they may do more complex projects such as structure refinement, quantification, or determination of crystal perfection.

REFERENCE NUMERIC DATA BASES
FOR POWDER DIFFRACTION ANALYSIS

Many philosophers have described the impact of the computer on society as being not the "Age of Computers" but rather the "Age of Communication and Information Transfer." Now that the cost of data storage is no longer the most expensive component in the computer system, vast amounts of data are being stored for direct access by the user. Actually, the two primary diffraction oriented databases both have their roots outside computer influences. The *Powder Diffraction File* (PDF) began in the late 1930s following the publication of the material that became the first set by the Dow Chemical Company (Hanawalt et al., 1938). *Crystal Data* was first published by Donnay et al. (1954) as a Memoir of the Geological Society of America. Both projects have continued; the PDF under the auspices of what is now the JCPDS-International Centre for Diffraction Data (ICDD) and *Crystal Data* under what is now the National Institute for Science and Technology, formerly the National Bureau of Standards. The PDF derives its financial support entirely from the sales of the PDF. *Crystal Data* is partly supported by sales and partly by funding from the U.S. Government. Detailed information on the present status of the many crystallographic databases may be found in Allen et al. (1987).

The Powder Diffraction File

The *Powder Diffraction File* (PDF) contains diffraction data primarily in the form of tables of d's and I's coupled with derived crystallographic information and documentation on the sample used, the experimental conditions, and on the source of information (Jenkins and Smith, 1987). Much of the information has been derived from the open literature and reviewed and edited by a group of experts in diffraction analysis. The literature information has been augmented by experimental projects sponsored by the ICDD to collect accurate data on the most important compounds and by calculation of theoretical powder diffraction patterns using crystal structures determined by single-crystal methods. The editing procedure selects the best data for each compound where there are multiple data sets reported; thus, the PDF is continually being revised to keep it as accurate and as current as possible. The present PDF contains data for over 50,000 X-ray diffraction patterns. Minerals are carefully edited for conformity with the nomenclature of the International Mineralogical Association Commission on New Minerals and Mineral Names, and indexes are provided for both mineral names and crystal-chemically classified mineral groups.

The PDF was initially published on 3"x5" cards, and this format is still the image used for data presentation in all hard copy forms of the PDF currently produced. All information on these card images is now in computer-readable form, and there are two products, known as PDF-1 and PDF-2, being distributed by the ICDD on various computer media. A third product, presently designated PDF-3, containing the full diffraction data trace along with all the derived and documenting information is under development.

PDF-1**. The first computer-readable version of the PDF contained only the d's, I's, the chemical formula and name, and the PDF number. These data were used by the developers of the phase-identification programs that compared the d-I sets to experimental data. These SEARCH/MATCH programs will be described in the section on the analysis of d-I data. The PDF-1 version was the basis for on-line search services as well as being distributed on magnetic tape for installation at individual computer facilities. Currently, PDF-1 is the basis for most APD search packages.

There are several routines available to convert the d-I data into a graphical image for visual comparison to the experimental data. Such simulations allow the user to view the subtleties of the "pattern" as a whole and usually assist the user in confirming the identification of the sample under study. Simulations may use single bars, isosceles triangles, or analytical profiles to represent each diffraction peak. Simulations may be made to any scale on either a graphics terminal or a hard-copy device and have even been made on transparent bases for direct comparisons with diffractometer charts. Film images have also been simulated, complete with the line density and curvature characteristic of the Debye-Scherrer pattern. Some routines even allow the user to combine several single patterns to simulate a mixture.

Programs that provide simulations include:

DISPLAY	Goehner and Garbauskas (1983)
SIMUL/COMBIN*	Smith
PEAK**	Sonnefeld and Visser (1975)
PPDP*	Okamoto and Kawahara
XRAYPLOT*	Canfield
DEBYE*	Blanchard
DEBYE*	Millidge
DEBYE	Bideaux
DEBYE	Morton

PDF-2**. The weakness of the PDF-1 version is that it only contains the key data for identification but none of the supporting information. Supporting information must be obtained from a hard copy of the PDF. In the days of limited storage on the computer, this situation was the only reasonable alternative. With the advent of inexpensive storage, it is now feasible to have all the data in computer-readable form. PDF-2 contains all the information that appears on a PDF card image, so that the user can display these data on the computer screen after determining the desired phase. Not only are the crystal unit cell and the Miller indices available, the symmetry, some physical property information, and the reference are also accessible. Most of the information can be directly searched to locate compounds by means other than through the d-I search procedures.

PDF-2 is available from the ICDD on several computer media. Probably the most interesting medium to the independent user is the CD-ROM, which can be installed on most PC's and on the microVAX2[1] series of computers. This medium is here to stay, and it will replace the hard-copy forms of the PDF in most laboratories in the not-too-distant future. The CD-ROM does not replace PDF-1 as a search database because data retrieval from the CD-ROM is slow, but it can provide all the necessary d-I information. It is ideal, though, for other types of searches that index on specific data. Searches by chemistry, symmetry, color, etc. are feasible and provide many new ways of using the PDF effectively. Perhaps the most important information that can be located is the reference to the source of the information for a given compound. The PDF is an excellent entry into the literature.

There are two major CD-ROM-directed programs in the public domain that are available through the ICDD. These programs are by Holomany and Jenkins and by Toby, Harlow and Holomany (1989). Many commercial systems also exist including Philips; Materials Data, Inc.; Socabim; INEL; and A. Wassermann.

PDF-3**. With the growing importance of utilizing the full digitized diffraction trace, the ICDD is researching the feasibility of providing a database of diffraction patterns. The value of such a database will become clear in the discussions below. PDF-3 is presently envisioned to contain all the information in PDF-2 along with the digitized trace. The current storage capabilities are no longer a limitation. Once a standard file structure that meets the needs of the user is agreed upon, a pilot project will begin. One of the problems is setting up a universal file transfer system, and considerable attention is being given to JCAMP-DX (McDonald, 1988). This program was devised by the optical spectroscopists for establishing communication between instruments and among users. Prototype PDF-3 type databases are now in use for some of the programs to be described in later sections.

---

[1]MicroVAX is a trademark of Digital Equipment Corporation.

NIST Crystal Data File**

The *Crystal Data File* (CDF) is a compendium of information on crystalline materials for which the crystal unit cell is reported (Stalick and Mighell, 1986; Mighell et al., 1987). Along with the unit cell, many other properties of the crystal are reported including the chemistry, status of the crystal structure knowledge, and the source of the information. The philosophy of editing data for the CDF is different from that used for the PDF in that all literature is retained either as an independent entry or through cross referencing from related compounds. Consequently, the CDF is useful for locating literature references as well as for identifying and characterizing materials; it currently has over 136,000 entries and is available in two forms: on magnetic tape or on CD-ROM. Both are available from the ICDD.

To use the CDF effectively with powder diffraction information, one must determine the unit cell from the d-spacings. With these data and a knowledge of the rules for defining cells, one can search the CDF for possible matching cells without regard to any chemical information. Either the compound of interest or an isostructural compound might be located. These topics are discussed in the section on "Unit-cell Identification."

NIST/Sandia/ICDD Electron Diffraction Database**

The *Electron Diffraction Database* (EDD) is a new database that has been derived from the PDF and CDF specifically for use with electron diffraction data (Carr et al., 1989). By using the crystal cell data in both the PDF and the CDF, d-spacings have been calculated for all the compounds present. In electron diffraction, the intensities are not as useful as in X-ray diffraction because of sensitivity to the instrument conditions and multiple diffraction, so the emphasis for identification is placed on the d-spacings of the first 10 lines along with chemical information. One advantage of electron diffraction is that the region of the sample producing the pattern is more likely to be single phase than for the larger sample used in X-ray diffraction.

The main responsibility for developing this database is with the Crystal Data group at the National Institute for Science and Technology. Details on the use of this database will be given in the section on "Unit-cell Identification." There are over 70,000 entries in this database.

FIZ-4 Inorganic Crystal Structure Database**

The *Inorganic Crystal Structure Database* (ICSD) is a product of earlier independent databases developed by G. Bergerhoff and I.D. Brown (1987). It is presently based at the Fachsinformationzentrum-4 in Karlsruhe in conjunction with the Gmelin Institute in Frankfurt, FRG. This database contains over 24,000 entries of crystal structure information for inorganic compounds other than metals. In addition to the crystal cell and chemistry, all the atom locations are given. Thus, from this information, it is possible to calculate the powder diffraction pattern that can then be compared with experimental data. Many programs for calculating patterns will be described in the section on "Crystal Structure Analysis." This file is also an entry into the literature. At present this file is available only through on-line subscription, but it will soon be available on CD-ROM also. Programs are available to search on various names and chemistry, and it is possible to display the structures graphically.

NRCC Metals Structure Database**

The *Metals Structure Database* (MSD) is similar to the ICSD in purpose (Rodgers and Wood, 1987). It contains the same type of information as is found in the ICSD except for metals and their corrosion products. It is based at the National Research Council of Canada in Ottawa. It augments the information in the ICSD, so between them, the mineralogist will find most of the minerals whose structures are known. It also provides data for the calculation of powder diffraction patterns. The MSD also may soon be available on CD-ROM.

Mineral Data databases**

There are many new PC-based databases whose goal is to provide complete descriptive information for all known minerals. MINERAL by Nichols and Nickel (ALEPH) is derived from the Fleischer (1983) compendium and contains crystal information along with other data. This file provides an entry to the PDF by references to the PDF file numbers, and the crystal cell information is included so the user can calculate d-spacings. Many other such databases are appearing now that PC systems are so common.

ANALYSIS OF d-I DATA

There are many routines that use discrete d-I data for various purposes. The d-I data may be determined by the diffractionist from the film or chart or provided by the APD. Prior to use for interpretive purposes, the data should be calibrated against an internal standard, if used, or by external calibration. Then the data are ready for study.

By far the most common use of powder diffraction data is the confirmation of materials or the identification of unknowns. Identification usually involves the interrogation of the PDF database using experimental d's and I's for comparisons. Such programs are described in the section on "Identification of Mineral Phases." If identification is not successful by this approach, the d's may be used in an indexing program to attempt to find a unit cell that is compatible with the observed d-spacings using one of the routines described in the section on "Indexing of Unknown d-spacing Data." A successful indexing may provide a cell that can be matched against the CDF database for an identification. Cell reduction programs may be necessary to locate a match.

There are other applications for discrete d-I data. Accurate d-values can be refined to provide accurate unit cell parameters that are sensitive to subtle differences in composition or structural ordering, so cell refinements are often used in the studies of materials as described in the section on "Unit-cell refinements." Discrete intensity data may be used for applications such as quantitative phase analysis, texture analysis, and crystal structure determination as discussed in succeeding sections.

Processing of d-I data sets

In order to achieve accuracy with diffraction data, it is necessary to eliminate or minimize systematic errors in the experimental data. There are several important steps in achieving accurate data, the most important of which is alignment of the instrument. Analysis of the systematic errors is then possible. By far the most common method of reducing systematic errors in experimental data is to include a known phase in the sample and to use the known phase to calibrate the measurement scale for the data. Common calibrators are Si, W, Ag, CaF_2, $MgAl_2O_4$, SiO_2, fluorophlogopite, and LaB_6 (Hubbard, 1980, 1983a,b). Si, fluorophlogopite, and LaB_6 are available from the Office of Standard Reference Materials (OSRM), National Institute of Standards and Technology, Washington, DC, USA. Adding an internal standard with known lattice parameters permits the correction of all the systematic errors at the same time. Several routines make this correction:

        INTCAL   Snyder
        PPLP   Gabe et al.
        CALIBR   Hubbard

These programs use the precise angle values from the carefully calibrated standards to correct the angular diffractometric measurements prior to the calculation of the d-spacings. Careful calibration can provide data accurate to 0.005 degrees 2θ.

Some comments on the preparation and use of internal standards are appropriate here. For internal standards other than those available from the OSRM, the user will usually have to calibrate their own supply against one of these certified standards. Quartz is nearly ubiquitous in most geological samples, and its unit cell varies little from sample to sample. The variation is less than the accuracy necessary in many experiments, so it may be employed without calibration in many studies. Spinel is easily made by firing a stoichiometric mixture of components to 1400-1600°C with regrindings and pelletizing. Do not use a commercially grown single-crystal boule as a source of spinel because they are always off stoichiometry. Chemically pure CaF_2 can be fired at 1000°C to improve crystallinity. Corundum, rutile, zincite, and some other commonly used standards are usually better for intensity references rather than as d-spacing standards because they usually have small crystallite sizes and exhibit line broadening. Any standard prepared by the user should be calibrated by preparing a mixture with one of the certified standards, and refining the corrected d-spacing measurements using one of the routines mentioned in the section on unit-cell refinements.

<u>Identification of mineral phases</u>

Diffraction data are not a collection of independent d's and I's but a very systematic set of interrelated values that reflect the lattice and the crystal structure of the material. Consequently, the set of d's and I's are a pattern of values characteristic of the material. Thus, it is possible to use this pattern of data as a means of identification, and many routines have been developed to achieve this end. Most of these routines were developed on the large main-frame computers, and some have seen over 20 years of development (Johnson and Vand, 1967; Frevel et al., 1976; Edmonds, 1980). Now, there are many routines on PC-based systems. The early limitations on the developments were the storage restrictions that prevented the database being directly accessible, and magnetic tapes were used to run through the database. Ingenious search schemes were employed to locate candidate answers. Direct access to the database is now the norm, and the search schemes are much more straightforward. PC-based search programs are also now feasible, and several are available.

The philosophy of identification of unknowns using a database of known phases is to ask the question whether the pattern of the known is present in the pattern of the unknown. The diffraction pattern of a mixture is a composite of the patterns of each phase in the mixture, so the approach is to determine a figure-of-merit for each known and then to rank the known phases for user interpretation. The database of knowns may be pre-screened using chemical information or other information such as the knowledge that the specimen under study is a mineral. Prescreening creates a much smaller set of knowns that needs to be examined. The search process then takes considerably less time to execute, and the results are usually much easier to interpret.

Although the computer is a tireless comparer of numbers, the results are only as good as the data allow and are highly dependent on the algorithm used. The major problem in searches is the degree of accuracy of both the unknown and the known data, and how well the known data represent the sample under study. Mineral identification can be very difficult when the mineral exhibits extensive solid solution. Ionic substitutions can alter both the lattice dimensions and the characteristic intensities of the mineral, and it is unlikely that an exact match is present in the reference database for such minerals. Thus, it is usually up to the user to recognize the probable answer from the short list of possible phases provided by the algorithm. Logic and an understanding of the geologic environment are often the key to selecting the right answer.

Diffraction patterns of mixtures are generally more difficult to unravel than those for single phases. The computer can be very useful for mixtures, but it still has limitations where the quantity of the phase is small and its pattern is only weakly represented in the experimental data. Logic, additional information such as chemical or optical data, perseverance, and luck are needed to identify such phases.

Mini- and main-frame routines. All of the early search programs were developed on large computers. Most of these programs interrogate the PDF as the primary database. Most also modify the primary data to structure the data files for efficiency according to the algorithm used. Consequently, it is necessary to acquire these files or the ability to build them when implementing one of the programs. Because the PDF is a private database, it is also necessary to obtain lease permission from the ICDD prior to using it for identification.

Mini- and main-frame programs include:

SEARCH/MATCH	GJohnson and Vand (1967)
PDIDENT/PDMATCH	Goehner and Garbauskas (1983a,b)
XRDQUAL	Clayton
SEARCH	Toby et al. (1989)
FSRCH	Carr et al. (1986)
ZRD-SEARCH	Frevel et al. (1976)
FAZAN	Burova and Schedrin
SEARCH/MATCH	Lin et al. (1983)
SEARCH/MATCH	Marquart et al. (1979)

Some of these are available from ICDD, but most should be obtained from the authors.

PC-based search routines. All of the PC-based search programs are commercial. They follow similar philosophies to the mini and main frame programs that are now possible with new data storage devices designed for PC's. These programs include:

μ-PDSM**	Marquart (1986)
Micro-ID**	QJohnson
SEARCH**	O'Connor

These programs must be obtained from the authors.

<u>Indexing of d-spacing data for an unknown</u>

Where one has absolutely no knowledge of any crystallographic information on a material and is unable to identify the phase using the search procedures mentioned above, the next step is to try to determine a cell that will explain the d-spacing data observed. This procedure requires the material to be single phase and the experimental data to be very accurate. The procedure is usually known as *indexing a pattern*.

Indexing procedures use only the position variables of the pattern and try to find a solution to the relations:

$$d(hkl) = f(h, k, l, a, b, c, \alpha, \beta, \gamma) ,$$

where h, k, and l are the Miller indices and a, b, c, α, ß, γ are the parameters that describe the lattice. The form of the equation in real crystal space is complex for the triclinic crystal system and quite simple for the cubic system where it takes the form:

$$Q(hkl) = \frac{1}{d^2(hkl)} = \frac{h^2 + k^2 + l^2}{a^2} .$$

Most indexing procedures use the equation in reciprocal crystal space that takes the form:

$$d^{*2}(hkl) = (h^2 + k^2 + l^2)a^{*2}$$

in the cubic system and:

$$d^{*2}(hkl) = h^2 a^{*2} + k^2 b^{*2} + l^2 c^{*2} + 2hk a^* b^* \cos\gamma^* + 2hl a^* c^* \cos\beta^* + 2kl b^* c^* \cos\alpha^* ,$$

or

$$Q = h^2 A + k^2 B + l^2 C + hkD + hlE + klF$$

in the general case of the triclinic system. This approach linearizes the problem to determining six unknowns, A, B, C, D, E, and F, that have integer coefficients in every d-spacing equation. Finding the proper values for the lattice parameters so that every observed d-spacing satisfies a particular combination of Miller indices is the goal of indexing. It is not easy even for the cubic system, but it is very difficult for the triclinic system. Because one knows nothing about the material, one does not even know which form of the equation to use.

There are two general approaches to indexing (Shirley, 1980). One can assume in the beginning that the material is cubic and increase the a value until a solution is found to the above equation. That such a solution can be found is not proof of success, e.g. if the a value is ridiculously large, then it is apparent that it is not acceptable. One then tries the equations for hexagonal and tetragonal trying to determine reasonable values for a and c. Beyond orthorhombic, this approach is difficult even for a computer. This approach has been termed 'exhaustive indexing' or 'trial and error.'

The other approach is analytical indexing where the crystal is assumed to be triclinic, and a final solution is approached by finding sub-solutions called zones using part of the data and then trying combinations of the sub-solutions to find the full solution. This method is known as the ITO method (Ito, 1950; deWolff, 1957, 1962). Once a lattice is found that satisfies the experimental d-spacing data, the geometry of the lattice is examined for possible higher metric symmetry (*International Tables*, Vol. I, pp. 530-533, 1965).

Both of these approaches require very accurate d-spacing data. The smaller the errors, the easier it is to test solutions because there are often missing data points due to intensity extinctions related to the symmetry or the structural arrangement or due to lack of resolution of the d-spacings themselves. The earliest approaches were of the exhaustive type and were done by graphical fitting or numerical table fitting. One of the earliest computer routines was by Goebel and Wilson (1965), who programmed the cubic and hexagonal/tetragonal equations for trial and error testing. The method used the Table of Quadratic Forms (*International Tables*, Vol. II, pp. 124-146, 1965) and assumed the first observed d-spacing was among the lowest allowed Miller indices. The method was extended to the orthorhombic system by Louer and Louer (1972) and to monoclinic by Smith and Kahara (1975). Related routines were developed by Louer and Vargas (1982), Shirley and Louer (1978), Werner (1964), Werner et al. (1985), and Taupin (1968); and routines by these latter authors are commonly used today.

The ITO analytical approach was used by Visser (1969). This routine is the basis of other derivative programs available today. Whereas the exhaustive approaches are not limited by data sets with few values in the high symmetry cases, the analytical approach is. At least 20 data points are required to be successful and more are desirable. However, it is usually not advisable to use too many data points because it becomes harder to assign hkl's to specific points in the high-angle part of the data set due to the larger number of acceptable assignments. The practical limit is about 40.

Indexing programs that were identified include:

    CUBIC/HEX-TET   Goebel/Wilson (1965)
    CUBIC   Snyder
    ITO   Visser (1969)
    TREOR   Werner (1964)

DICVOL	Louer and Louer (1972)
POWDER	Taupin (1968)
CELL	Takeugi et al.
KOHL	Kohlbeck and Hoerl (1976)
INDEXING	Paszkowicz (1989)
AIDED	Setten and Setten (1979)

PC versions include:

PC-ITO*	Visser
UNITCELL	Garvey (1986a)
Micro-INDEX**	QJohnson and Snyder

Indexing programs usually provide the user with several answers and a figure-of-merit that is based on the lattice fit. It is up to the user to decide on the best answer. There are two figures-of-merit used: M_{20}, deWolff (1968, 1972) and F_N, Smith and Snyder (1979).

$$M_{20} = \frac{Q_{20}}{2 |\Delta Q|_{20}},$$

where Q_{20} is the Q value for the 20th observed line and N_{20} is the number of possible lines that could be observed out to Q_{20}.

$$F_N = \frac{1}{|\Delta 2\Theta|} \cdot \frac{N_{obs}}{N_{poss}},$$

where N_{poss} is the number of possible lines out to the Nth observed line. There is some controversy concerning which of these proposed figures-of-merit is the most appropriate. Shirley (1980) claims that M_{20} is more sensitive to the correctness of indexing and F_N is more a measure of the accuracy of the diffraction data. Both figures-of-merit are now included in all PDF data sets that are indexed.

Any lattice with an acceptable figure-of-merit needs to be examined further to determine if it explains all the data including physical property measurements on the material. Several suggested solutions may actually be the same because there are different ways to describe the same lattice. At this stage it is necessary to reduce the cell by the methods described in the next section and then examine it for possible higher symmetry using the 43 Niggli matrices (*International Tables,* Vol. I, pp. 530-534, 1965; Azaroff and Buerger, 1958).

<u>Cell reduction and lattice evaluation</u>. As indicated above, once a unit cell is found by an indexing procedure, it must be examined for other possibilities of different metric symmetry. For any given lattice there are many ways to describe the periodicity by selecting any combination of three non-coplanar vectors whose unit-cell volume does not enclose another lattice point. The problem is to determine if there is a better choice of lattice vectors. The procedure follows two steps. The first step is to determine what is called the *reduced cell* (Azaroff and Buerger, 1958). This cell is unique for every lattice and is found by following specific rules in the selection of the lattice vectors. This process is known as cell reduction. The cell edges of this cell are usually the three shortest noncoplanar vectors in the lattice with the interaxial angles defined in a specific manner. This cell is usually of low metric symmetry. (Metric symmetry means the symmetry of the shape of the unit cell only without regard to the positions of other lattice points.) Once the reduced cell is found, it is straightforward to test relationships between the vector magnitudes and directions to seek nonreduced cells with higher metric symmetry. Unfortunately, the metric symmetry may not be the true symmetry. The true symmetry cannot be confirmed without further tests such as physical property measurements, single crystal X-ray diffraction, or structure determination.

There are several programs other than the indexing programs that allow the user to determine the reduced cell. These programs include:

Cell Reduction	Lawton
NBS*LATTICE	Himes and Mighell (1985)
PC-R-TRUEBLOOD	Blanchard
CREDUC	Gabe, Lee and LePage
PC-LEPAGE	Spek
NEWLAT	Mugnoli (1985)

There is a related routine that looks at the position coordinates in a crystal structure description to determine higher symmetry that is:

MISSYM	Gabe, Lee and LePage

Once the cell is selected, there are two further steps toward identification of the material that are refinement of the cell and then lattice matching against a database.

Unit-cell refinements

To determine the best unit cell, initial cell parameters are usually refined using the entire set of experimentally measured and corrected d-spacings. The most common procedure is to use a least-squares technique that minimizes the squares of errors between the calculated and measured data. As done in indexing, this calculation is usually done in $Q = (1/d^2)$ form because it linearizes the equations and simplifies the mathematics.

$$Q(hkl) = \frac{1}{d^2} = \frac{4(\sin^2\Theta)}{\lambda^2} .$$

It is the sum of the squares of Q(obs)-Q(calc) that is minimized in the calculation. A good general discussion of refinement may be found in Klug and Alexander (1974) and Wilson (1963).

Unit-cell refinement is usually done with weighting factors based on the experimental reliability of the measurement of the data. Weighting factors may include contributions from the ability to detect the peak usually based on the intensity of the peak and the ability to resolve the peak both from adjacent peaks including the α_2 component. Where resolution is impossible at low angles, the appropriate wavelength must be used in the refinement. Systematic error functions may also be included for instrumental and sample errors that could not be eliminated experimentally. The method for this error analysis is known as the Cohen (1936a,b) method and involves added terms in the Q equation for position compensation functions. This method is described in Cullity (1978).

There are several routines available for cell refinement, some with systematic errors. These routines require that you assign hkl's to each data point and select the symmetry and error functions prior to the refinement. The routines include:

ARGONNE	Mueller and Heaton (1961)
LCLSQ	Burnham
PODEX	Foris
FINAX	Hovestreydt (1983)
PC-CELRF	Prewitt
INDEXING	Novak and Colville (1989)

There is another approach that may be used during refinement which is to start with a cell as input parameters and allow the program to determine the hkl's to be assigned to each data point. The input cell must be close to the correct cell or the results may stray, but this approach allows the user to determine a cell from the low angle low-accuracy peaks and to use this cell as the starting point. The calculation usually follows the sequence: (1) Use the starting cell to index points within a defined error and 2θ range; (2) refine the accepted points to yield a better cell; and (3) reduce the defined error, open the 2θ range and repeat step one. This procedure usually converges in less than 10 cycles if it is going to converge at all. It is possible to force the program to use specific hkl's for some points, and this assignment is often necessary when cells are pseudosymmetric. Because of the nature of the algorithm, it is not possible to refine systematic errors at the same time, but weighting factors are used.

All the programs that use this technique have evolved from the routine written by Appleman et al. (1973). Presently available routines include:

APPLEMAN	Appleman et al. (1973)
NBS*LSQ85	Hubbard
PPLP	Gabe et al.
PC-APPLVANS	Benoit
PCPIRAM	Werner and Brown
LSUCRIPC	Garvey (1986b)

Because of the importance in determining the best data for inclusion in the PDF and CDF databases, one of the steps in the evaluation of the experimental data includes refinement of the unit cell and the comparison of the measured and calculated 2θ values and their figures-of-merit. There is a routine that is used specifically for this purpose that includes a modified APPLEMAN component along with many other checks. This routine is known as NBS*AIDS83 that was prepared under the direction of Mighell et al. (1983). In addition to refining the unit cell and calculating the figures-of-merit, it warns the user of poorly fitting data values and the presence of systematic errors. Any laboratory planning to prepare data for either of the databases or for publication should use this program to screen the data for errors prior to submission.

There are many applications for refined cell data. Comparing refined unit cells is the most sensitive measure of subtle differences between similar materials. All data that are to be published in any form should be refined. Unit-cell data may be the best way to study systematic changes in compounds with crystallochemical substitutions or with environmental changes such as temperature, pressure and/or atmosphere. If nothing else is known about the material, the cell may also be the basis for identification by comparing it to one of the crystallographic databases.

<u>Unit-cell identification</u>

Where the compound under study is not represented in the PDF, d-I pattern searching is usually not successful unless there is an isostructural compound present. All is not lost at this point, as it is very effective to use the CDF database and search the cell directly against the cells in this database. NBS*LATTICE, developed by Himes and Mighell (1985), uses the reduced cell and derivative cells formed by doubling or halving input parameters to locate possible cells in the database that are within specified error limits. This database has cell information for over 70,000 inorganic compounds that is a large portion of the known inorganic materials. This large selection of unit cells provides a good chance of finding a cell match and thus an identification. Along with the identification comes additional information on the compound including its composition, true symmetry, and the reference to the paper from which the cell was abstracted.

Because of the large amount of information in the CDF database, other programs have been written to access the file other than through the cell. BOOLEAN/LATTICE, written by Harlow et al., allows searches on symmetry, chemistry, authors, journals, density, optics, and many other properties that are included in the data entries. This type of search is useful for examining the coverage of the CDF database. It allows the user to select compounds that may show specific properties, e.g., piezoelectricity, that are symmetry-sensitive. It also allows the user to locate a specific compound to obtain the unit cell for generation of allowed d-values prior to an experiment.

Although designed for electron diffraction, the *Electron Diffraction Database* (EDD) is a useful source for the X-ray diffractionist. It was produced to allow identification where the intensity values are generally very unreliable and the emphasis is on the d-value measurements only. The database contains the first 10 lines generated from the unit cell, and it is the d-values of these lines that are searched in the D-SPACING program (Carr et al., 1989). There is a lattice-matching routine available for this database also known as UNIT-CELL (Himes and Mighell, 1987). It works like NBS*LATTICE described above.

Quantitative analysis

There are other uses for d-I data in the analysis of materials. One use is for the quantitative analysis of the phases present in a mixture. The diffraction pattern of a mixture is a composite of the individual patterns of the component phases. The integrated intensities from each phase in the mixture are proportional to the amount present in the mixture modified by absorption effects due to the phases in the mixture. The theory of analysis developed by Alexander and Klug (1948) is the basis for several techniques that use the intensities of specific peaks from the diffraction pattern usually after creating a calibration suite of similar samples. The fundamental equation is

$$I_{iJ} = \frac{K_{iJ} x_J}{\rho_J \mu^*} ,$$

where i represents the ith line and J the Jth phase. The weight fraction of the Jth phase is x_J, and its density is ρ_J. The factor μ^* is the mass absorption coefficient of the whole sample. Thus, the intensities are not independent of the other phases in the mixture. Methods have been devised to get around this problem most commonly by including a known amount of an added phase as an internal standard or by preparing calibration mixtures. No details will be given here because this topic is the subject of Chapter 5 of this volume. The calculations are straightforward, so that most programs are subsets of the general routines within the system, but there is one program for the PC worth mentioning:

        PC/PEAKS   Hill and Foxworthy.

Tests for preferred orientation

One of the problems with most diffraction experiments involving intensities of powdered samples is that the crystallites in the sample are not randomly oriented, i.e., the crystallites do not assume all possible orientations with equal probability. Considerable effort must be expended in sample preparation to achieve the random state, and it is often difficult to determine when randomness is attained unless one has a reference set of intensities that are known to represent the random sample. However, if one does have such a reference, it is possible to test the experimental intensity measurements and to determine the type of orientation that is present in the sample. This procedure is known as texture analysis. The most common procedure for determining texture is to tilt the specimen and rotate it while following the variation in intensity of a single diffraction peak. Such an experiment involves elaborate equipment to quantify the measurements. When the texture is only needed qualitatively, the standard powder pattern may be examined for its intensity deviations from the random sample values.

Preferred orientation in a sample is usually due to the shape of the particles in the sample and to the lack of sufficient numbers of particles to allow randomness to be attained. This situation is often referred to as "particle statistics." The shape is usually due to crystallite shapes or to cleavages when the sample is being crushed. Where there is one well-developed cleavage, the peaks corresponding to the cleavage plane will be enhanced and all others diminished. Where several symmetry-related cleavages occur, the orientation becomes more complicated. If two cleavages result in a needle shaped particle, the orientation effect is similar to the plate effect. There are several ways to describe the orientation effect on intensity values by applying an analytical function to calculated intensities; these are reviewed by Dollase (1987). The program PREF/SUPREF has been written by Snyder to evaluate preferred orientation in a sample by comparing the experimental intensities to the expected intensities from that material. POWD12 by Smith et al. (1983) includes orientation functions to model preferred orientation. Orientation functions are an option in most Rietveld programs.

Structure analysis using integrated intensities

True structure analysis, i.e., the determination of atomic coordinates in the unit cell for an unknown structure, is rarely done today using only powder diffraction data. Usually the structures are too complex, and it is not possible to resolve the individual diffraction peaks sufficiently to measure the needed integrated intensities. Peaks must be considered as clusters. Most of the intensity modeling programs were written to calculate the intensities from a proposed structural arrangement to test the model. One program that uses the individually resolved intensities from a powder pattern for structure analysis is

POWDER   Rossel and Scott

Any single-crystal structure analysis program may be used when the intensity values can be fully resolved. Usually the individual intensities cannot be fully resolved, and the methods in the next section must be employed.

ANALYSIS OF DIGITIZED DIFFRACTION TRACES

With the availability of modern computer-controlled diffractometers, many of the experiments utilize the digitized diffractometer trace directly in the data analysis. It is now routine to collect the full trace and determine the peak positions and intensities by algorithmic processes rather than by visual estimation. Such processes lead to more accurate measurements and improve the utility of the experiment considerably. Some experiments use these derived values with the analytical programs already described above. For these studies, the d's and I's are often extracted by routines that are part of the APD system software, but there are some non-APD programs available also. These programs will be discussed in the next two sections. Subsequent sections are devoted to programs that use the digitized trace either directly or only slightly modified from the raw data, e.g., removal of the background and/or the α_2 component. Although it is not the intent of this paper to discuss APD systems, programs by Rogers and Lane (1987) and Madsen et al. (1988) for controlling a diffractometer using a PC should be mentioned. They provide the do-it-yourself user with a limited budget a means to control the data collection system.

Processing of raw digitized data

There are several programs designed to process raw diffraction traces and extract d-I information. The processing includes options for smoothing of the data and for stripping out the background and the α_2 component of the diffraction peak and then locating the peaks and intensity values. Experience has shown that there is no perfect algorithm for peak recognition due to the complexities of the shapes. It is usually necessary for the diffractionist to examine, preferably using display graphics, the selections and to add or subtract peaks where necessary. Most APD packages allow the user to test the positions with the screen cursor and edit the data file as necessary. Background stripping is usually accomplished by selecting points on the trace that are the lowest and fitting either a polynomial or spline function to these points. Steenstrup (1981) discussed the use of orthogonal polynomials for background fitting. Fitting problems often occur at the ends of the 2θ range when using analytical functions. Linear fits within selected regions are also used. Routine

        BKGRD   Smith

is available as a separate program for these operations. The smoothing is either done with a Savitsky and Golay (1964) procedure or by Fourier filtering (Armstrong, 1989) The α_2 stripping uses the Rachinger (1948) method that uses one-half the α_1 profile to image the α_2 shape at its proper position or the Ladell (1965) algorithm that uses experimentally determined profiles for each component.

Programs for the reduction of digitized data include:

    ADR   Mallory and Snyder (1979)
    PC/DRX   Vila1
    ISIP   Rogers and Lane (1987)

There are also programs that digitize and reduce the data on film patterns, including

    GUFI*   Dinnebier and Eysel
    SCAN/SCANPI**   Werner and Brown

<u>Location of peak positions and intensities</u>

There are two general procedures used to determine the positions of peaks in a diffraction trace: derivative methods and fitting methods. Both have their advantages and their problems. The derivative method is very sensitive to the noise due to counting statistics; peak fitting is slower and sensitive to the peak shape assumed.

In the derivative method, both the first and second derivatives are determined using the techniques of Savitsky and Golay (1964). The first derivative changes sign whenever the diffraction trace forms a peak or valley. The sign of the second derivative for the same point identifies which are peaks. The second derivative also locates inflection points in the diffraction trace that indicate half-widths and shoulders of peaks. The noise effects are minimized by the smoothing step, but in the process, some detail is lost and peaks are missed or misplaced by the algorithm. These missed peaks must be reentered by the user usually employing a interactive graphical display. Once the peaks are located, the intensities are determined from the raw trace by either selecting the intensity at the peak point or by fitting a parabola near the peak to define its position and maximum. Some attempts are made to integrate peak intensities that are usually unsatisfactory for experiments requiring accurate measurements. General programs that use the derivative method include

    ADR   Mallory and Snyder (1979)
    PEAK**   Sonneveld and Visser (1975)
    POWDER PATTERN   Pyrros and Hubbard (1983)

Each diffraction peak is a *convolution* of many components that can be classified into three groups. The first component is the source profile that is dependent on whether the source is a sealed X-ray tube, a synchrotron, or a nuclear reactor. The sealed tube produces the most complicated spectral distribution, being composed of several different wavelengths each yielding peaks whose shape is empirically close to a Lorentzian (Cauchy)-like profile with long tails and a sharper than Lorentzian peak. The α_1 and α_2 peaks are dominant but there is a small α_3 peak that also must be considered when detailed studies are done. This source profile is modified by the aberrations of the diffractometer, known as the instrumental profile, and by the sample profile that depends on the perfection of the sample. Ideally, one would like to remove the contributions of the source and the instrument and isolate the sample profile that contains the desired information. Achieving this goal requires extremely accurate data over the whole profile because of the large amount of intensity in the long tails. Several mathematical methods have been devised to deconvolute diffraction peaks (Klug and Alexander, 1974), but the details are beyond the scope of this book. Deconvolution methods are not used for d-I determination. They are used for determining crystallite size and strain.

Profile-fitting to locate peak positions shows considerable promise to unravel complex peaks and yield accurate locations and intensities, both peak height and integrated. To use profile-fitting, one must assume an analytical peak shape and fit all three wavelength components using the known dispersion and known intensity ratios and a changing breadth over the 2θ range. After finding the best fit for each peak in the diffraction pattern, the centroid of the profile is used to determine the d-value and the area or peak height is used for the measure of the intensity.

A mathematically simple approach to peak finding is to fit an inverted parabola to several points around each peak position. This parabola then indicates a peak (centroid) position, a peak intensity, and a probable half-width. The number of points to be used in the fitting procedure depends on the data collecting strategy (the 2θ interval) and the half-widths of the experimental peaks. All the points in the top half of the profile should be used for this method to be successful. This method has great difficulty with overlapped peaks where weaker peaks appear as shoulders on the sides of stronger peaks.

In the more elaborate profile-fitting methods, there are several analytical functions that have been used to represent the experimental profile. Some fit each peak with profiles using many parameters. Others use the same profile throughout the whole pattern varying only the half-width and intensity. These profile models and fitting procedures have been evaluated by Howard and Snyder (1983). Profiles that have been employed for X-ray studies include: the Lorentzian (exponent = -1), the Intermediate Lorentzian (exponent = -1.5), the Modified Lorentzian (exponent = -2), Voigt and pseudo-Voigt functions, the Pearson VII function, and the Edgeworth series. The Gaussian profile commonly is used for neutron diffraction. The functional forms of these analytical profiles are listed in Table 1. The selected function is then best-fit to the diffraction trace, usually by a least-squares method; and all the peak positions and intensities are recorded. Note that this process *is not deconvolution*; it is *decomposition*.

The intensities determined by profile-fitting are usually applicable to all experiments requiring accurate intensities. If time permits, profile-fitting is the best procedure for determining pattern parameters. However, profile refinement is hampered by the lack of any *a priori* information on the expected peak positions and intensities. Any profile-fitting procedure that is not constrained by the crystallography of the material under study may generate peaks that are not real or might miss peaks, and all located peaks must be examined to confirm that they are allowed by the unit cell. Routines that are available for profile-fitting include:

SHADOW	Howard
Micro-SHADOW**	Howard and QJohnson
PROFIT**	Sonneveld and Visser (Langford et al., 1986)
PRO-FIT*	Toraya et al. (1983)
REGION	Hubbard and Pyrros
XRAYL	Zhang and Hubbard
Pi'oPiliPa'a	Jones
EDINP	Pawley (1980, 1981)
LAT1	Tran and Buleon (1987)

Chapter 8 of this volume examines profile-fitting in more detail.

Table 1. Analytical functions used in representing diffraction profiles.

NAME	FUNCTION
Gaussian	$A_1 \exp(-\dfrac{x^2}{k_1^2})$
Lorentzian	$A_2(1 + k_2^2 x^2)^{-1}$
Modified Lorentzian	$A_3(1 + k_3^2 x^2)^{-2}$
Intermediate Lorentzian	$A_4(1 + k_4^2 x^2)^{-1.5}$
Edgeworth Series	$\dfrac{A_5}{P}(1 - \dfrac{x}{Q})$
Voigt	$A_6 \int_{-\infty}^{\infty} L(x')G(x-x')dx'$
Pseudo-Voigt	$A_7[wL(x) + (1-w)G(x)]$
Pearson VII	$A_8[1 + 4(2^{1/m} - 1)\dfrac{x^2}{k_8^2}]^{-m}$

where $m = k'_8 + k''_8/2\theta + k'''_8/(2\theta)^2$

$x = 2\theta_i - 2\theta_o$

k_n are constants related to the half-width

A_n are normalizing constants

L and G are Lorentzian and Gaussian functions

P and Q are polynomials containing only even exponential terms

Crystal structure analysis

The determination of the arrangement of atoms in a crystal structure is the ultimate goal of some diffraction experiments, so the researcher can relate physical or chemical properties to the structure. Powder diffraction is rarely used in the determination of a totally new structure because it is difficult to separate the diffraction intensities into the individual contributions with sufficient accuracy to employ the standard structure-solving techniques. Most structure types are known now, and it is possible to test structure models by comparing the theoretical diffraction intensities with experimentally measured values. Two main approaches are used; one is to start with a model of a structure, make the substitutions and adjustments in parameters expected, and calculate the theoretical pattern for comparison to an experimental diffraction pattern. Structural parameter changes are guessed by examining the contributions of individual atoms to the intensities, and a new model is calculated to see if the fit is improved. The other is to refine the structure parameters using the differences between observed and calculated powder intensity data to calculate parameter shifts for the next cycle. Both techniques are extensively used.

Structure modeling. Programs to calculate integrated intensities essentially evaluate the equation

$$I(hkl) = \alpha m Lp A |F_t(hkl)|^2 ,$$

where α is a scale factor, m is the multiplicity factor, Lp is the Lorentz-polarization factor, A is the absorption factor, and F_t is the temperature-corrected structure factor that describes the types and positions of atoms in the unit cell.

$$F(hkl) = \sum_{j=1}^{N} f_j \, e^{2\pi i(hx_j+ky_j+lz_j)},$$

where f_j is the atomic scattering factor and x_j, y_j, and z_j are the position coordinates of the N atoms in the unit cell. The calculation is straightforward. Input consists of the symmetry and size of the unit cell and the atom types and coordinates. Starting with a given structure model, one can substitute other atoms in selected sites and examine the predicted intensity values. Position parameters or occupancy factors can be adjusted to achieve improvement in the match between experimental and calculated intensities until the user is satisfied with the new model. Models do not have to be close to the true structure for the calculations to be useful. A series of patterns for completely hypothetical structures will provide data for determining which member of the series might be the true one. For example, the 12 possible one-layer chlorite structures of Brown and Bailey (1962) have been calculated for comparison with patterns from natural specimens.

The following programs are available for calculating powder intensities:

POWD12	Smith et al. (1983)
Micro-POWD**	QJohnson
LAZY PULVERIX	Yvon et al. (1977)
DISPOW	Gabe et al.
ENDIX	Hovestreydt and Parthe (1988)
PC/POWIN	Prewitt
PC/XDPG	Russ
EDDA	Gerward and Olsen (1987)

Of these programs, only POWD and its derivatives use the integrated intensities to simulate the diffractometer trace. Each diffraction peak is distributed over one of the selected analytical profiles from Table 1 so that the area conforms to the integrated intensity. Both wavelength components are included, and the half-widths are adjusted with the diffraction angle. All the peaks thus created are then added to generate a single intensity trace that is to be compared with the experimental diffractometer pattern. This trace can be scanned for peak-height intensities that are the values usually incorporated as "C" patterns in the Powder Diffraction File. The POWD family of programs list both the peak-height intensities and the integrated intensities as output data.

Clay minerals and other two-dimensional structures are usually studied in oriented diffraction mounts, so that only the 00l reflections are observed. The programs mentioned above are not appropriate for this type of structure. The following programs:

MOD2**	Reynolds
INTER	Vila1 and Ruiz-Amil (1988)

are designed specifically for this calculation, and the details of the modeling are discussed in Chapter 6.

Structure refinement. Structure refinement is the refinement of the structural parameters of a model, that is close to the true structure, by comparing the calculated and experimental intensities of all the reflections in a pattern at the same time. Discrepencies are used to determine the parameter changes. Pattern parameters and unit cell values are refined along with the structure parameters. There are two basic techniques: the profile-fitting technique and the Rietveld technique.

The profile-fitting structure analysis programs for refinement of crystal structures use one of the analytical profiles from Table 1 for each peak and minimize the fit to the pattern on a peak by peak basis. The minimization procedure is similar to conventional crystal structure analysis except for adjustments to proportionate the observed intensities in complex clusters of peaks based on the profile fits. The equations are essentially the same as for intensity modeling. Each intensity is distributed over the selected profile, and the collection of peaks are summed on the 2θ scale and matched to the experimental pattern. Programs that fit into this category include:

POWLS	Will (1979)
EDINP	Pawley (1980)
SCRAP	Cooper et al. (1981); Rouse et al. (1981)
PROFIT	Scott
WPPF*	Toraya
PROF	Rudolf and Clearfield

The Rietveld (1967,1969) technique examines the pattern on a point-by-point basis. It is a very powerful method and is becoming very popular to refine the details of specific crystal structures. It requires accurate data collected over a wide angular range. The technique was first developed for neutron diffraction and then adapted to X-ray analysis. An excellent review of the problems of X-ray Rietveld methods is given by Young et al. (1977, 1980). The Rietveld technique is not a profile-fitting method; it fits the whole pattern considering each digital point on the intensity trace as a separate data point. The function that is minimized is

$$M = \Sigma w_j [y_i(obs) - k y_i(calc)]^2 ,$$

where $y_i(obs)$ is the observed intensity at step i in the pattern, $y_i(calc)$ is the calculated intensity and w_i is a weighting factor usually related to the counting statistics. Each y_i may have several peaks contributing to its intensity, so that

$$y_i(calc) = \sum_{hkl} I_i(hkl) .$$

The analytical profile may be any of the profiles listed in Table 1. Depending on the 2θ range attributed to a peak profile, a dozen or more peaks may have a contribution to any given intensity point. The details of this technique are presented in Chapter 9, so they will not be given here. The programs that have been used for X-ray refinements include:

DBW3.2	Wiles and Young (1981)
PC/WYRIET	Schneider
RIETVELD	Toby and Cox
RIETVELD	Baerlocher and McCusker
RIETVELD	Prince and Finger
CCSL	RAL
LANCE-GSAS	Larson and Von Dreele (1987)
LHPM7	Hill and Howard
RIETVELD	Nurmela and Sourtti
RIETAN*	Izumi
DBW4.1	Howard
FIBLS	Busing

FIBLS is a new program designed for the refinement of two-dimensional structures, especially polymers. It may have applications to clays, disordered mica structures and asbestos-like minerals. It is expected to be available by the end of 1989.

Two interesting programs that supplement the above refinement routines are

DLS-76	Baerlocher et al. (1976)
DLS-78	Baerlocher et al.

These programs do not use intensity data. They examine structure models by constraining the bond distances and minimizing the geometric variations in the arrangement of the atoms. They have been used primarily with zeolites, but may be useful for other framework-type structures including tetrahedral and octahedral linkages.

Quantification of crystallite parameters

More common applications of diffraction analysis determine sample parameters such as phase identification or cell dimensions. Identification and cell refinements are usually done using derived d-I data, although the whole pattern may be used in some instances. Information requiring accurate intensities or profile shapes, e.g., phase analysis, determining percent crystallinity, and measuring crystallite perfection, usually is more accurate when the digitized data are used directly in the analysis.

Quantitative phase analysis. X-ray diffraction has the unique ability to be sensitive to both the elements present and the physical state of a sample. Compounds with the same structural arrangement but with different elements yield similar patterns of peaks with different intensities. Compounds with the same chemistry but with different structures yield distinctly different diffraction patterns. Consequently, X-ray diffraction techniques were developed for determining the phase composition of a multicomponent mixture. These techniques have been considered in detail in Chapter 5.

The programs that are used for quantitative analysis fall into three categories. The first category integrates selected ranges of the diffraction pattern to determine an intensity value for each range that is then used in one of the procedures mentioned in Chapter 5. The second category consists of programs that fit the digital pattern on a point-by-point basis with weighted individual patterns of the component phases from a database of measured reference patterns. Corrections are usually based on the reference-intensity-ratio method. The third category also fits patterns on a point-by-point basis using the Rietveld method to calculate the patterns of the component phases. Quantitative analysis using the integrated intensities of a few characteristic peaks of each of the phases present in the mixture is most applicable in systems involving simple high-symmetry materials that create few peak overlap problems in the diffraction pattern.

The basic theory was well defined by Alexander and Klug (1948). Where some of the peaks do overlap, the extension of Copeland and Bragg (1958) is applicable that considers a cluster of peaks as a single measurement. DeWolff and Visser (1964) showed how to scale intensities to the relative-absolute scale using intensity ratios to a standard material, and Chung (1974a,b), Hubbard et al. (1976), and Davis (1986) incorporated this reference-intensity ratio into quantitative analysis. Programs that apply these methods are:

QUANT85   Snyder and Hubbard
PC-RIMPAC**   Davis (1986)

Quantitative analysis using the whole diffraction trace has considerable advantages over the independent-peak method when the patterns are very complex such as those obtained from feldspars and clay minerals. The limitations are the need to have pure samples of the analytes to produce the necessary experimental patterns for the reference database. These samples not only need to be pure, they must represent the crystallinity and structural state of the phases in the sample to be analyzed. There is some forgiveness in the fitting procedure, however, that compensates for mismatch of profiles and for preferred orientation effects because the procedure averages over the whole pattern. In some instances, reference patterns can be extracted from a mixture if the patterns of the other phases are known by pattern subtraction methods, but this approach is of limited usefulness. Programs that use the whole-trace approach include:

GMQUANT   Smith, et al. (1987)

PFLS   Toraya (1986)

Up to 10 phases have been analyzed by this method. Smith et al. (1989) extended the method to allow constraints on the weight fractions based on chemical information that considerably improved the accuracy of the weight-fraction determinations.

Quantification by the Rietveld method is particularly advantageous with complex mixtures and can be accomplished with mixtures of crystalline materials of any complexity. The method will not explicitly treat clay minerals exhibiting two-dimensional diffraction or amorphous components. Most Rietveld programs allow more than one structure to be refined at the same time, and many have been adapted to provide quantitative estimates of the several phases present. Knowledge of the structures and their abundances permits the calculation of the probable absorption coefficient of the bulk sample allowing conversion of the relative intensities to provide weight ratios. Examples of this approach were presented in Chapter 5, and the general problem of structure refinement was covered previously, so details will not be presented here. In addition to the general Rietveld programs, the following Rietveld programs are specifically designed for quantitative analysis:

QPDA**   Hill and Masden
PC/QXRD**   Hill
DBW4.1   Howard and Bish (Bish and Howard, 1988)

Crystallinity. The term crystallinity has several interpretations in crystallography, but it usually is some measure of the fraction of a sample that is three-dimensionally organized in a matrix of disorganized amorphous material. The absolute value reported by the many proposed methods is often questionable, but the relative values of a series of related samples is useful when determining which have progressed further in a reaction. Crystallinity indexes are usually based on some ratio of a crystalline diffraction pattern to the scattering intensity of the amorphous portion. This ratio may be aa area measurement or some function of peak heights and valleys. Crystallinity in geologic systems is applicable to graphite in coal, crystalline SiO_2 to amorphous SiO_2, clay minerals, and fibrous materials, such as asbestos. Most measurements are taken directly from a graphical display of the diffraction pattern. One program that is available that attempts to determine the ratio of crystalline fraction to amorphous fraction:

POLYMER   Wims et al.

Crystallite size and strain. In the introduction it was mentioned that the information on crystallite size and perfection was in the shape of the diffraction peaks. Note that "crystallite size" is not synonymous with "particle size," and X-ray diffraction is sensitive to crystallite size. Even perfect appearing crystals may be composed of microdomains that act as the coherent diffracting unit. Strained crystals often are composed of elastically strained microdomains separated by plastically-deformed regions of high defect density.

Crystallite domain size and strain are two measures of crystallite perfection. Very small crystallites are considered more imperfect than large ones because the surface interactions affect a larger volume fraction of the crystal. Crystallite size causes the peaks to broaden as the size gets smaller than ~2000 Å. Strain causes the peaks to shift from their ideal location for the unstrained state. Usually, there is a distribution of strain in a sample that results in a distribution of shifts that distorts the peak shape considerably. Strain is also invariably accompanied by the breakup of the macrocrystallites into microdomains of different strain values that complicates the diffraction profile by superimposing broadening from the domains on the range of shifts caused by the range of strain values. The analysis of this strained state is not easy. Strain is defined as "change in length per unit length" and is measured as the change in d-spacing of a strained sample compared to the unstrained state usually expressed as

$$\varepsilon = \frac{d_{strained} - d_{unstrained}}{d_{unstrained}}.$$

Strain measurements are usually converted to a stress using the physical properties of the material. The conversion is complicated because of the dependence on the sample material and configuration and direction of measurement with respect to the sample. Even a cursory discussion is beyond the scope of this presentation. Stress analysis is discussed in Cullity (1978) and Klug and Alexander (1974). It is more commonly used in metallurgy than in mineralogy.

In the absence of significant strain, such as in a loose aggregate of clay particles, a reasonably good estimate of the crystallite size is possible. The most commonly employed relation is the Scherrer equation:

$$D = \frac{K\lambda}{b \cos \Theta},$$

where K is the shape constant (approximately 0.9), and b is the half-width of the diffraction profile due to the sample (measured breadth minus the instrumental breadth) in radians. For equant crystallites of a single size, the equation works quite well. When there is a distribution of sizes or crystallite shapes that are tabular or elongate, the equation is less accurate, and more elaborate methods similar to those used to determine the strain must be employed. The diffraction peaks must be separated into the components due to the various sources of distortion. This process is called *deconvolution* and is an involved operation. There are several programs that are available to deconvolute diffraction data for estimates of the crystallite size and strain.

WARREN-AVERBACH	Cohen
UNFOLD	Roof
DECON	Wiedemann et al. (1987)
CRYSIZE**	Zhang et al.
LWL/SIZEDIST	Louer and LeBail

Profile-fitting may also be used to obtain crystallite size and strain measurements (Langford et al., 1986). The method is not strictly deconvolution, but because the fitting function is analytical, the desired crystallite parameters can be modeled, or the best-fit profile can be deconvoluted with a profile fit to data from a "perfect" sample that provides the instrumental and spectral contributions. This approximation to deconvolution of the true experimental data yields parameters usable for evaluating the crystallite size and strain. Profile-fitting programs that have been used for determining crystallite size and perfection include

PROFIT	Sonneveld and Visser (Langford et al. 1986)
SHADOW	Howard
Micro-SHADOW**	Howard and QJohnson
WPPF*	Toraya

Most Rietveld programs use adjustable parameters for broadening that may be interpreted in terms of crystallite size, and deviations of the crystal lattice parameters may be interpreted as strain if they are not due to chemical variations. Asymmetry parameters determined by the Rietveld method are more dependent on instrumental aberrations and are difficult to interpret in terms of strain distributions.

<u>File building and transfer</u>

Now that digitized data are utilized for so many purposes in modern diffraction analysis, it is necessary to be able to read and transfer these data from one system or one laboratory to another. There are several routines that allow the transfer of files from APD's to other computer systems by creating standard file structures for communication. These routines include:

JCAMP-DX**	McDonald and Wilks (1987)
GMQUANT	Smith et al. (1986)
Micro-PEAK**	QJohnson
VAXCONV	Zhou and Snyder

JCAMP-DX is designed to be a universal digitized-file transfer system that has been adopted by the spectroscopists and is being introduced to diffraction patterns.

STRUCTURE DISPLAY PROGRAMS

Once a structure is solved, the results are often displayed graphically especially for publications. There are many routines to accomplish this task.

ORTEP	CJohnson (1970)
PRETEP	Izumi
PLORTEP	Bandel and Sussman (1983)
PLOTMD	Luo et al. (1989)
NAMOD	Beppu
MODEL	Smith
SCHAKAL88	Keller (1989)
PLOTMOL	Gabe, Lee and LePage
STRPLT	Fischer (1985)
PLUTO	Motherwell
PLUTO78	Crennell and Crisp (1984)
PLORTEP	Bandel and Sussman (1983)
SNOOPI	Davies
CRYES**	Vila2 and Vegas
B&S**	Muller
CHEM-X/MITIE**	Chemical Design Co.
Mol. Plot	Radhakrishnan (1982)
ALCHEMY	Tripos

The best known program ORTEP that plots a ball-and-spoke image with the option to represent the atoms as ellipsoids showing the thermal motion. PRETEP is a setup program for ORTEP, which is the main difficulty in using ORTEP. PLOTMD simplifies the labelling of an ORTEP plot. MODEL is a setup program for NAMOD. Most of these programs plot ball-and-spoke type pictures, and most have options to remove hidden lines, so the picture looks clean. STRPLT is designed specifically for plotting polyhedral images commonly used to represent framework silicates in particular and many minerals with octahedral coordination as well.

Mineralogists are also interested in displaying the external morphologies of crystals. Program

    SHAPE**   Dowty (1988)

is available for this purpose. Construction of ball-and-spoke models is also of interest. Program

    DRILL   Smith and Clark

will calculate projection coordinates in orthogonal space for all the atoms in the unit cell and drilling angles for all the bonds around the atoms for a drilling machine using spherical coordinates.

ANALYTICAL PACKAGES

Several laboratories have assembled many of the above programs into a package that allows data to be transferred from one program to another with a minimum of data entry. In addition to communicating internally, each of these packages can accept data files from at least one APD system. These packages are generally system-dependent. Some of the packages are complete substitutions for the analytical component of most APD systems, except for the data collection activities. Packages for powder diffraction analysis include:

SPECPLOT	Goehner and Garbauskas (1979)
AUTO	Snyder et al. (1982)
ALFRED	Mallory and Snyder (1979)
ZEOPAK**	GJohnson and Smith
NRC/VAX	Gabe, Lee and LePage
PC/NRC	White
FAT-RIETAN**	Rigaku

Some packages designed for single-crystal analysis have routines useful to powder analysis. These single-crystal packages include:

SHELXTL**	Sheldrick
XRAY88**	Stewart
XTAL**	Stewart and Hall (1985)

MISCELLANEOUS ROUTINES

There are many other programs that are available to the user that do not fit into the previous categories. Some of these programs will be mentioned here.

Almost every diffraction laboratory will have some routine for generating d-spacings for a given unit cell. These programs are known collectively as D-GEN's, and the calculations can be programmed on pocket calculators. Such programs are indispensable in the analysis of a diffraction pattern. There are too many versions to list here.

Although the *International Tables of Crystallography* are usually available in every diffraction laboratory, the program

CRYSET   Larson

is very useful to generate the symmetry positions from the space symmetry information. Any space group symbol can be interpreted, but the program only allows settings with the center of symmetry at the origin. Program

TABLES   Abad-Zapetro and O'Donnell (1987)

will allow the plotting of the symmetry related points for viewing and manipulation in real time to see how the points interact.

Occasionally, Laue patterns will be used for determining crystal orientation or for distinguishing specific directions in a crystal. Programs

STEREOPL	Scientific Software
STLPLT	Canut-Amoros (1970)
LAUE	Goehner and Garbauskas
Laue	Laugier and Folhol (1983)

will plot both transmission and back-reflection Laue patterns for a given unit cell and symmetry. Such plots save the time of finding a crystal of known orientation and making a reference photograph.

LOTUS-123** should be mentioned as a very useful way of preparing data reduction and comparison tables of diffraction data. Spread sheets are also a good way of presenting data and interpretations to requestors of sample analyses and for publication. McCarthy (1986) has developed a series of templates for various diffraction applications. Novak and Colville (1989) use d-templates to refine the unit cells for each of the crystal systems.

GENERAL COMMENTS AND DISCLAIMER

Over 150 programs for the analysis of diffraction data have been identified in this paper. Many of these programs duplicate the calculations of other routines in the same category, so the diffractionist certainly does not need to acquire every program to have a complete system of routines. However, because there are some differences in programs for similar purposes, it is usually advisable to have more than one routine from a given category. Given a specific set of unit-cell parameters, all D-GENS will calculate the same d-spacings. On the other hand, search/match and indexing programs may not give identical results because they use different strategies on experimental data that are not free of errors. Selection of programs for one's own use depends on the types of experiments commonly performed and on the type of computer that is available.

For the diffractionist who has an APD, the selection of programs is usually to augment the APD software package for a specific application. This paper will help identify the individual who can supply that program. For the diffractionist who is starting from nothing, it is a different matter. Perhaps the best way to start would be to contact one of the authors who has a collection of programs already integrated into a basic package. Additional programs can then be obtained later. The advantage of a package of programs is that usually that the data entry is relatively simple compared to that for a collection of individual programs. Once data are entered or transferred from an APD, the data can be reformatted for entry to all the programs in the package.

This author makes no claim with respect to the completeness in the coverage of programs described in this paper or their success in doing the calculations that are reported. It has proved to be a formidable task to assemble this list, and I wish to thank the many contributors who supplied information on programs with which I had no direct experience. I hope I have described each routine properly, but without first-hand use of each program, I have probably misclassified some routines. Obviously, there is an emphasis on those programs that have been in use in North America, but I have made an attempt to collect programs from overseas where possible. This list has been submitted to the International Union on Crystallography Commission on Powder Diffraction, where it is hoped that it will be maintained for future reference.

SOURCES OF PROGRAMS

Most of the programs listed in this paper are available from the author(s) or as part of one of the analytical packages of programs. Where the author list has a date attached, the entry will be found in the section on REFERENCES. If a date is not given, check for one of the authors in this section. I have tried to indicate the level of price that might be charged for a copy of the specific program by the (*) and (**) marks. The single mark (*) indicates a nominal charge and the double mark (**) indicates a commercial program. Prices may include only the supply of the medium for the copy, and sometimes there is an additional small copying charge. PC programs seem to be more commonly commercialized than main-frame programs; however, there are many PC freeware programs available. Even when requesting freeware, a requestor should supply the medium and ask if a small remittance would be appropriate.

ALEPH
Aleph Enterprises, Inc.
P. O. Box 213
Livermore, CA 94550 USA

Benoit
Paul Benoit
Department of Geology
31 Williams Hall, Lehigh University
Bethlehem, PA 18015

Beppu
Yoshitaka Beppu
Laboratory of Theoretical Biophysics
Department of Physics
Nogoya University
Nogoya 464, Japan

Bideaux
Richard A. Bideaux
710 W. Bangalor Drive
Oro Valley, AZ 85704

Bish
David L. Bish
ESS-1, Geology/Geochemistry
MS D469
Los Alamos National Laboratory
Los Alamos, NM 87545

Blanchard
Frank N. Blanchard
Department of Geology
University of Florida
Gainesville, FL 32611

Burnham
Charles W. Burnham
Hoffman Laboratory
Harvard University
20 Oxford Street
Cambridge, MA 02138

Busing
William R. Busing
Chemistry Division
Oak Ridge National Laboratory
P. O. Box 2008
Oak Ridge, TN 37831

Canfield
Dennis V. Canfield
Univ. of Southern Mississippi
Department of Polymer Science
Hattiesburg, MS 39406-0076

Clayton
William Rex Clayton
Miller Brewing Company
3939 West Highland Blvd.
Milwaukee, WI 53208

Clearfield
Abraham Clearfield
Department of Chemistry
Texas A&M University
College Station, TX 77843

Cohen
Jerome B. Cohen
Technical Institute
Northwestern University
Evanston, IL 60201

Cooper
M. J. Cooper
Materials Physics Section
AERE
Harwell, Oxfordshire, OX11 0RA, UK

Davies
Chemical Designs, Ltd., Unit 12
7 West Way
Oxford OX2 0JB England

Davis
Briant L. Davis
Institute for Atmospheric Research
South Dakota School of Mines
Rapid City, SD 57701-3995

Dowty
Eric Dowty
196 Beechwood Avenue
Bogota, NJ 07603

Eysel
Walter Eysel
Mineralogische und Petrographisches Institut
Universität Heidelburg
Im Neuerheimer Feld 236
D-6900 Heidelburg, FRG

Fischer
Reinhard X Fischer
Mineralogishes Institut
Universität Wurzburg
D-8700 Wurzburg, FRG

Forrest
Allister M. Forrest
Department of Physics
Paisley College of Technology
High Street, Paisley
Renfrewshire, PA1 2BE, UK

Foris
C. M. Foris
Central Research and Development
E. I. DuPont de Nemours & Co.
Experimental Station, Bldg. 356
Wilmington, DE 19898

Frevel
　Ludo K. Frevel
　1205 West Park Drive
　Midland, MI 48640

Gabe
　Eric J. Gabe
　Chemistry Division
　National Research Council
　Ottawa, Ontario, K1J 8L9 Canada

Garbauskas
　Mary F. Garbauskas
　General Electric Company
　Corporate Research and Development
　P. O. Box 8, K-1 2C35
　Schenectady, NY 12301

Garvey
　Roy G. Garvey
　Department of Chemistry
　North Dakota State University
　Fargo, ND 58105-5516

Goehner
　Raymond A. Goehner
　Siemens Analytical X-ray Instruments
　6300 Enterprise Lane
　Madison, WI 53719-1173

Harlow
　Richard L. Harlow
　Central Research and Development
　E. I. DuPont de Nemours & Co.
　E228/316 D
　Wilmington, DE 19898

Hill
　Roderick J. Hill
　Division of Mineral Chemistry
　C. S. I. R. O.
　P. O. Box 124
　Port Melbourne, Victoria 3207, Australia

Holomany
　Mark A. Holomany
　JCPDS-ICDD
　1601 Park Lane Road
　Swarthmore, PA 19081

Hovestreydt
　Eric R. Hovestreydt
　Institut für Kristallographie
　Universität Karlsruhe
　D-7500 Karlsruhe, FRG

Howard
　Scott A. Howard
　Department of Ceramic Engineering
　University of Missouri - Rolla
　Rolla, MO 65401

Hubbard
　Camden R. Hubbard
　Oak Ridge National Laboratory
　P. O. Box 2008
　Oak Ridge, TN 37831

INEL
　INEL
　J. P. Duchemin
　Avenue de Scandinavie
　Z. A. de Courtaboeuf
　91953 Les Ulis Cedex, France

Izumi
　F. Izumi
　National Institute for Research
　　in Inorganic Materials
　Namiki 1-1, Tsukuba-Shi, 305, Japan

Jenkins
　Ron Jenkins
　JCPDS-ICDD
　1601 Park Lane Road
　Swarthmore, PA 19081

CJohnson
　Carroll K. Johnson
　Oak Ridge National Laboratory
　P. O. Box 2008
　Oak Ridge, TN 37831

GJohnson
　Gerald G. Johnson, Jr.
　164 Materials Research Laboratory
　The Pennsylvania State University
　University Park, PA 16802

QJohnson
　Quintin C. Johnson
　Materials Data, Inc.
　P. O. Box 791
　Livermore, CA 94550

Jones
　Rolland C. Jones
　Department of Agronomy
　University of Hawaii
　1910 East West Road
　Honolulu, HI 96822

Kawahara
　A. Kawahara
　Faculty of Science
　Okayama University
　3-1-1, Tsoshimanaka
　Okayama 700, Japan

Keller
　Egbert Keller
　Kristallographisches Institut
　　der Universität
　Hebelstrasse, 25
　D-7800 Frieburg, FRG

Larson
  Allen C. Larson
  LANSCE
  Los Alamos National Laboratory
  Los Alamos, NM 87545

Lawton
  Stephen L. Lawton
  Mobil Research Center
  Paulsboro, NJ 08096

Louer
  Daniel Louer
  Laboratorie de Cristallochimie
  Université de Rennes
  Avenue du General Leclerc
  35042 Rennes-Cedex, France

Marquart
  Fein-Marquart Associates, Inc.
  7215 York Road
  Baltimore, MD 21212

MDI
  Materials Data Inc.
  P. O. Box 791
  Livermore, CA 19550 USA

McCarthy
  Gregory J. McCarthy
  Department of Chemistry
  North Dakota State University
  Fargo, ND 58105-5516

Mighell
  Alan D. Mighell
  National Institute of Standards
    and Technology
  A209, MATLS
  Gaithersburg, MD 20899

Millidge
  H. Judith Millidge
  Crystallography Unit
  University College, London
  Gower Street
  London, WC1E 6BT, UK

Mueller
  Melvin H. Mueller
  IPNS Building 360
  Argonne National Laboratory
  9700 Cass Avenue
  Argonne, IL 60439

Morton
  Roger D. Morton
  Department of Geology
  University of Alberta
  Edmonton T6G 2H7, Canada

Motherwell
  William D. S. Motherwell
  Automation Office
  Cambridge University Library
  West Road
  Cambridge CB3 9DR, UK

Muller
  Norbert Muller
  Michael Hainischstrasse 2/12
  A-4040 Linz, Austria

Nichols
  Monte C. Nichols
  Aleph Enterprises
  P. O. Box 213
  Livermore, Ca 94550

Nurmela
  Pekka Suortti
  Department of Physics
  University of Helsinki
  Siltavuorenpenger 20 D
  SF-00170 Helsinki 17, Finland

O'Connor
  Brian H. O'Connor
  WAITEC Consulting Services
  Kent Street, Bentley
  Western Australia 6102, Australia

Parthe
  Erwin Parthe
  Laboratoire de Cristallographie
    aux Rayons X
  24, Quai Ernest-Ansermet
  CH 1211 Geneve 4, Switzerland

Pawley
  G. Stuart Pawley
  Department of Physics
  University of Edinburgh
  Mayfield Road
  Edinburgh EH9 3JZ, Scotland, UK

Philips
  Philips Electronic Instruments
  85 McKee Drive
  Mahwah, NJ 07430

Prewitt
  Charles T. Prewitt
  Geophysical Laboratory
  2801 Upton Street
  Washington, DC 20008-3898

Prince
  Edward Prince
  Reactor Radiation Division
  National Institute for Standards
    and Technology, Bldg. 235 A106
  Gaithersburg, MD 20899

RAL
  Rutherford-Appleton Laboratory
  Chilton, Didcot
  Oxon OX11 0QX, UK

Reck
  Guenter Reck
  Zentralininstitut für Physikalische Chemie
  Akademie der Wissenshaften der DDR
  Rudower Chausse 5
  DDR-1199, Berlin, GDR

Reynolds
  Robert C. Reynolds
  Department of Earth Sciences
  Dartmouth College
  Hanover, NH 03755

Rigaku
  Rigaku Corporation
  Segawa Building
  2-8 Kandasurugadai
  Chiyoda-ku, Tokyo, Japan

Rose
  Charles E. Rose
  Helios Software
  P. O. Box 44-1215
  West Somerville, MA 02144

Rossell
  Henry J. Rossell
  Division of Materials Science
  C. S. I. R. O.
  Normanby Road
  Clayton, Victoria 3168, Australia

Russ
  John C. Russ
  School of Engineering
  North Carolina State University
  P. O. Box 5995
  Raleigh, NC 27650

Scientific
  Scientific Software Services
  3497 School Road
  Murraysville, PA 15668

Scott
  Henry G. Scott
  C. S. I. R. O. Division of Materials Science
  Normanby Road
  Clayton, Victoria 3168, Australia

Schneider
  Julius Schneider
  Institut für Kristallographie
    und Mineralogie der Universität
  Theresienstrasse 41
  D-8000 Munchen 2 FRG

Shchedrin
  Boris M. Shchedrin
  Moscow State University
  Computing Center
  Leninskiye Gory
  Moscow 117234, USSR

Sheldrick
  George M. Sheldrick
  Institut für Anorganische Chemie
  der Universität
  Tammannstrasse 4
  D-3400 Goettingen, FRG

Smith
  Deane K. Smith
  Department of Geosciences
  239 Deike Building
  The Pennsylvania State University
  University Park, PA 16802

Snyder
  Robert L. Snyder
  College of Ceramics, Alfred University
  Alfred, NY 14802-1296

Socabim
  Socabim
  Julian Nusinovici
  9 Bis Villa du Bel-Air
  Paris 75012, France

Sonneveld
  Eduard J. Sonneveld
  Technische Physiche Dienst, TNO-TH
  Stieltjesweg 1, P.O. Box 155
  2600 AD Delft, The Netherlands

Spek
  Anthony L. Spek
  Lab. voor Kristal en Structuurchemie
  Universite Utrecht
  Padualaan 8, P.O. Box 80050
  3508 TB Utrecht, The Netherlands

Stewart
  James M. Stewart
  Department of Chemistry
  University of Maryland
  College Park, MD 20742

Suortti
  Pekka Suortti
  Department of Physics
  University of Helinski
  Siltavuoropenger 20 D
  SF-00170 Helinski 17 Finland

Takagi
  Yutaka Takagi
  Osaka Kyoiku University
  Minamikawahoricho 43
  Tennouji-ku, Osaka 543, Japan

Toby
  Brian H. Toby
  Department of Materials Science
  University of Pennsylvania
  Philadelphia, PA 19104-6272

Toraya
 Hideo Toraya
 Ceramic Engineering Research Center
 Nagoya Institute of Technology
 Asahigaoka, Tajimi 507, Japan

Tripos
 Tripos Associates, Inc.
 1699 South Hanley Road
 Suite 303
 St. Louis, MO 63117

Vila1
 A. Vila
 Instituto de Ciencia de Materiales CSIC
 Serrano 113
 28003 Madrid, Spain

Vila2
 SciSoft
 Attn: Mr. Arias
 c/ Providencia 27
 E-08800 Vilanova i La Geltru
 Barcelona, Spain

deVillers
 Johann P. R. deVillers
 MINTEK
 Private Bag X3015
 Randsburg, South Africa

Wassermann
 A. Wassermann
 Kaufbeurerstrasse 4, Postfach 2631
 D-8960 Kempten, FRG

Werner
 Per-Erik Werner
 Arrhenius Laboratory
 University of Stockholm
 S-106 91 Stockholm, Sweden

White
 P. S. White
 Chemistry Department
 University of New Brunswick
 Fredericton, N.B. E3B 5A3 Canada

Will
 Georg Will
 Mineralogisches Institut
  der Universität
 Poppelsdorfer Shloss,
 D-5300 Bonn, FRG

Wims
 Andrew M. Wims
 Research Center
 General Motors Technical Center
 30500 Mound Road
 Warren, MI 48090

Young
 Ray A. Young
 School of Physics
 Georgia Institute of Technology
 Atlanta, GA 30332

REFERENCES

Abad-Zapetro, C. and O'Donnell, T.J. (1987) TABLES, A program to display space-group symmetry information in three dimensions. J. Appl. Cryst. 20, 532-535.

Alexander, L.E. and Klug, H.P. (1948) Basic aspects of X-ray absorption in quantitative diffraction analysis of powder mixtures. Anal. Chem. 20, 886-889.

Allen, F.H., Bergerhoff, G. and Sievers, R., Editors (1987) "Crystallographic Databases." International Union of Crystallography, Chester, UK.

Appleman, D.E. and Evans, H.T., Jr. (1973) U.S. Geological Survey Computer Contribution 20. U.S. National Technical Information Service Doc. PB2-16188.

Armstrong, E.E. (1989) The use of fourier self-deconvolution to resolve overlapping X-ray powder diffraction peaks. Powd. Diff. (in press).

Azaroff, L.V. and Buerger, M.J. (1958) "The Powder Method in X-ray Crystallography." McGraw-Hill, New York, NY, USA.

Baerlocher, Ch., Hepp, A. and Meier, W.M. (1977) DLS-76, A program for the simulation of crystal structures by geometric refinement. Institute of Crystallography and Petrology, ETH, Switzerland.

Bandel, G. and Sussman, J.L. (1983) PLORTEP: a computer program to translate PLUTO instructions into those of ORTEP. J. Appl. Cryst. 16, 650-651.

Barker, T.V. (1922) "Graphical and Tabular Methods in Crystallography as a Foundation of a New System of Practice. London, UK.

Bergerhoff, G. and Brown, I.D. (1987) Inorganic crystal structure database. In Crystallographic Databases, Allen, Bergerhoff and Siever, Eds. International Union of Crystallography, Chester, England, p. 77-95.

Bish, D.L. and Howard, S.A. (1988) Quantitative phase analysis using the Rietveld method. J. Appl. Cryst. 21, 86-91.

Brown, B.E. and Bailey, S.W. (1962) Chlorite polytypism: I. Regular and semi-random one-layer structures. Amer. Mineral. 47, 819-850.

Canut-Amoros, M. (1970) STLPLT: Calcomp plot of crystallographic projections of Laue photographs. Computer Phys. Comm. 1, 293-305.

Carr, M. Chambers, W.F. and Melgaard, D. (1986) A search/match procedure for electron diffraction data based on pattern matching with binary bit maps. Powd. Diff. 1, 226-2340.

Carr, M., Chambers, W.F., Melgaard, D., Himes, V.L., Stalick, J., and Mighell, A.D. (1989) NIST/Sandia/ICDD Electron Diffraction Database: a database for phase identification by electron diffraction. J. Res. NIST 94, 15-20.

Chung, F.H. (1974a) Quantitative interpretation of X-ray patterns of mixtures: I. Matrix-flushing method of quantitative multicomponent analysis. J. Appl. Cryst. 7, 519-525.

Chung, F.H. (1974b) Quantitative interpretation of X-ray patterns of mixtures: II. Adiabatic principle of X-ray diffraction analysis of mixtures. J. Appl. Cryst. 7, 526-531.

Cohen, M.U. (1936a) Elimination of systematic errors in powder photographs. Z. Kristallogr. 94, 288-298.

Cohen, M.U. (1936b) Calculation of precise lattice constants from X-ray powder photographs. Z. Kristallogr. 94, 306-310.

Cooper, M.J., Rouse, K.D. and Sakata, M. (1981) An alternative to the Rietveld profile refinement method. Z. Kristallogr. 157, 101-117.

Copeland, L.E. and Bragg, R.H. (1958) Quantitative X-ray diffraction analysis. Anal. Chem. 30, 196-201.

Crennell, K.M. and Crisp, G.M. (1984) An enhanced PLUTO78 for molecular display. J. Appl. Cryst. 17, 366-368.

Cullity, B.D. (1978) Elements of X-ray Diffraction. 2nd Ed., Addison-Wesley, Reading, MA, USA.

Dana, J.D. (1837) System of Mineralogy. 1st Ed., New Haven, CT, USA.

Davis, B.L. (1986) Reference Intensity Method of Quantitative X-ray Diffraction Analysis. South Dakota School of Mines, Rapid City, SD, USA.

Dollase, W.A. (1986) Correction of intensities for preferred orientation in powder diffractometry: Applications of the March model. J. Appl. Cryst. 19, 267-272.

Donnay, J.D.H., Nowacki, W. and Donnay, G. (1954) Crystal Data. Memoir 60, Geological Society of America, Boulder, CO, USA.

Dowty, E. (1988) Shape software for drawing crystals on personal computers. J. Appl. Cryst. 21, 211.

Edmonds, J.W. (1980) Generalization of the ZRD-SEARCH-MATCH program for powder diffraction analysis. J. Appl. Cryst. 13, 191-192.

Fischer, R.X. (1985) STRUPLO84, A Fortran plot program for crystal structure illustrations in polyhedral representation. J. Appl. Cryst. 18, 258-262.

Fleischer, M. (1987) Glossary of Mineral Species. Mineralogical Record, Tuscon, AZ, USA.

Frevel, L.K. (1982) Structure-sensitive search-match procedures for powder diffraction analysis. Anal. Chem. 54, 691-697.

Frevel, L.K., Adams, C.E. and Ruhberg, L.R. (1976) A fast search/match program for powder diffraction analysis. J. Appl. Cryst. 9, 199-204.

Garvey, R.G. (1986a) UNITCELL finds unit cell parameters from a powder diffraction pattern. Powd. Diff. 1[2], 89.

Garvey, R.G. (1986b) LSUCRIPC, Least squares unit-cell refinement with indexing on the personal computer. Powd. Diff. 1[1], 114.

Gerward, L. and Olsen, J.S. (1987) EDDA, a program for producing energy-dispersive powder diffraction spectra. J. Appl. Cryst. 20, 324.

Goebel, J.B. and Wilson, A.S. (1965) Indexing Program for Indexing X-ray Diffraction Powder Patterns. Batelle Northwest Laboratories, Report BNWL-22.

Goehner, R.P. (1979) SPECPLOT, An interactive data reduction and display program for spectral data. Adv. X-ray Anal. 23, 305-311.

Goehner, R.P. and Garbauskas, M.F. (1983a) Computer-aided qualitative X-ray powder diffraction phase analysis. Adv. X-ray Anal. 26, 81-86.

Goehner, R.P. and Garbauskas, M.F. (1983b) PDIDENT, A Set of Complete Programs for Powder Diffraction Phase Identification. General Electric Technical Information Series 83CRD062, Schenectady, NY, USA

Hanawalt, J.D., Rinn, H.W. and Frevel, L.K. (1938) Chemical analysis by X-ray diffraction--Classification and use of X-ray diffraction patterns. Ind. Eng. Chem. Anal. Ed. 10, 457-512.

Himes, V.L. and Mighell, A.D. (1985) NBS*LATTICE: A Program to Analyze Lattice Relationships. NIST Crystal Data Center, Gaithersburg, MD, USA.

Himes, V.L. and Mighell, A.D. (1987) NSB Crystal Data: NBS*SEARCH: a program to search the database. In Crystallographic Databases, Allen, Bergerhoff and Siever, Eds. International Union of Crystallography, Chester England, p. 144-155.

Howard, S.A. and Snyder, R.L. (1983) Evaluation of some profile models used in profile-fitting. Adv. X-ray Anal. 26, 73-80.

Hovestreydt, E. (1983) FINAX: A computer program for correcting diffraction angles, refining cell parameters and calculating powder patterns. J. Appl. Cryst. 16, 651-653.

Hovestreydt, E. and Parthe, E. (1988) ENDIX: a program to simulate energy dispersive X-ray powder diffraction diagrams. J. Appl. Cryst. 21, 282-283.

Hubbard, C.R. (1980) Standard reference materials for quantitative analysis and d-spacing measurements. In Accuracy in Powder Diffraction, Block and Hubbard, Eds. NBS Special Publication 567, p. 489-502.

Hubbard, C.R. (1983a) New standard reference materials for X-ray powder diffraction. Adv. X-ray Anal., 26, 45-51.

Hubbard, C.R. (1983b) Certification of Si powder diffraction Reference Material 640a. J. Appl. Cryst. 16, 285-289.

Hubbard, C.R., Evans, E.H. and Smith, D.K. (1976) The reference intensity ratio, I/Ic, for computer simulated powder patterns. J. Appl. Cryst. 9, 169-174.

Ito, T. (1950) X-ray Studies on Polymorphism. Maruzen, Tokyo, Japan.

International Tables for X-ray Crystallography, Vols. 1 & 2 (1965, 1959). Kynoch Press, Birmingham, England.

Jenkins, R. and Smith, D.K. (1987) Powder Diffraction File. In Crystallographic Databases, Allen, Bergerhoff and Siever, Eds. International Union of Crystallography, Chester, England, p. 158-177.

Johnson, C.K. (1970) ORTEP: A Fortran Thermal-Ellipsoid Program for Crystal Structure Illustrations. Report ORNL-3794. (See also ORTEPII Report ORNL-5138) Oak Ridge National Laboratory, TN, USA.

Johnson, G.G., Jr. and Vand, V. (1967) A computerized powder diffraction identification system. Ind. Eng. Chem. 59, 19-26.

Keller, E. (1989) Some computer drawings of molecular and solid-state structures. J. Appl. Cryst. 21, 19-22.

Klug, H.P. and Alexander, L.E. (1974) X-ray Diffraction Procedures for Polycrystalline and Amorphous Materials. 2nd Ed., J. Wiley, New York, NY, USA.

Kohlbeck, F. and Hoerl, E.M. (1976) Indexing program for powder patterns especially suitable for triclinic, monoclinic and orthorhombic lattices. J. Appl. Cryst. 9, 28-33.

Ladell, J., Zagofsky, A. and Pearlman, S. (1975) CuKα_2 elmination algorithm. J. Appl. Cryst. 8, 499-506.

Langford, J.I., Louer, D., Sonneveld, E. and Visser, J.W. (1986) Applications of total pattern fitting to a study of crystallite size and strain in zinc oxide powders. Powd. Diff. 1, 211-221.

Larson, A.C. and Von Dreele, R.B. (1987) GSAS, Generalized Crystal Structure Analysis System. LAUR-86-748, Los Alamos National Laboratory, Los Alamos, NM, USA.

Laugier, J. and Folhol, A. (1983) An interactive program for the interpretation and simulation of Laue patterns. J. Appl. Cryst. 16, 281-283.

Lin, T.H., Zhang, S.Z., Chen, L.J. and Cai, X.X. (1983) An improved program for searching and matching of X-ray powder diffraction patterns. J. Appl. Cryst. 16, 150-154.

Louer, D. and Louer, M. (1972) Methode d'essois et erruers pour l'indexation automatique des diagrammes poudre. J. Appl. Cryst. 5, 271-275.

Louer, D. and Vargas, R. (1982) Indexation automatique des diagrammes de poudre per dichotomies successives. J. Appl. Cryst. 15, 542-545.

Luo, J., Ammon, H.L. and Gilliland, G.L. (1989) PLOTMD, An interactive program to modify molecular plots on a graphics terminal. J. Appl. Cryst. 22, 186.

Mallory, C.L. and Snyder, R.L. (1979) The Alfred University Powder Diffraction System, Technical Paper No. 144. New York State College of Ceramics, Alfred University, Alfred, NY, USA.

Marquart, R.G., Katsnelson, I., Milne, G.W.A., Heller, S.R., Johnson, G.G., Jr. and Jenkins, R. (1979) A Search-match system for X-ray powder data. J. Appl. Cryst. 12, 629-634.

Marquart, R.G. (1986) (8Ml (8U-PDSM: Mainframe search/match on an IBM PC. Powd. Diff. 1, 34-39.

Masden, J.C., Skov, H.J. and Rasmussen, S.E. (1988) An inexpensive automation of a powder diffractometer. Powd. Diff. 3, 91-92.

McCarthy, G.J. (1986) LOTUS 1-2-3 Templates for XRD and XRF. Powd. Diff. 1, 89.

McDonald, R.S. and Wilks, P.A., Jr. (1988) JCAMP-DX: A standard form for exchange of infrared apectra in computer-readable form. Appl. Spectros. 42, 151-162.

Mighell, A.D., Hubbard, C.R., Stalick, J.K and Holomany, M.A. (1983) NBS*AIDS83: A Manual Describing the Data Format Used in NBS*AIDS83. JCPDS-International Centre for Diffraction Data, Swarthmore, PA, USA.

Mighell, A.D., Stalick, J.K. and Himes, V.L. (1987) NBS Crystal Data: Database description and applications. In Crystallographic Databases, Allen, Bergerhoff and Siever, Eds. International Union of Crystallography, Chester, England, p. 134-143.

Mueller, M.H. and Heaton L. (1961) Determination of Parameters with the Aid of a Computer. ANL-6176. Argonne National Laboratory, Argonne, IL, USA.

Mugnoli, A. (1985) A microcomputer program to detect higher lattice symmetry. J. Appl. Cryst. 18, 183-184.

Niggli, P., Ewald, P.P., Fajans, K. and von Laue, M. (1931) Structurbericht Ergoenzungsband der Z. Kristallogr. 1, 818.

Novak, G.A. and Colville, A.A. (1989) A practical interactive least-squares cell-parameter program using an electronic spreadsheet and a personal computer. Amer. Mineral. 74, 488-490.

Paszkowicz, W. (1989) INDEXING, a program for indexing powder patterns of cubic, tetragonal, hexagonal and orthorhombic substances. J. Appl. Cryst. 22, 186-187.

Pawley, G.S. (1980) EDINP, The Edinburgh powder refinement program. J. Appl. Cryst. 13, 630-633.

Pawley, G.S. (1981) Unit-cell refinements from powder diffraction scans. J. Appl. Cryst. 14, 357-361.

Pyrros, N.P. and Hubbard, C.R. (1983) POWDER PATTERN: A system of programs for processing and interpreting powder diffraction data. Adv. X-ray Anal. 26, 63-72.

Rachinger, W.A. (1948) A correction of the $\alpha_1-\alpha_2$ doublet in the measurement of the widths of X-ray diffraction lines. J. Sci. Instr. 25, 254-255.

Radhakrishnan, R. (1982) A molecular plotting program. J. Appl. Cryst. 15, 135-136.

Rietveld, H.M. (1967) Line profiles of neutron powder-diffraction peaks for structure refinement. Acta Cryst. 22, 151-152.

Rietveld, H.M. (1969) A profile refinement for nuclear and magnetic structures. J. Appl. Cryst. 2, 65-71.

Rodgers, J.D. and Wood, G.H. (1987) NRCC Metals Crystallographic Data File. In Crystallographic Databases, Allen, Bergerhoff and Siever, Eds. International Union of Crystallography, Chester, England, p. 96-106.

Rogers, K.D. and Lane, D.W. (1987) A simple system for enhanced data collection from a powder diffractometer. Powd. Diff. 2, 227-229.

Rouse, K.D., Cooper, M.J. and Sakata, M. (1981) SCRAP, a program for the analysis of powder diffraction patterns. Report AERE-R9718, AERE, Harwell, England.

Savitsky, A. and Golay, M.J.E. (1964) Smoothing and differentiation of data by simplified least squares procedures. Anal. Chem. 36, 1627-1639. [Correction: Steiner, J., Termonia, Y. and Deltour, J. (1974) Comments on smoothing and differentiation of data by simplified least squares procedures. Anal. Chem. 44, 1906-1909.]

Setten, A. and Setten, R. (1979) AIDED, A program for the automatic indexing of epitaxial derivatives; applications to the graphite lamellar compounds. J. Appl. Cryst. 12, 147-150.

Shirley, R. and Louer, D. (1978) New powder indexing programs for any symmetry which combine grid-search with successive dichotomy. Acta Cryst. A34, S382.

Shirley, R. (1980) Data accuracy for powder indexing. In Accuracy in Powder Diffraction Block and Hubbard, Eds. NBS Special Publication 567, Gaithersburg, MD, USA.

Smith, D.K., Johnson, G.G., Jr., Kelton, M.J. and Anderson C.A. (1989) Chemical constraints in quantitative X-ray powder diffraction for mineral analysis of the sand/silt fraction of sedimentary rocks. Adv. X-ray Anal. 32, (in press).

Smith, D.K., Johnson, G.G., Jr., Schieble, A., Wims, A.M., Johnson, J.L. and Ullmann, G. (1987) Quantitative X-ray powder diffraction method using the full diffraction pattern. Powd. Diff. 2, 73-77.

Smith, D.K., Nichols, M.C. and Zolensky, M.E. (1983) POWD10, A Fortran IV Program for Calculating X-ray Powder Diffraction Patterns. The Pennsylvania State University, University Park, PA, USA.

Smith, G.S. and Kahara, E. (1975) Automated computed indexing of powder patterns, the monoclinic case. J. Appl. Cryst. 8, 681-683.

Smith, G.S. and Snyder, R.L. (1979) F N : A criterion for rating powder diffraction patterns and evaluating the reliability of powder-pattern indexing. J. Appl. Cryst. 12, 60-65.

Snyder, R.L., Hubbard, C.R. and Panagiotopoulos (Pyrros), N.P. (1982) A second generation automated powder diffraction control system. Adv. X-ray Anal. 25, 245-260.

Sonneveld, E.J. and Visser, J.W. (1975) Automatic collection of powder data from photographs. J. Appl. Cryst. 8, 1-7.

Stalick, J.K. and Mighell, A.D. (1986) Crystal Data, Version 1.0, Database Specifications. NBS Technical Note 1229, Gaithersburg, MD, USA.

Steenstrup, S. (1981) A simple procedure for fitting a background to a certain class of measured spectra. J. Appl. Cryst. 14, 226-229.

Stewart, J.M. and Hall, S.R. (1985) XTAL, A program system for crystallographic calculations. J. Appl. Cryst. 18, 263.

Taupin, D. (1968) Une methode generale pour l'indexation des diagrammes de poudres. J. Appl. Cryst. 1, 178-181.

Toby, B.H., Harlow, R.L. and Holomany, M.A. (1989) The POWDER SUITE: Computer programs for searching and accessing the JCPDS-ICDD powder diffraction data base (I.). Powd. Diff. (in press).

Toraya, H. (1986) Whole-powder-pattern fitting without reference to a structural model: Application to X-ray powder diffraction spectra. J. Appl. Cryst. 19, 940-947.

Toraya, H., Yoshimura, M. and Somiya, A. (1983) A computer program for the deconvolution of X-ray diffraction profiles with the composite of Pearson type VII functions. J. Appl. Cryst. 16, 653-657.

Tran, V. and Buleon, A. (1987) Diffraction peak shape: A profile refinement method for badly resolved powder diagrams. J. Appl. Cryst. 20, 430-436.

Vila, E. and Ruiz-Amil, A. (1988) Computer program for analysing interstratified structures by Fourier transform methods. Powd. Diff., 3, 7-11.

Visser, J.W. (1969) A fully automatic program for finding the unit cell from powder data. J. Appl. Cryst. 2, 89-95.

Werner, P.-E. (1964) Trial and error computer methods for the indexing of unknown powder patterns. Z. Kristallogr. 120, 375-387.

Werner, P.-E., Eriksson, L. and Westdahl, M. (1985) TREOR, a semi-exhaustive trial-and-error powder indexing program for all symmetries. J. Appl. Cryst. 18, 367-370.

Wiedemann, K.E., Unnam, J. and Clark, R.K. (1987) Computer program for deconvoluting powder diffraction spectra. Powd. Diff. 2, 137-145.

Wiles, D.B. and Young, R.A. (1981) A new computer program for Rietveld analysis of X-ray powder diffraction patterns. J. Appl. Cryst. 14, 149-151.

Will, G. (1979) POWLS: A powder least-squares program. J. Appl. Cryst. 12, 483-485.

Wilson, A.J.C. (1963) Mathematical Theory of X-ray Powder Diffractometry. Philips Technical Library, Eindoven, The Netherlands.

deWolff, P.M. (1957) On the determination of unit-cell dimensions from powder patterns. Acta Cryst. 10, 590-595.

deWolff, P.M. (1962) Indexing of powder diffraction patterns. Adv. X-ray Anal. 6, 1-17.

deWolff, P.M. (1968) A simplified criterion for the reliability of a powder pattern indexing. J. Appl. Cryst. 1, 108-113.

deWolff, P.M. (1972) A definition of the indexing figure of merit, M20. J. Appl. Cryst. 5, 243.

deWolff, P.M. and Visser, J.W. (1964) Absolute Intensities--Outline of a Recommended Procedure. TPD Technical Report 641-109. Reprinted (1988) Powd. Diff. 3, 202-204.

Young, R.A.(1980) Structural analyses from X-ray powder diffraction patterns with the Rietveld nethod. In Accuracy in Powder Diffraction, Block and Hubbard, Eds. NBS Special Publication 567, Gaithersburg, MD, USA, p. 143-163.

Young, R.A., Mackie, P.E. and Von Dreele, R.B. (1977) Application of the pattern-fitting structure-refinement method to X-ray powder diffractometer patterns. J. Appl. Cryst. 10, 262-269.

Yvon, K., Jeitschko, W. and Parthe, E. (1977) LAZY PULVERIX: A computer program for calculating X-ray and neutron diffraction powder patterns. J. Appl. Cryst. 10, 73-74.

8. Profile Fitting of Powder Diffraction Patterns

S.A. Howard & K.D. Preston

INTRODUCTION

The primary objective of profile fitting is to fit a numerical function, specifically referred to as a profile-shape-function (PSF), to a measured diffraction line. Several of the more common PSFs used in profile-fitting, normalized to generate a line with unit area, are listed in Table 1. These PSFs are typically described by three parameters: (1) the line position, $2\theta_k$, (2) a peak or integrated line intensity, I_0, and (3) the line width expressed as the full-width at half-maximum intensity, FWHM or H_k. An optimization algorithm is employed to adjust the PSF's parameters until the differences between the measured and the calculated lines are minimized.

Figure 1 shows the results of fitting a pair of Lorentzian PSFs to a Cu $K\alpha_1$-$K\alpha_2$ doublet measured from a Si specimen. The apparent distortions in the calculated line shapes, i.e., the asymmetry at the top of the α_1 line and the truncated top of the α_2 line, result from the peak maxima lying between two measured points. The intensity values for the refined pattern and profiles were generated at the same 2θ values at which the pattern was collected. The refined pattern and profiles would have appeared very smooth, with the maxima lying at the anticipated positions, had they been generated with a much smaller increment between points. This illustrates how profile-fitting can be used to obtain precise and accurate line positions and intensities from step-scanned diffraction data. The width and integrated areas of the reflections are also easily obtained from the refined profile parameters.

The subtleties involved in profile fitting commonly relegate the technique to the realm of art rather than science. For example, the selection of the PSF must consider both the physical characteristics of the specimen and the instrument on which the diffraction data are collected. The problem of refining the overlapping lines shown in Figure 1 is trivial in comparison to the problem of separating the complex groups of lines commonly encountered in specimens of interest, e.g., Figure 2. The proximity of the lines in such groups creates problems due to the interdependence of the PSF parameters. In the refinement of the data shown in Figure 2, the best results were achieved by determining the α_2 line positions and intensities from those of the α_1 while constraining the values of the line widths to have the same value. Further difficulties are encountered if the shapes of the lines are dissimilar due to (1) anisotropic physical characteristics of a single phase, e.g., anisotropic crystallite size, or (2) the lines are from different phases. Perhaps the worst case is when the number of lines in a group is unknown; it is often possible to achieve "acceptable" results by using either a lesser or greater number of lines than are actually present!

Table 1 - Normalized Profile-Shape-Functions and their Designations in the Text.

$$I_{i,k} = \frac{2\sqrt{\ln(2)}}{H_k} \exp\left[\frac{-4\ln(2)}{H_k^2}(2\theta_i - 2\theta_k)^2\right] \quad \text{Gaussian (G)}$$

$$I_{i,k} = \frac{\sqrt{4}}{\pi H_k} \left[1 + \frac{4(\sqrt{2}-1)}{H_k^2}(2\theta_i - 2\theta_k)^2\right]^{-1} \quad \text{Lorentzian (L)}$$

$$I_{i,k} = \frac{\sqrt{[4(2^{3/2}-1)]}}{2\pi H_k} \left[1 + \frac{4(2^{2/3}-1)}{H_k^2}(2\theta_i - 2\theta_k)^2\right]^{-1.5} \quad \text{Intermediate Lorentzian (IL)}$$

$$I_{i,k} = \frac{2\sqrt{[4(\sqrt{2}-1)]}}{\pi H_k} \left[1 + \frac{4(\sqrt{2}-1)}{H_k^2}(2\theta_i - 2\theta_k)^2\right]^{-2} \quad \text{Modified Lorentzian (ML)}$$

$$I_{i,k} = \frac{2\sqrt{m}\,\Gamma(2^{1/m}-1)}{\pi\,\Gamma(m-0.5)\,H_k} \left[1 + \frac{4(2^{1/m}-1)}{H_k^2}(2\theta_i - 2\theta_k)^2\right]^{-m} \quad \text{Pearson VII (PVII)}$$

$$I_{i,k} = \beta_g^{-1} \operatorname{Re}\left\{\Omega\left[\frac{\sqrt{\pi}}{\beta}|2\theta_i - 2\theta_k| + i\frac{\beta_c^2}{\beta_g^2 \pi}\right]\right\} \quad \text{Voigt (V)}$$

$$I_{i,k} = \eta L_{i,k} + (1-\eta) G_{i,k} \quad \text{pseudo-Voigt}$$

Key: $I_{i,k}$, intensity at i^{th} point in pattern due to k^{th} line; $2\theta_k$, Bragg angle; H_k, full-width at half-maximum intensity; β_c & β_g, integral-breadths of Lorentzian and Gaussian components, respectively; η, mixing parameter; Ω, complex error function; Re, real part of function.

Table 2 - Use of Refined Profile Parameters in Various Types of Analyses.

Analytical Technique	Bragg Angle	Peak Intensity	Integrated Area	Line Shape
Search/match	x	x		
Indexing	x			
Lattice-parameter refinement	x			
Quantitative analysis		x	x	
Solid-solution analysis	x	x	x	
Order/disorder evaluation	x	x	x	x
Crystallinity		x	x	x
Strain-isotropic	x			
Strain-anisotropic	x			x
Crystallite size				x
Two-step structure refinement	x	x	x	x

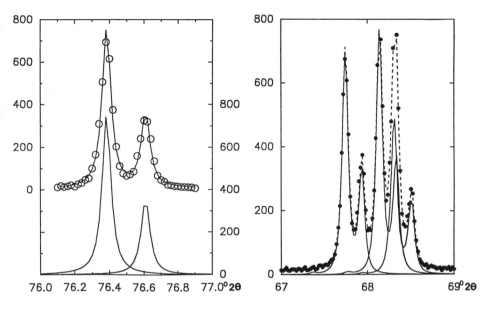

Figure 1 (left). The open circles represent a portion of a pattern collected from a Si specimen showing a Cu $K\alpha_1$-$K\alpha_2$ doublet. The dashed line through the points are the sum of two refined Lorentzian profiles, shown in the bottom half of the figure, and a constant to accommodate the background.

Figure 2 (right). The circles represent a portion of a pattern collected from a quartz specimen containing three Cu $K\alpha_1$-$K\alpha_2$ doublets. The dashed line through the points is the sum of the three refined profile-shape-functions, shown with solid lines, and a constant to accommodate the background. Each profile-shape-function is the sum of two modified Lorentzian functions, one for each of the α_1-α_2 lines in a doublet. The position and intensity of the α_2 component is determined from the corresponding α_1 line's parameters.

The organization of this chapter is as follows. In the first part of the chapter, examples of the advantages of using profile fitting are presented. The literature is also reviewed to acquaint the reader with the historical development of the technique and to illustrate applications, subtleties, and pitfalls. One point that becomes apparent in reviewing the literature is the extent to which line-shape analysis is involved in most powder diffraction techniques including X-ray, neutron and synchrotron. To limit the length of this manuscript, the Fourier line analysis techniques are not discussed. The Fourier techniques, being the traditional methods of analysis, have a rich history which can be found in many texts.

The second part of this chapter details the considerations involved in profile fitting. A series of examples and figures is presented to illustrate the effects of various parameters on the refined lines. An attempt has been made to confine the data used in these examples to a set of synthesized data representing the "five-finger" region of a quartz pattern so that the values of the refined parameters can be compared to their "true" values. The final section contains a summary and indicates some of the areas involving profile-analysis that generate the most activity.

Advantages of using refined line parameters

Table 2 lists some of the applications that take advantage of the high quality of the parameters obtainable through the profile-fitting of step-scanned data. In most cases, a significant increase in the precision of the line parameters, e.g., line positions, can be obtained which, in turn, may diminish many problems associated with random errors. One should note that systematic errors are not necessarily reduced since accuracy is influenced by such things as the alignment of the diffractometer and preparation of the specimen. As an illustration of the nature of the problem associated with systematic errors, consider a set of data collected from a specimen that was intentionally displaced from the focusing plane. The profile-fitting algorithm may estimate the precision of the refined line positions to be on the order of 0.001° 2θ whereas the accuracy of the positions may be on the order of 0.1°. The estimated precision of the line positions from a properly mounted specimen may be on the same order, yet the accuracy may be better than 0.01°. The estimates of precision obtained from profile-fitting algorithms should be regarded only as an estimate of how well the values of the parameters could be determined for a particular set of data and not as an indication of how close the parameter values are to the true values.

Perhaps one of the easiest mistakes one can make is to over-estimate the precision of the refined line intensities. The measured line intensities are strongly influenced by particle-counting statistics, i.e., the number of crystallites favorably oriented for diffraction. Since line intensities are most often used in quantitative analysis, further discourse will follow in the corresponding section below. The following sections describe the advantages of using refined line parameters in several of the more common applications.

Qualitative phase analysis. Perhaps the most routine use of diffraction data is for phase identification. The reliability of search/match algorithms is primarily determined by the accuracy and precision in the measurement of the line positions and intensities (Frevel, 1965; Johnson and Vand, 1967; Frevel and Adams, 1968; Schliephake, 1970; Fiala, 1976). Recently, Schreiner et al. (1982a,b) expressed the matching of patterns in terms of probabilities due to the inexactitudes in peak positions and intensities. The search/match problem is two-fold: 1) there is a significant error in many of the d-spacings recorded in the JCPDS data base, and 2) there is difficulty in obtaining good measurements in the laboratory. Snyder et al. (1978) evaluated the errors associated with the d-spacings in the JCPDS data base and showed that significant improvements occurred as the data-collection technique shifted away from powder cameras and towards the use of the powder diffractometer. Errors associated with the indexable cubic patterns in sets 1-24 showed an average error of 0.091° in 2θ. Using profile fitting, Brown and Edmonds (1980) reported average $\Delta 2\theta$ values in the 0.001° range whereas Parrish (1980) achieved values in the range of 0.0001°. Even routine accuracy and precision on the order of 0.01° would significantly improve the performance of search/match algorithms. Snyder (1985) suggested that accurate and precise determination of line positions, coupled with sophisticated indexing techniques, may allow for identification of the phase purely from the lattice parameters using the Crystal Data database!

Indexing. Indexing procedures are also sensitive to the lack of precision and accuracy in line positions or derived d-spacing values, regardless of the technique employed. The most frequent reason for failure to obtain a correct indexing is inaccuracy of the input data. Visser (1969) and Taupin (1973) recommended a maximum error in 2θ of 0.03° whereas Louer and Louer (1972) suggested a maximum of 0.02°. Kohlbeck and Horl (1976) are more tolerant in their requirements with their low-symmetry indexing program; errors in d on the order of 0.3% can be accommodated. Although line positions within the suggested limits of error may be obtained with careful measurements using strip-chart recorder output, successful indexing also depends heavily on the number of lines that can be input to the algorithm. In many cases even 20 lines is too few. Obtaining 20 lines within the indicated errors in patterns with a large number of overlapping reflections often may only be possible by profile fitting.

Accurate lattice parameter determination. Accurate lattice parameter values are used in a number of analyses including the determination of thermal expansion behavior, densities, interatomic distances, and compressibilities. Relatively accurate determination of lattice parameters can be obtained in a routine way using extrapolation techniques, but they are generally limited to higher symmetry crystal systems (e.g., Beu, 1965; Klug and Alexander, 1974). These extrapolation techniques accommodate the systematic errors associated with the measurement conditions, e.g., flat specimen, specimen displacement, axial divergence, and specimen transparency, by plotting the value of the lattice parameter(s) as a function of the angular dependence of the expected systematic errors (e.g., Alexander, 1954; Jenkins, 1984). Random errors, therefore, will play the dominant role in determining the accuracy of the derived lattice parameters. The raw precision of the line positions obtained from profile-fitting allows for increased accuracy when extrapolation techniques are used.

An interesting example of the precision that may be obtained using profile fitting techniques involves Hubbard's (1983) certification of the Si standard reference material SRM 640a. He reported that the standard deviations associated with the peak positions, based on data from 12 specimen loadings, ranged from 0.0025° to 0.005° 2θ. The result, based on 26 refined line positions from each specimen, which were corrected using two internal standards, was a value for the lattice parameter with an estimated error of approximately 1 part in 500,000. Estimates of the precision for the individual specimen's lattice parameter was approximately 1 part in 100,000.

Quantitative phase analysis. Quantitative analysis is accomplished using either the peak intensity values or, preferably, integrated line intensities since they are less affected by the shape of the line. Overlapping lines will probably cause the most difficulty since it is often necessary to separate their contributions from the group. If the entire set of lines in a group can be used in the analysis, then the Copland-

Bragg (1958) internal standard method will allow for using the net area under the group obviating the need for line separation. However, if good estimates of the individual component line areas are necessary, then profile fitting provides the most accurate means of determining them.

As indicated above, it is easy to over-estimate the precision of refined line intensities. For example, Klug and Alexander (1974) described the effects of different crystallite size distributions on the integrated line areas and their estimated standard deviations. The results were based on 10 loadings of quartz powders held stationary, i.e., not rotated during measurement. The particle size ranges and mean percentage deviation of the line areas were given as: 15-50 μm, 18.2%; 5-50 μm, 10.1%; 5-15 μm, 2.1% and <5 μm, 1.2%. In each case the error associated with the measurement at each point in the measured profiles was kept below 1%. (The specimens having relatively large crystallites had a smaller fraction meeting the Bragg condition and thus their intensities varied from loading to loading.) Had profile fitting been used with these data, the esd's associated with each size fraction would almost certainly have been equal to one another. Repeating a statement made earlier, the estimates of precision reflect the optimization algorithm's ability to determine the best set of parameters for a particular data set and, in general, do not reflect their deviation from the true values.

If one needs to obtain precise line intensities from powders with a particle size larger than ~5 μm, the specimen should be spun during measurement. Parrish and Huang (1983), for example, showed that spinning the specimen will reduce the relative error in intensity measurements to approximately 0.5% on powders having crystallite sizes as large as 30 μm. These intensity variations are also a function of the absorption coefficients of the phase and the roughness of the specimen surface and can be accommodated to some degree with corrections for microabsorption (e.g., Gonzalez, 1987; Hermann and Ermrich, 1987; Kimmel and Schriener, 1988). The effects of microabsorption may play a significant role in quantitative analysis, particularly in the *ab-initio* methods that do not employ the measurement of standard mixtures (Bish and Howard, 1988). In general, microabsorption may manifest itself as an angular dependent deviation of the measured line intensities from their expected values with the deviation increasing with an increase in crystallite size. The uniformity of intensity values achieved from specimen spinning and keeping the crystallite size small aids in justifying the application of profile fitting to obtain precise and accurate line intensities.

Development of the profile-shape-functions used in profile and pattern fitting

Perhaps the earliest routine use of directly fitting profile-shape functions (PSFs) to powder data was in Rietveld's (1967,1969) pattern-fitting structure-refinement algorithm. The choice of PSF and its broadening as a function of angle were based on the optics of the neutron powder diffraction system. The positions and relative integrated intensities of the PSFs were based on the crystal structure of the phase. These constraints quite effectively limited the number of parameters associated with the relatively large number of measured profiles in the diffraction pattern. Since several of the numerical methods developed and employed by Rietveld are used in many techniques involving profile fitting today, Rietveld's methodology will be described in moderate detail.

In contrast to the comprehensive set of profile constraints employed by the Rietveld algorithm, few, if any, are typically employed in profile fitting. This lack of constraints makes refinements difficult due to the proximity of overlapping lines. In many cases, some constraints on the profile parameters must be established in order to ensure that realistic parameter values are obtained. For example, it may be necessary to assume that the shape of all lines in a group are the same. These problems emphasize the necessity of using a PSF that accurately matches the observed line shape.

There has recently been interest in whole-pattern-fitting (WPF) techniques that do not require an explicit structural model. The nature of the WPF technique is intermediate between simple profile-fitting and the more sophisticated Rietveld pattern-fitting structure-refinement techniques. In WPF, the line positions are based on a chosen crystal system and set of lattice parameters -- perhaps obtained from an indexing program. To reduce further the number of simultaneously refined parameters, constraints are placed on the angular dependence of the parameters associated with the shape of the profiles. The large number of permissible lines, particularly if there are no extinction conditions imposed, puts a great demand on using the correct PSF.

Although analytical descriptions of the X-ray powder diffraction profile based on the instrumental optics have been explored, no such model is in routine use. In a summary of classic studies, Alexander (1954) suggested that the description of the aberrations arising from the instrument was fairly complete. By generating profiles based on the convolution of seven specific functions, the author obtained sufficiently good fits to observed diffraction lines to confirm the various contributions to the line shape. The relatively large number of parameters, the correlation between parameters, and the minimal contribution from some of the sources to the profile shape makes such an analytical description difficult. In addition, the contributions from the sources vary with the instrumental configuration and with the specimen. For these reasons, it has been found to be more practical to measure rather than analytically describe profiles that represent the instrument contributions.

Regardless of the algorithm employed, the desired attributes of a PSF are essentially the same. The PSF should accurately fit the measured diffraction lines, accommodate asymmetry, and should be applicable over the entire angular range of an instrument. A PSF should utilize angle-dependent shape-related parameters that vary smoothly as a function of angle and these parameters should reflect the characteristics of the instrument and/or the specimen. The shape-related parameters should allow the profile width to be expressed in terms of a full-width at half-maximum to afford compatibility with previous profile and specimen characterization, i.e., crystallite size and strain. One consideration that has diminished in importance is that the PSF be numerically simple to generate. As computers continue to become more powerful, fewer approximation techniques are being relied upon. This is perhaps best illustrated by the increased use of the convolution products as PSFs. Table 1 lists a number of PSFs referred to in the following sections which have been normalized for unit area.

<u>The Rietveld method of pattern-fitting structure-refinement.</u> The profile-shape-function began to play a considerable role in the area of neutron powder diffraction after the publication of two landmark papers by Rietveld (1968, 1969). Prior to the development of the Rietveld method, one of the major problems with using powder data for structure refinement was the overlapping of the diffraction lines. If the patterns were simple enough, the integrated intensity of the lines could be measured and used to derive the values of $|F_{hkl}|^2$. In patterns with significantly overlapped lines, the line intensities could not be determined with sufficient precision to allow for use of the well-established single crystal structure-refinement techniques.

Rietveld developed a technique for effectively extracting the line intensities by fitting all the lines in a pattern with a PSF. Rietveld employed an analytical model for the diffracted line intensity and profile-shape for neutron data:

$$I_{i,k} = t\, S_k^2\, j_k\, L_k\, \frac{2\sqrt{ln2}}{H_k}\, \exp\left[-4ln2\left(\frac{2\theta_i - 2\theta_k}{H_k}\right)^2\right], \qquad (1)$$

where: $I_{i,k}$ = the intensity at the ith point in the pattern due to the kth line,
  t = the step width of the detector,
  S_k^2 = $F_k^2 + J_k^2$, the nuclear and magnetic structure factors,
  j_k = the multiplicity of the kth reflection,
  L_k = the Lorentz factor,
  $2\theta_k$ = the calculated Bragg angle,
  $2\theta_i$ = the angle at the ith point in the pattern, and
  H_k = the full-width at half-maximum intensity for the reflection.

Now, following Rietveld, new variables are defined:

$$I_k = t\, S_k^2\, j_k\, \frac{2\sqrt{ln2}}{H_k}, \qquad (2)$$

$$b_k = \frac{4\, ln2}{H_k^2}, \qquad (3)$$

and substituted to yield the profile-shape-function:

$$I_{i,k} = I_k \exp[-b_k(2\theta_i - 2\theta_k)^2] \quad . \tag{4}$$

The PSF is thus a Gaussian function having an integrated area, I_k, calculated from the structure. The shape of the Gaussian PSF very effectively matches the observed reflection. The position of the lines are fixed by the lattice parameters.

The slight asymmetry of the line shape was accommodated with a modifying term accounting for the vertical divergence of the neutrons from the source. This asymmetry factor was applied to each profile:

$$I_{i,k} = I_k \exp[-b_k(2\theta_i-2\theta_k)^2] \left[1 - P(2\theta_i-2\theta_k)^2 \frac{S}{\tan\theta_k} \right] , \tag{5}$$

where: P = the asymmetry parameter, and
S = +1, 0, or −1 if $(2\theta_i - 2\theta_k)$ was positive, 0, or negative, respectively.

Use of this factor does not affect the profile's integrated intensity but will change the apparent peak position. This factor is often employed in profile fitting of X-ray and synchrotron data.

The broadening of the neutron diffraction line profile as a function of scattering angle and beam collimators had previously been investigated by Caglioti et al. (1958). Rietveld incorporated their line broadening function in his PSF description:

$$H_k^2 = U \tan^2\theta_k + V \tan\theta_k + W \quad . \tag{6}$$

The angular dependence of the width is thus a function of the parameters U, V and W. This expression is also commonly used in the profile-fitting of diffraction data obtained from other sources, although such use is not analytically correct.

The primary interest in using the Rietveld technique is to obtain values for the atom-related parameters, i.e., positions, occupancies and thermal parameters. Considerable effort has been expended in evaluating profile-shape functions for use in Rietveld algorithms since the values of the atom-related parameters effectively depend on obtaining good measurements of the integrated line intensities, and these in turn depend on the quality of the fit of the PSF to the measured diffraction lines. The results obtained from these evaluations influence the choice of PSF used in profile-fitting and the constraints imposed on the angular dependence of the PSF parameters, and so they are reviewed here.

The PSF in Rietveld structure refinement algorithms. A review of the PSFs used in Rietveld type refinements was given by Young and Wiles (1982). Their review provided results from a test of six analytic functions on X-ray data obtained from seven different specimens with differing degrees of line broadening. The six PSFs tested were the Gaussian, Lorentzian, modified Lorentzian, Intermediate Lorentzian, Pseudo-Voigt, and the Pearson VII. Comparison of the results was based on difference plots and the two residuals R_{wp} and R_B. The authors stressed the need for a profile that has a physical foundation as well as just fitting the observed diffraction lines. This is important since the adjustable parameters in the PSFs appear in the normal equation matrix with the parameters representing the crystal structure. Refinement of both sets of parameters is simultaneous and, hence, the failure of the PSF to describe the observed lines may perturb the structure parameters.

In agreement with Sakata and Cooper (1979), Sabine (1977), and Rotella et al. (1982), Young and Wiles (1982) pointed out the need for variable parameters representing background, as visual estimates are subject to unknown certainty, especially in patterns exhibiting heavily overlapping diffraction lines. An uncertainty in background is most strongly related to the rate of decay of the tail of the PSFs. For example, too high a background estimate may produce residuals favoring those PSFs with more rapidly decaying tails. An *a priori* knowledge of the profile-shape would reduce the uncertainty associated with a variable background.

Poorly fitting PSFs most strongly affect the occupancy factors and thermal parameters for the structure (Rotella et al., 1982; Malmros and Thomas, 1977), but in general, the positional parameters are not significantly affected, at least not at the 3σ level, by the choice of PSF. Thermal parameters and occupancy factors are no doubt influenced by crystallite size and strain, which, in turn must be accommodated by the PSF employed.

The R_{wp} is the most statistically meaningful indicator of calculated-to-observed pattern fit because the numerator is the residual minimized during refinement. However, its value is inflated by the statistical fluctuations in the background and is strongly affected by the misfit of the PSF to the diffraction line, especially in the region of the tails. In pattern-fitting structure refinement, the quantity of interest is the integrated line area since it is directly related to the structure parameters. In terms of structure refinement, the residual of interest is therefore the difference between the calculated and observed integrated line intensities, i.e., the R_B residual. If we consider the best fit of two PSFs having different shapes to a diffraction line, the value of R_B would be zero for the PSF that refined to the same area as the observed line. The value for R_{wp} would refine to some value based on the misfit of PSF shape and, in particular, the misfit in the tails. Therefore, in order to obtain accurate line areas, it is necessary to employ a well fitting PSF since the R_{wp} error is the criterion for refinement.

Malmros and Thomas (1977) applied the Rietveld technique to X-ray film intensity data digitized using an automatic microdensitometer. Data were obtained from Bi_2O_3 using a Guinier-Hagg focusing film camera which afforded good resolution and the complete elimination of the $K\alpha_2$ component through the use of an incident-beam monochromator. The refinement was carried out using Gaussian, Lorentzian, and Modified Lorentzian PSFs. The final choice of the profile model was based on the model yielding the lowest residual R_p, R_{wp} and R_B values. The authors, disregarding the inherent profile asymmetry, indicated that the best results were obtained using the Modified Lorentzian function and that the values of the parameters resulting from the use of either of the three PSFs were essentially the same. Comparison of the results of this study to those obtained from single crystal X-ray data showed a satisfactory agreement between the refined parameters, generally within 3σ of one another. However, negative values for the isotropic temperature factors were obtained from the powder refinements. Attention was called to the fact that the estimated standard deviations (esds) obtained from the powder refinement were on an average of three times larger than those obtained from the single crystal work.

Young et al. (1977) used conventional powder diffractometer data in their assessment of the applicability of the Rietveld technique to X-ray data. In this case, the reflections could be represented by either a Gaussian or Lorentzian PSF. Five specimens were examined, each having a structure in a different space group. R_{wp} values ranged from approximately 9% to 20% and the R_B values ranged from approximately 2% to 12%. The authors stated that the results obtained from the powder refinement, even in the simplest case, were not as good as expected. Poor fit of the PSFs to the diffraction lines was cited as the cause. The tails of the Gaussian functions fell too rapidly whereas the Lorentzian function fell too slowly. Both were accepted as an equally good representation of the profile function. No statistical analysis of which profile function provided a better fit was performed. The esds observed in this study were 3 to 10 times higher than those obtained in single crystal studies. The relatively large esds obtained from the refinement of the more complex structures were attributed to the limited angular range over which the diffraction data were collected. The scans were generally limited to the front-reflection region thereby neglecting the high-angle information with inherently lower error in d-spacing associated with two-theta measurements. Although their study showed a dependence of the profile width parameters U,V,W on the physical characteristics of the specimens, need was cited for adjustable parameters characterizing the crystallite size distribution and strain.

Khattak and Cox (1977) applied the Rietveld technique to X-ray powder diffraction data using Lorentzian, intermediate-Lorentzian, modified-Lorentzian, and Gaussian PSFs (Table 1). The patterns were collected using an X-ray diffractometer equipped with a diffracted-beam pyrolytic graphite monochromator tuned for the single $K\beta$ radiation line. A standard Si specimen was used to test the fits of the PSFs and a $La_{0.75}Sr_{0.25}CrO_3$ specimen was used to assess the structure-refinement capabilities. The authors indicated that, for the instrumental configuration used, the Gaussian function was not adequate in any case and that either the intermediate-Lorentzian or modified-Lorentzian functions were preferred over the Lorentzian function. In the neutron studies using a Gaussian PSF, the tails were assumed to extend no more than $1.5 \times H_k$ from the peak maximum. The authors found it necessary, due to the slower rate of decay of the Lorentz-type PSFs used with the X-ray data, to increase this distance to $3 \times H_k$ since the tails were most influential in the fit. It was noted that the Gaussian function was too low in peak intensity and the H_k too wide, whereas the Lorentzian-type functions proved to have peak intensities and tails too high and H_k too narrow. This study also included an evaluation of the equation of Cagliotti et al. (1958), relating H_k to the adjustable parameters U,V,W. It was found that this expression, at least in the case involving the well-separated lines of Si, could be simplified to $H_k^2 = V \tan\theta_k + W$.

The profile asymmetries observed by Khattak and Cox (1977) were notably less than those observed by Malmros and Thomas (1977) so that inclusion of the Rietveld asymmetry correction yielded only a slightly lower residual error after refinement. Again most parameters resulting from this study agreed within 2σ of results obtained from neutron refinements, but refinement of the thermal parameters yielded values somewhat higher than those obtained from neutron study. The authors suggested that the esds in both cases were unrealistically low. The intermediate-Lorentzian PSF was preferred based on the residual after refinement. The application of the Rietveld asymmetry correction resulted in no significant change in structure parameters.

<u>Profile-Shape-Functions in profile refinement of X-ray data.</u> The resolution of the X-ray powder diffractometer is generally much better than that of a neutron powder diffractometer. When there is a strong smearing from many sources, as is the case with a neutron powder diffractometer, the PSF will tend to a symmetrical Gaussian function. The instrument contributions to the width and shape of the diffraction line tends to mask those contributions from the specimen. In contrast, the better resolution of the X-ray powder diffractometer means more of the individual contributions to the line shape are better experimentally observed. Depending on the nature of the specimen, i.e., crystallite size and lattice imperfections, the X-ray diffractometer's contributions to the profile shape may be less than those of the specimen. This means that fitting the X-ray powder diffractometer line shape is more difficult due to its stronger asymmetry and the angular dependence of its shape.

It became apparent early in the X-ray Rietveld studies (Malmros and Thomas, 1977; Young et al., 1977; Khattak and Cox, 1977) that simple PSFs, e.g., Gaussian, Lorentzian, modified-Lorentzian or intermediate-Lorentzian functions, would not adequately model diffraction lines. It is obvious that for routine X-ray powder diffraction work, the Gaussian profile is inappropriate (Malmros and Thomas, 1977; Young et al., 1977; Khattak and Cox, 1977; Howard and Snyder, 1983; Young and Wiles, 1982). Evaluations of the modified-Lorentzian and intermediate-Lorentzian PSFs indicated that a Lorentz-type function might be a likely possibility.

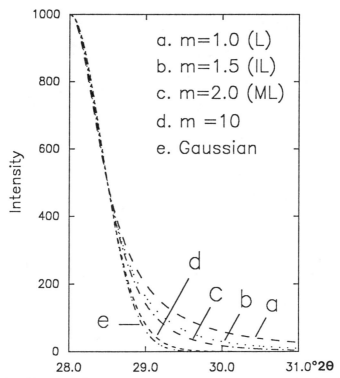

Figure 3. A series of Pearson VII profiles generated with the same full-width at half-maximum but with different values of the exponent m. A Gaussian profile is also shown for comparison. Depending on the value of m, the function replicates the Intermediate (IL), Modified (ML), and pure (L) Lorentzian profiles. The shape of the profile is essentially Gaussian when m exceeds approximately 10.

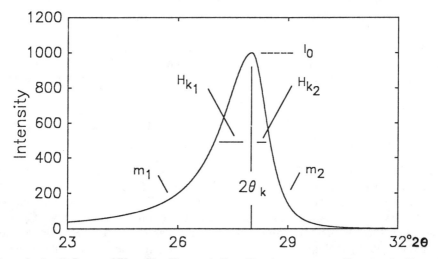

Figure 4. A split-Pearson VII profile. The two half profiles share a common Bragg angle, $2\theta_k$, and peak intensity I_0. Their different full-widths at half-maximum, H_k, and exponents, m, allow the profile to model an asymmetric line.

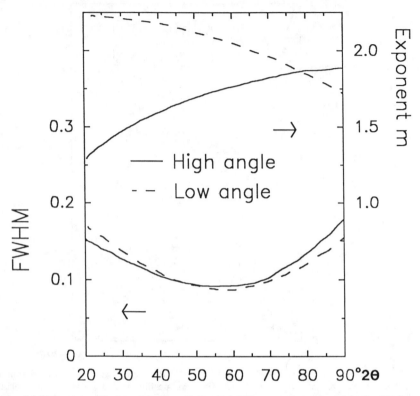

Figure 5. The measured values for the split-Pearson VII parameters H_k and m for the low- and high-angle sides of $Pb(NO_3)_2$ obtained from a Guinier camera by Brown and Edmonds (1980).

Brown and Edmonds (1980) evaluated the Pearson VII function which can range from a basic Lorentzian to a Gaussian function. Due to its flexibility, it is sometimes referred to in the literature as a "variable Lorentzian". When the exponent m = 1, 1.5, and 2, the function takes the form of the basic, intermediate, and modified Lorentzian, respectively (Table 1). Not so apparent is that when m approaches infinity the Pearson VII function becomes equivalent to the Gaussian function. The value of m is most strongly related to the shape of the tails of the profiles as illustrated in Figure 3. In practice, m = 10 is sufficient to approximate the Gaussian function.

Using two half-profiles with a common Bragg angle, $2\theta_k$, and peak intensity, I_0, allows for the modeling of asymmetric line shapes. Such a PSF is termed a split profile function. In the case of a split-Pearson VII PSF, the values of m and H_k on the low- and high-angle sides can be allowed to vary independently. Figure 4 illustrates the asymmetric nature of the split Pearson VII PSF. Brown and Edmonds (1980) evaluated the values of m and H_k for the low and high sides of the profiles as a function of angle using a sample of $Pb(NO_3)_2$ (Fig. 5). Grinding of the sample increased values for both H_k and m, but an analysis of the broadening was not carried out.

An evaluation of seven PSFs on a single data set using the traditional Gauss-Newton, the Marquardt (1963), and the non-linear Simplex (Nelder and Mead, 1965) optimization algorithms was given by Howard and Snyder (1983). The authors concluded that of the basic, intermediate, and modified Lorentzian line shapes, the basic Lorentzian PSF consistently provided the best fit to their data. Overall, the profile giving the most acceptable results was the split-Pearson VII with the position and intensity of the α_2 line determined from the position and intensity of the α_1 line. The constraints on the α_2 lines circumvented a problem which manifested itself most strongly when refining profiles below approximately 70 degrees two-theta, i.e., one line would account for the bulk of the doublet area whereas the second line would become an appendage to account for the "bump" on the side due to the α_2.

The analytical convolution product of a Lorentzian and Gaussian function is the Voigt function. One of the most significant features of the Voigt function is that it allows for the determination of the fractions of Lorentzian and Gaussian nature in a diffraction line (Brown and Edmonds, 1980; Langford, 1978; Suortti et al., 1979; de Kiejser et al., 1982; Delhez et al., 1982, Asthana and Kiefer, 1982). Due to the high interest in this PSF, a detailed description of the Voigt function is given later in this chapter. Widespread use of this profile is described in the spectroscopic literature, for example Sundius (1973), Armstrong (1967), de Vreede et al. (1982), and Asthana and Kiefer (1982).

The use of direct convolution functions

One could quite easily argue that Fourier line shape analysis, mainly used for assessing line broadening mechanisms, preceded the development of the Rietveld technique and therefore should be regarded as the earliest algorithms employing profile-fitting. In an effort to narrow the scope of this manuscript, the authors are making use of a very fine distinction between the profiles used in the Fourier techniques and those discussed here. The Fourier techniques are essentially deconvolution techniques. In other words, a line from a standard is mathematically deconvoluted from another measured line profile. The techniques discussed here involve the fitting of PSFs directly to the observed diffraction line. Detailed discussions regarding the deconvolution techniques can be found elsewhere, e.g., Klug and Alexander (1974), Warren and Averbach (1950), and Delhez et al. (1982).

The PSFs discussed here are either analytical convolution products or their approximations. The principle is analogous to the Fourier techniques except that the convolution process is used to fold two profiles together rather than unfolding them as in the deconvolution process. Generally, the parameters in the components of the convolution product, or in the convolute PSF, are varied to obtain the best fit to the observed data. With either technique, it is necessary to know precisely the nature of contributions from both the instrument and the specimen in order to avoid ambiguities in the interpretation.

Since the analytically based PSFs do not always provide acceptable fits to measured lines, numerical approximations have been used. One principle relied upon is that the PSF representing the instrument can be obtained by measuring a set of lines from a specimen free of crystallite-size broadening and lattice defects. Also, the mean particle size must be sufficiently small and the size distribution sufficiently narrow to minimize orientation effects without having particles so small as to introduce line broadening. With these considerations in mind, a discussion of direct convolution products is presented.

<u>Direct convolution products.</u> Two of the first studies involving direct convolution products as PSFs were reported by Taupin (1973) and Parrish et al. (1976). Both authors used a multiplet of Lorentzians to approximate the diffraction line profile. The multiplet consisted of three Lorentzian functions for each of the α_1 and α_2 lines and a seventh profile for the α_3 component. The profile parameters were evaluated as a function of angle so that interpolation between the standard lines could be performed. The values of the parameters obtained from specimens free from size and strain effects were used to determine the contributions from the radiation source and the instrument. Using W to denote the radiation source contributions, i.e., the α_1, α_2, and α_3 spectral components, and G to represent the instrumental contributions, the resulting function can be written as:

$$F' = W * G \quad , \tag{7}$$

where the (*) represents the convolution operation and the F' denotes the ideal profiles or pattern.

Lorentzian functions have a particular advantage in convolution relations as the mathematics are relatively simple. The convolution product of two Lorentzians is another Lorentzian; the intensity is equal to the product of the two component intensities whereas the value of H_k is the simple sum of the components. Although the use of multiple Lorentzians is not theoretically justified, such a composite can be evaluated rapidly.

The first step in the procedure is to determine the wavelength distributions in the X-ray source. This is done by measuring a line at low angle and stripping the wavelengths other than α_1 from the line, for example by the Rachinger method (Rachinger, 1948). This line approximates the geometrical and instrumental function G. Then a line is measured at high angle where the separation between wavelength components is substantial. Three profiles representing the component wavelengths are convoluted with the line shape calculated from the low angle measurement. The parameters of the wavelength profiles are refined until the best fit is obtained. The components are thus separated and the wavelength function W determined.

A series of standards is then chosen and the narrowest and most intense lines, dispersed throughout the angular range to be used later, are measured. The data are collected with a fine step width, e.g., 0.01°, so that the shape of the profiles can be precisely determined. Typically, three or four Lorentzian functions are used to represent the geometric and instrumental aberrations, G. The G function is convoluted with the wavelength distribution function W and the parameters associated with the G function are adjusted until the best fit between the resultant convolute profile and measured line is obtained. The parameters associated with the G function are then evaluated as a function of angle to obtain a set of curves to be used for interpolation.

Once the G and W functions are obtained, it is now possible to consider the relation of Jones (1938), stated here as:

$$F = (W*G) * S \quad , \tag{8}$$

where S represents the sample contributions to the observed line. The PSF is, again for convenience, assumed to be a Lorentzian function. In practice, since W*G has been previously determined, it is only necessary to refine three parameters: the position and intensity of the convolution product and the full-width at half-maximum of the S function.

The combination of mathematical ease with which the Lorentzian functions can be convoluted and the result of having only three parameters required to represent the specimen contributions make for a remarkably quick refinement algorithm. Additionally, it is no longer necessary to gather the data with as fine a step-width as required previously since the profile shape can be defined by the W*G convolution product. The precision associated with the refined angles and intensities is also significantly improved. Another merit of this technique is that the integrated line intensities can be calculated more accurately as the areas under the Lorentzian profiles are evaluated analytically. This is important since the tails of the profiles, which have a significant area beneath them, may extend a substantial distance from the peak position and may not be recorded or are difficult to separate from the background.

Computational improvements to the algorithm were made and use of the technique extended to data from X-ray energy- and wavelength-dispersive systems by Parrish et al. (1976). The power of this technique was demonstrated using back-reflection X-ray powder diffraction data from topaz (Huang and Parrish, 1975). A total of 144 doublets were found and fitted in a matter of approximately 1.1 minutes.

Some points deserve attention regarding this type of convolution method, which we will refer to as the direct convolution method. Often the W and G functions are not determined separately but as (W*G) directly from the measured

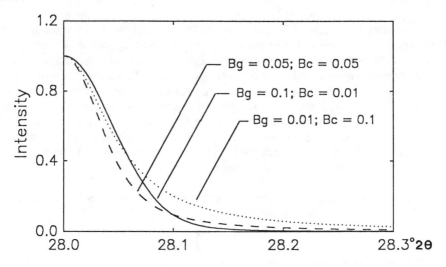

Figure 6. Three Voigt profiles normalized to unit intensity. The parameters B_g and B_c are the integral-breadths of the Gaussian and Lorentzian components, respectively.

lines of the standards. However, the determination of the W function is sometimes desirable. Indeed, the apparent wavelength values will vary depending on the setting of the monochromator. Because most systems have the α_2 component in the spectrum, there is a real need for an *a priori* knowledge of the wavelengths and their relative intensity ratios in profile fitting (Parrish et al., 1976; Howard and Snyder, 1983). Secondly, no modeling was done for crystallite size and strain broadening in the profiles. Combinations of these sample characteristics may change the profile shape to be different from that of the Lorentzian PSF assumed here. It is also interesting to note that Parrish has not indicated a need for another PSF to represent the specimen contributions in spite of the practice of modeling strain broadening by a Gaussian function.

<u>The Voigt and pseudo-Voigt function.</u> The Voigt function, the analytical convolution product of a Lorentzian and Gaussian function, is of considerable interest for its potential ability to (1) model size and strain line-broadening contributions simultaneously, and (2) allow for separation of instrumental and specimen contributions to the diffracted line profile. Due to this interest, the function is described in detail below. In addition, the Voigt profiles shown in Figure 6 have been generated to illustrate the relative contributions of the Lorentzian and Gaussian components to the overall line shape. The parameters β_g and β_c refer to the integral-breadths of the Gaussian and Lorentzian components, respectively. The Gaussian-Lorentzian convolution product also illustrates a general problem with convolution products; since an analytical convolution product is generally not available, the approximation to the solution often requires numerical methods which tend to be computationally intensive.

The Lorentzian function may be defined as (Langford, 1978):

$$I_c(x) = I_c(0) \frac{\omega_c^2}{\omega_c^2 + x^2} , \qquad (9)$$

and the intensity distribution for the Gaussian function as:

$$I_g(x) = I_g(0) \exp(-x^2/\omega_g^2) , \qquad (10)$$

where: $2\omega_g = 2\beta_g (\ln 2)^{1/2}/\pi^{1/2}$
$\omega_c = \beta_c / \pi$, the half-width at half-maximum intensity,
$I_g(0), I_c(0)$ = the respective peak intensities,
β_g and β_c = the integral breadths of the distributions and
$x = |2\theta_i - 2\theta_k|$.

The convolution of these functions is arrived at by solving

$$I(x) = \int_{-\infty}^{+\infty} I_c(u) I_g(x-u) \, du . \qquad (11)$$

The solution of interest is the real part of a complex expression, i.e.,

$$I(x) = \beta_c I_c(0) I_g(0) \, \mathrm{Re} \left\{ \Omega \left[\frac{\pi^{1/2} x}{\beta_g} + iy \right] \right\} , \qquad (12)$$

where: $y = \beta_c / \pi^{1/2} \beta_g$, and
Ω = complex error function, and
Re = indicates the real part of the complex function.

The Voigt function changes from a Gaussian function at $y = 0$ to a Lorentzian function as y approaches ∞. For $y = 3.9$, the value of β_c is about seven times greater than β_g. Asthana and Kiefer (1982) showed that as y varies from 4.0 to 15.0, the percentage Lorentzian character varies by only approximately 7%, from 92.3 to 99.4%.

The following relations apply for the area, A, peak intensity, I(0), and integral breadth, β, of the Voigt distribution:

$$A = \beta_c \, \beta_g \, I_c(0) \, I_g(0) \, , \tag{13}$$

$$\begin{aligned} I(0) &= \beta_c \, I_c(0) \, I_g(0) \, \Omega[iy] \\ &= \beta_c \, I_c(0) \, I_g(0) \, \exp(y^2) \, \mathrm{erfc}(y) \, , \end{aligned}$$

$$\beta = \frac{\beta_g \, \exp(-y^2)}{1 - \mathrm{erfc}(y)} \, . \tag{14}$$

The full-width at half-height of the Voigt function, 2ω, is related to y as (Langford, 1978):

$$\mathrm{Re}\left\{ \Omega\left[\frac{\pi^{1/2}\omega}{\beta_g} + iy \right] \right\} = \frac{1}{2} \Omega[iy] = \frac{\beta_g}{2\beta} \, . \tag{15}$$

The ratio of half-width to integral-breadth, ω/β, is related through the following (Ahtee et al., 1984):

$$\mathrm{Re}\left\{ \Omega\left[\frac{\pi^{1/2}\omega}{2} \frac{2\omega/\beta}{\exp(y^2) \, \mathrm{erfc}(y)} + iy \right] \right\} = \frac{1}{2} \exp(y^2) \, \mathrm{erfc}(y) \, . \tag{16}$$

Approximations are typically used to obtain β and 2ω so that evaluation of the error functions is not necessary. These quantities are underestimated approximately by 5% and 2%, respectively, through the following approximations (Langford, 1978):

$$\beta^2 \approx \beta_c \, \beta + \beta_g^2 \, , \tag{17}$$

$$(2\omega)^2 \approx \frac{4\beta_g^2}{\pi}(1 + y^2) \approx \frac{(2\omega_g)^2}{\ln 2} + (2\omega_c)^2 \, . \tag{18}$$

The quantities β and $2\omega/\beta$ for the F profile, accurate to ~1%, are obtained by the following approximations (de Keijser et al., 1982):

$$\frac{\beta_g}{b} = \frac{1}{2} k \pi^{1/2} + \frac{1}{2}(\pi k^2+4)^{1/2} - 0.234 \ k \ \exp(-2.176k) \ , \tag{19}$$

$$\frac{2\omega}{\beta} = \left[\frac{1+k^2}{\pi}\right]^{1/2} [-k\pi^{1/2} + (\pi k^2+4)^{1/2}] - 0.1889 \exp(-3.5k) \ . \tag{20}$$

Ahtee et al. (1984) gave an approximation for $2\omega/\beta$ accurate to better than 0.16% (at y=0.15) as:

$$\frac{2\omega}{\beta} = 2\left[\frac{\ln 2}{\pi}\right]^{1/2} \frac{1 + ay + by^2}{1 + cy + dy^2} \ , \tag{21}$$

where a = 0.90309645, b = 0.7699548, c = 1.364216 and d = 1.136195.

A parameter of interest, particularly with the Voigt function, is $2\omega/\beta$. Since

$$\beta = A / I(0) \ , \tag{22}$$

the parameter $2\omega/\beta$ contains the H_k, area, and peak intensity. This parameter has a value of 0.6366 for the Lorentzian function and 0.9395 for the Gaussian function, and the Voigt function has values ranging between those for the Lorentzian and Gaussian functions. Furthermore, the value of this parameter can be used to determine the fractional Lorentzian and Gaussian components in the convolute. The simplest expressions from which the fractional components above can be evaluated, while still retaining approximately 1% accuracy, are given by Asthana and Kiefer (1982) as:

$$\beta_c/\beta = a_0 + a_1 p + a_2 p^2 \ , \tag{23}$$

$$\beta_g/\beta = b_0 + b_{1/2}\left[p\frac{2}{\pi}\right]^{1/2} + b_1 p + b_2 p^2 \ , \tag{24}$$

where: $p = 2\omega/\beta$ $a_0 = 2.0207$ $a_1 = -0.4803$ $a_2 = -1.7756$
$b_{1/2} = 1.4187$ $b_0 = 0.6420$ $b_1 = -2.2043$ $b_2 = 1.8706$.

Further consideration of the Voigt function used as a PSF leads one to suggest using this function to represent both instrumental and sample lines or both size and strain contributions to line broadening. Denoting the observed line profile as H, the instrumental profile G, and the sample profile as F, the Lorentzian and Gaussian components of the convoluted line profile are given as:

$$H_l = G_l * F_l \qquad (25)$$

and

$$H_g = G_g * F_g , \qquad (26)$$

and the integral breadths for the F profile are given by:

$$\beta_{l\text{-}F} = \beta_{l\text{-}H} - \beta_{l\text{-}G} \qquad (27)$$

and

$$(\beta_{g\text{-}F})^2 = (\beta_{g\text{-}H})^2 - (\beta_{g\text{-}G})^2 . \qquad (28)$$

An excellent summary of the mathematical properties of the Voigt function is given by Armstrong (1967). Because the complex error function Ω can not be solved in closed form, approximations for the values of this function are typically obtained using numerical techniques. A review of the techniques for generating values approximating those of the Voigt function may be found in Sundius (1973).

Brown and Edmonds (1980) reported approximating the Voigt function using the Pearson VII function. Pairs of Pearson VII functions having various values of the exponent, m, and half-widths, b, were numerically convoluted and the products compared to observed diffraction lines. Results indicated that the sample lines could be suitably fit if one profile used m = 1 and the other m = 10 while their respective values for b were allowed to vary. (These values of m, of course, lead to a Lorentzian and a close approximation to a Gaussian function, respectively.) Experimentally then, Pearson VII functions for which 1 ≤m≤ 10 are hybrids that represent the convolution of a Lorentzian and Gaussian function, i.e., a Voigt function. It was observed that the value of m for a Pearson VII function fit to the convolution products varied smoothly as a function of the ratio of the widths of the m = 1 and m = 10 components, b(L)/b(G), as shown in Figure 7. The ratio of the convolute half-width to that of the input Gaussian, b(H)/b(G), was also given. A predominant Lorentzian nature of the diffraction lines was observed in their study,

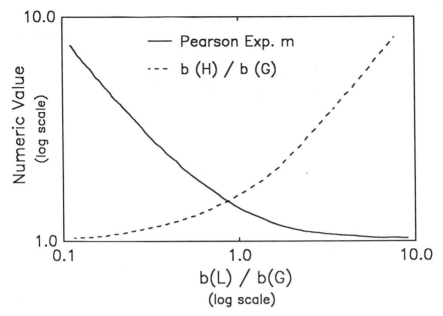

Figure 7. The value of the Pearson VII exponent obtained from fitting the convolution product of two Pearson VII profiles, one with an exponent m = 1 (L) and the other with an exponent m = 10 (G), as a function of the ratio of the half-widths of the respective components. The ratio of the convolute half-width, b(H), to that of the input (G) function is also given. (Brown and Edmonds, 1980)

i.e., the value of m was in the range of 1.2 to 2.3 corresponding to values of b(L)/b(G) of 1.4 and 1/(2.1), respectively.

The pseudo-Voigt function, Table 1, is often used to approximate the Lorentzian/Gaussian convolution product by using the simple addition of the two functions. The mixing parameter, η, varies the function from Lorentzian to Gaussian for $\eta = 1$ to $\eta = 0$, respectively. Hastings et al. (1984) gave the following approximations for the un-normalized PSF:

$$H_{k\text{-}G}/H_k = (1 - 1.10424n + 0.05803n^2 + 0.04622n^3)^{1/2} \tag{29}$$

and

$$H_{k\text{-}L}/H_k = 1.07348n - 0.06275n^2 - 0.01073n^3 , \tag{30}$$

where $H_{k\text{-}G}$, $H_{k\text{-}L}$ and H_k are the full-width at half-maximum values for the Gaussian, Lorentzian, and pseudo-Voigt PSFs, respectively. The errors in these approximations were less that 4×10^{-3}. Thompson et al. (1987) gave a series expansion relating η to the H_k of the Lorentzian and Gaussian components of the pseudo-Voigt PSFs as:

$$\eta = 1.36603(H_{k\text{-}G}/H_{k\text{-}L}) - 0.47719(H_{k\text{-}G}/H_{k\text{-}L})^2 + 0.11116(H_{k\text{-}G}/H_{k\text{-}L})^3. \quad (31)$$

They also used a set of numerically convoluted profiles to obtain the series approximation for the full-width at half-maximum of the pseudo-Voigt profile:

$$H_k = [\,(H_{k\text{-}G})^5 + 2.69269(H_{k\text{-}G})^4 H_{k\text{-}L} + 2.42843(H_{k\text{-}G})^3(H_{k\text{-}L})^2 +$$

$$4.47163(H_{k\text{-}G})^2(H_{k\text{-}L})^3 + 0.07842 H_{k\text{-}G}(H_{k\text{-}L})^4 + (H_{k\text{-}L})^5\,]^{1/5}. \quad (32)$$

It is often advantageous to refine the values of $H_{k\text{-}G}$ and $H_{k\text{-}L}$ separately, particularly when these terms are used to account for the strain- and size-broadening contributions to the diffraction line profile, respectively. In this case, their angular dependencies may be constrained according to the following:

$$H_{k\text{-}G} = V \tan\theta \quad (33)$$

and

$$H_{k\text{-}L} = X / \cos\theta \,. \quad (34)$$

<u>Particle-size and strain analysis via direct- or approximated-convolution relations</u>

Modeling of the convolution relations between instrumental and sample PSFs allows separation of the sample characteristics from those of the instrument. This principle has been used to evaluate particle size and strain using X-ray powder diffraction by Langford (1978), Nandi and SenGupta (1978), Delhez et al. (1982), deKeijser et al. (1982) and using neutron powder diffraction by Suortti (1979). Although many Fourier techniques exist for evaluating the crystallite or domain size (τ) and the strain (ϵ) (e.g., Delhez et al., 1982; Nandi and SenGupta, 1978; Klug and Alexander, 1974), rapid approximations may be obtained using the integral breadths of the specimen-related profile.

The most rapid and least accurate approximations to these parameters may be obtained using the width of a single diffraction line. Typically, the contribution from the specimen is estimated from the difference between a line from the specimen under consideration and the same line from an ideal specimen of the same composition. However, what really should be measured is the breadth of the specimen line after the instrumental line profile has been deconvoluted from it. A profile from a standard of the same composition may be broadened from the intrinsic physical characteristics of the specimen, e.g, microstrain due to defects.

The breadth of the deconvoluted profile, since it yields only a single parameter depending on the PSF employed, might represent either strain or size broadening.

Profiles based on the direct convolution of an instrument profile (e.g., a split-Pearson VII multiplet) and a specimen function represented by a Lorentzian and Gaussian profile for size and strain analysis in a profile-fitting algorithm have been reported by Howard and Snyder (1985) and Yau and Howard (1989). The coefficients of the instrumental profile were based on those obtained from profile refinement of the National Institute of Standards and Technology Si reference material SRM 640a. They found that use of the Lorentzian specimen function yielded much better fits for all of their test specimens. In cases involving suspected strain-broadened materials, they found that convolutes employing Gaussian functions provided a significantly poorer fit than those employing the Lorentzian specimen profiles. Howard and Snyder (1989) and Yau et al. (1989) also reported the use of direct convolute functions in a Rietveld program. Again, strain-broadened materials were fit better with the specimen profile modeled with a Lorentzian function than with a Gaussian function. The angular dependence of the broadening was used to determine the relative contributions to line broadening from size and strain.

Enzo et al. (1988) and Benedetti et al. (1988) reported the use of a truncated exponential convoluted with a pseudo-Voigt to model the instrumental function. A second pseudo-Voigt was used to model the specimen function and was convoluted with the instrument function to account for line broadening. The latter authors compared the results of a traditional Warren-Averbach (1950) Fourier type of analysis with those obtained simply using the integral breadths of the specimen function. The authors found a generally good agreement between the results. It was noted that when the analysis was based upon the use of a single line, the results seemed to be less precise than when a multiple line analysis was employed, for example, when two orders of a reflection were examined.

Analysis using the Voigt function

A single Voigt function can, in principle, be used to assay both size and strain effects as it has both a Gaussian component to model the strain contributions and a Lorentzian component to model the size broadening contributions. However, to accommodate instrumental effects, the function must be convoluted with another PSF representing the instrument. Nandi and Sen Gupta (1978) used analytical convolutions of Gaussian and Lorentzian functions representing the strain and size effects, respectively. They used a Gaussian or Lorentzian function to represent the instrument function and stated that both resulting PSFs closely approached the true observed line profile. The PSF using the Gaussian function to represent the instrument produced more satisfactory results. Evaluation of the fit was based on visual comparison between the calculated (refined) PSF and the recorded lines. However, the authors did not investigate the likely combination of Gaussian and Lorentzian characteristics inherent in the instrumental function, nor was asymmetry modeled. Other authors have considered the instrumental and sample PSFs as

Voigtian and the observed line profile as a convolution of Voigt functions (Langford, 1978; de Keijser et al., 1982; Delhez et al., 1982; Suortti et al., 1979).

In an early use of the Voigt profile in crystallite size analysis, Langford (1978) used the single-line method to analyze an annealed submicron Ni powder. Comparison of the results obtained with a Voigt profile to those obtained with a Gaussian PSF showed the Voigt results to be 7-10% too high whereas, compared to results obtained from Lorentzian PSF, the Voigt estimates were 40-50% too low. The specimen line profile, corrected for instrumental contributions, had a value for $2\omega/\beta = 0.755$, intermediate between Gaussian and Lorentzian values. An interesting application in this article was the use of the 333 line from a Si crystal to diffract unfiltered Cu $K\alpha$ radiation. Evaluation of the α_1 line gave a value of $2\omega/\beta = 0.628(5)$ for the Voigt profile. This is close to, but slightly less than, the expected value for a Lorentzian wavelength distribution.

Suortti et al. (1979) investigated the use of the Voigt function in carrying out size and strain analyses using neutron and X-ray powder diffraction data. The authors argued that the integral-breadth of a diffraction line is determined by the average crystallite size and strain. Since most of the instrumental effects are limited to the proximity of the peak, the tails could be considered as the most accurate representation of the crystallite size. The fit of the Voigt to the neutron powder diffraction lines was indicated as being better than that obtained using a pure Gaussian. The fit was near perfect down to a fractional level of approximately $(10^{-2}$ to $10^{-3})I(0)$. The scattering at low angle was dominated by the Gaussian component, whereas lines at higher angles had a rapidly increasing Lorentzian component. The use of the Voigt profile with the X-ray data, however, resulted in higher tails in the region $0.1 \le |2\theta_i - 2\theta_k| \le 0.2$ than those actually observed.

The authors fitted the Voigt PSF to the X-ray data, corrected for instrumental broadening, using two separate refinement criteria. The first refinement was performed on the basis of predicting the correct peak behavior. The second test adjusted the PSF until the asymptotic behavior of the profile was accounted for while the integral breadth was constrained to be equal to that obtained in the first refinement. (This asymptotic behavior was reached at $|2\theta_i - 2\theta_k| > 0.7$.) Results indicated that when correct peak behavior was accounted for the estimates for the particle size were too large. Correct asymptotic behavior gave an incorrect strain contribution. Conclusions based on the comparison of the fitted profiles on Ni specimens throughout the angular range of the instrumentation were that the β_c and β_g parameters have little physical significance. In particular, β_c does not correspond to the average dimension of the crystal mosaics.

An evaluation of the errors associated with the determination of size, τ, and strain, ϵ, parameters using the Voigt function has been reported by de Keijser et al. (1982). Errors associated with the statistical fluctuation of the count rate, an incorrect ratio of α_1 to α_2, and the instrumental profile asymmetry were considered. Two examples were of particular interest. Let the α_1 to α_2 ratio be R and the error in R equal to dR. For a Voigt profile with $2\omega/\beta = 0.8$, R = 0.5 and dR = 0.03, then the errors associated with β_c and β_g are respectively 6% and 2.5%. Secondly, errors due to neglecting profile asymmetry can be large, particularly if asymmetry is pronounced and structural broadening is small. Neglect of asymmetry introduces fictitious size contributions to broadening and produces larger estimates of strain. Statistical analysis, considering only the statistical fluctuations in counts, show that errors for τ may be on the order of 10-20% whereas errors for ϵ are likely to be a factor of 2 lower.

Profile fitting of synchrotron diffraction data

Synchrotron X-ray diffraction is a relatively new technique and shows a great deal of promise. The greater intensity of the source, coupled with the narrower intrinsic diffraction profile, open avenues for investigation of symmetry changes, particle strain and size broadening effects, and the indexing of previously intractable patterns (Hastings et al., 1984; Will et al., 1987). In addition, the ability to select the incident wavelength has several advantages, including performing anomalous scattering experiments.

The parallel beam and the use of incident-beam monochromators eliminates many of the geometrical aberrations inherent in the traditional powder diffractometer and yields a more symmetric line shape. Line shapes are found to vary with the experimental configuration. In many cases, the intrinsic diffraction profile has been found to be nearly Gaussian. Will et al. (1987) found that the line shape could be suitably approximated using two Gaussian PSFs with a common peak position and the H_k and peak intensity ratios fixed at 2:1. In a similar study, Will et al. (1988) tested Gaussian, Lorentzian, Pearson VII, and pseudo-Voigt functions and, in agreement with Thompson et al. (1987), found the pseudo-Voigt provided a superior fit. Lehmann et al. (1987) reported using a Voigt function with success. Authors analyzing the profile shape typically report some asymmetry. An extended discussion of the use of synchrotron X-rays is given in a separate chapter in this volume by L. Finger.

Two-step structure refinement using a Whole-Pattern-Fitting algorithm. Perhaps one of the most exciting uses of high resolution synchrotron data, at least from the profile-fitting point of view, is its use in the two-step structure refinement technique using whole-pattern-fitting (WPF) (Pawley, 1984). The WPF technique is intermediate to that of a profile-fitting algorithm in which no reference is made to a structural model and the Rietveld technique which constrains almost all parameters to follow a structural model. In many cases, one desires to obtain a set of integrated intensities from a powder pattern which may be used, for example, in determining a set of $|F_{hkl}|$ values for structure determination. The WPF technique usually requires the generation of all lines that occur within the angular range of interest. The resulting high degree of line overlap induces significantly large correlations among the parameters undergoing refinement. An *a priori* knowledge of the profile shape would allow for refining fewer parameters and would diminish the likelihood of incorrect lines fitting an area of a pattern. In general, some constraints must be placed on the intensities, particularly where two or more lines share the same angle, e.g., symmetrically equivalent reflections.

Although not based on the use of a WPF algorithm, Will et al. (1983) used a two-step structure refinement process in which X-ray powder diffraction intensities were obtained using conventional profile-fitting techniques. The intensities and positions, along with a specimen-related broadening factor, were refined individually for each profile. The authors noted that although lines separated by one half-width could be refined without constraints on their positions, it was advantageous to calculate the line positions from the lattice spacings for those lines in closer proximity.

Toraya (1986) reported the use of the whole-pattern fitting algorithm on data obtained from an X-ray powder diffractometer. The first part of the paper reports a re-investigation of the angular dependence of the coefficients in a split-Pearson VII function. In this case, analytical curves were used to model the angular dependence of the low- and high-angle tails and an asymmetry parameter relating to H_k. The author suggested that it is also important to include corrections for line displacement. Combining adjacent lines into a single observation to minimize the possibility of an ill-conditioned matrix, and reducing the number of matrix elements due to the limited range in the pattern over which the profile are spread, were also addressed. Jansen et al. (1988) applied WPF to neutron powder data requiring the use of only a single Gaussian profile. The authors demonstrated the usefulness of WPF to texture and preferred orientation analysis.

In view of the large degree of correlation introduced by the overlapping of lines, it appears to be advantageous to employ a high resolution, monochromatic source such as an incident-beam monochromated X-ray powder diffractometer. Better still would be a synchrotron source due to the high resolution and high intensities obtainable. Several authors, including Lehmann et al. (1987) and Will et al. (1988), addressed the applicability of WPF using synchrotron data.

CONSIDERATIONS IN PROFILE FITTING

<u>General considerations in profile refinement</u>

The following sections outline several points that the analyst should consider prior to performing any profile refinements. The major considerations include the angular range of the pattern segment involved, the number of parameters, the profile model, and the type of optimization algorithm employed. There are, however, a number of aspects which are more esoteric to the numerical analyst, yet one should be aware of their existence and/or the role they will play in determining a solution. Examples of how incorrect solutions may result, even when using the correct technique, are given below. A set of illustrations, primarily related to instrumental considerations, has been given by Chipera and Bish (1988).

Most of the examples used throughout this section are based on profile fitting a region of a synthesized quartz pattern. The specific region lies between 66 and 70° 2θ which, when measured with Cu $K\alpha$ radiation, contains six overlapping lines consisting of three α_1-α_2 pairs. This region was chosen because a similar degree of overlap may be encountered in normal practice. The quartz pattern was calculated with Lorentzian profiles using the Rietveld program of Wiles and Young (1981). For these examples, a pattern was generated having a low signal-to-noise ratio (S/N) of 2, i.e., the maximum peak intensity was twice that of the background intensity. A second pattern was generated with a higher S/N ratio of 10. The files containing the calculated patterns were duplicated so that "noise" could be added to simulate real conditions. The noise added to the patterns was based on the Poisson distribution which governs X-ray counting statistics.

Two different PSFs were used in fitting the lines using the SHADOW profile-fitting program (Howard and Snyder, 1989; Howard, 1989). The first profile was a single-line PSF based on a normal Lorentzian function. Since a line was required for each of the α_1 and α_2 reflections, a total of six lines was refined. The second PSF was a compound profile (described below) requiring only one PSF for each of the three α_1-α_2 pairs. This PSF employed a split-Pearson VII function for each of the α_1 and α_2 reflections; the position and intensity of the α_2 lines were based on those of the α_1 lines. During refinement of this PSF, the shapes of the low-angle sides of the component lines were set equal as were those of the high angle sides. An additive background value was refined in addition to the profile parameters. The PSFs were refined both with and without constraints on their shapes. Constrained profiles were required to have the same shape throughout the segment while their intensities and position were allowed to vary independently. The shapes of the unconstrained profiles were free to vary independently.

Table 3 gives average values for the differences in positions, the relative peak intensities, and the integrated areas for each set of refinements as compared to the original calculated quartz pattern. The integrated areas of the constrained compound profiles without noise were assumed to be the same as those for the calculated pattern and all other integrated areas were referred to them. The weighted residual error is also given in the table for each refinement. Note that a smaller R_{wp} error does not necessarily indicate a better fit since a great deal of profile distortion may be present.

<u>Minimizing parameter correlation.</u> Parameter correlation can be illustrated by considering an α_1-α_2 doublet at a sufficiently low angle so that the α_2 causes only an inflection in the slope of the doublet on the high-angle side. An adjustment in the intensity of the α_1 line may require an adjustment in the intensity of the α_2 line. This may then require the adjustment of the α_1 line position which, in turn, might require an adjustment in the α_1 line intensity. In other words, the adjustment in any one of the parameters might require the simultaneous adjustment of one or

Table 3 - Average Errors in Refined Line Positions, Peak Intensities and Areas and the Weighted-Pattern Error Resulting from the Profile Fitting of a Synthesized Quartz Sextuplet.

Figure	Key	$\Delta 2\theta$	$\Delta I/I0$	ΔIntegrated area	Rwp (%)
8	W/N	0.0229	29.7617	7.1267	8.50
	O/N	0.0054	13.3833	3.9833	0.43
9	W/N	0.0175	43.8200	21.1150	8.31
	O/N	0.0041	25.9183	8.0633	0.42
12	W/N	0.0020	12.4400	36.3767	8.73
	O/N	0.0038	10.2433	3.0267	0.58
13	W/N	0.0086	71.16	10.8333	8.27
	O/N	0.0017	27.2567	34.3233	0.38
14	W/N	0.0712	25.2633	7.2717	6.02
	O/N	0.0051	10.8033	3.1683	0.90
15	W/N	0.0106	14.7550	3.6033	5.89
	O/N	0.1082	8.9650	2.7917	0.89
16	W/N	0.0051	15.9867	5.5900	6.12
	O/N	0.0009	0.4067	3.9500	1.24
17	W/N	0.0058	36.9867	18.4533	5.88
	O/N	0.0008	21.8000	12.3400	0.94
18	W/N	0.0238	8.9633	10.8767	15.09
	O/N	0.0023	0.7100	1.1233	0.66
19	W/N	0.0262	23.8533	1.4467	19.71
	O/N	0.0024	1.4267	1.0900	0.64
20	W/N	0.0029	29.0533	4.4167	11.30
	O/N	0.0024	5.8700	3.9133	1.08
21	W/N	0.0034	33.5367	4.8133	11.27
	O/N	0.0023	2.4767	4.1200	0.97
22	5P	0.0047	13.0100	23.0000	5.14
	11P	0.0143	25.2433	19.7533	3.12
23	5P	0.0028	12.4867	26.4400	8.28
	11P	0.0084	14.7333	15.3200	4.26
24	5P	0.0075	64.2933	21.6700	3.56
	11P	0.0082	36.4600	47.1000	2.18
25	5P	0.0086	71.1600	10.8333	5.88
	11P	0.0063	66.6567	1.7767	2.83

W/N : data with statistical noise;
O/N : data without noise;
5P : data smoothed with 5 point filter;
11P : data smoothed with 11 point filter.

more of the other parameters. This problem is particularly deleterious when the intensities of the lines are relatively low; the presence of statistical fluctuations in the intensities makes it difficult for the optimization algorithm to determine the best set of parameters.

A relatively small degree of parameter correlation can be accommodated by an optimization algorithm. When a high degree of correlation exists, the parameter values will tend to oscillate or, in the worst case, the optimization algorithm will fail altogether and the parameter values will diverge rather than converge. In the case where regions contain heavily overlapped lines, efforts must be taken to reduce the number of variable parameters. Possible ways of reducing the number of parameters include:

- Fix the α_2 line position and intensity based on those of the α_1.
- Choose a simpler, less flexible profile with fewer variable parameters.
- Constrain the shape of the lines to be the same.
- Fix the shapes of the lines based on previously established results.
- Fix positions and/or intensities of known lines.

<u>The least-squares error criteria for refinement.</u> The goal of the refinement algorithms used in profile-fitting is simply to obtain the set of parameters producing the least error between the calculated and measured profiles. The error criterion used is often called the "least-squares" error since it involves the sum of the squares of the differences between the measured intensities, I^{obs}, and the calculated intensities, I^{calc}. Either the weighted residual error, R_{wp}, or the unweighted residual error, R, may be used as the criterion for refinement. The R error is determined from the following expression:

$$R = \left[\frac{\Sigma [I(2\theta_i)^{obs} - I(2\theta_i)^{calc}]^2}{\Sigma [I(2\theta_i)^{obs}]^2} \right]^{1/2}, \quad (35)$$

where the subscript i runs over all points in the segment undergoing refinement. The R_{wp} error is the correct criterion to use with step-scanned data since the error associated with each intensity value is not constant but is a function of the number of counts. The R_{wp} error therefore uses the weighting factors $w(2\theta)$:

$$R_{wp} = \left[\frac{\Sigma [w(2\theta_i) \times (I(2\theta_i)^{obs} - I(2\theta_i)^{calc})^2]}{\Sigma [w(2\theta_i) \times (I(2\theta_i)^{obs})^2]} \right]^{1/2}, \quad (36)$$

where $w(2\theta_i) = 1/[\text{variance of } I(2\theta_i)] = 1/I(2\theta_i)$. Note that the effect of the weighting is to reduce the contributions to the error due to the misfit at the tops of the peaks. In contrast, the misfit in the regions near the tails will contribute significantly. Therefore, it is not unusual to observe discrepancies near the peaks and better fits near the tails of the profiles.

The value of the residual error should go to zero for a perfect match of calculated and observed intensities but, due to the statistical "noise" in the measurement of the intensities, the error can at best be minimized. Good results may yield an R_{wp} on the order of 2-10% whereas typical results may be 10-20%. The final error will be influenced most by the intensity of the diffraction line (i.e., the time spent counting X-rays) since the sum of the observed intensities is used in the denominator of the R and R_{wp} errors. To gauge the quality of the fit, one should compare the final R_{wp} value to the expected error value. The expected error is derived from the statistical error associated with the measured intensities:

$$R_{exp} = \left[\frac{N - P}{\Sigma\, I(2\theta_i)} \right]^{1/2} . \tag{37}$$

where: N = the number of observations, and
P = the number of variable parameters.

In order to use this statistic most effectively, the region should contain as few lines as possible so that R_{exp} reflects the error associated with individual lines or overlapped lines. Again the denominator contains the sum of the observed intensities; the greater the intensity in the range the lower the expected error. Note that there are no accommodations for any intrinsic misfit of the profile to the line shape.

The choice of optimization algorithm. The two optimization algorithms most frequently used are the Marquardt (1963) and the traditional Gauss-Newton techniques. Both use derivatives in determining a set of normal-equations for use in determining the adjustments for the parameters. The fundamental difference between the two involves the role that the off-diagonal elements in the normal equation matrix play in determining the new adjustments. The Marquardt algorithm may adjust the weighting of the off-diagonal elements thereby changing its nature from that of the Gauss-Newton technique to that of a gradient technique. This variability makes the Marquardt algorithm a bit more robust, but it performs slightly slower.

The non-linear SIMPLEX algorithm (Nelder and Mead, 1965), a "search" algorithm, is sometimes used instead of the Gauss-Newton and Marquardt algorithms. Given n parameters to refine, the SIMPLEX algorithm begins by selecting n+1 sets of n parameter values. The n+1 sets of parameters actually represent n+1 points in the n-dimensional solution space of the problem. The centroid of the points is calculated and the point with highest error, P_h, is determined. The point P_h then undergoes either a reflection through the centroid, contraction towards the centroid, or expansion away from the centroid, depending on the motion necessary to lower the associated error. A successful movement causes the new P_h to be located and reflected, contracted or expanded. The process is repeated until some criterion is met, e.g., a sufficiently small change in P_h after movement.

Howard and Snyder (1983) compared the performance of the Gauss-Newton, Marquardt, and non-linear SIMPLEX techniques using several different profile-shape-functions. The authors stated that the nature of the SIMPLEX technique tends to make the final parameters somewhat dependent on the initial parameters. To minimize this effect, a second application of the algorithm was used with the refined parameters from the first round as input. The SIMPLEX algorithm, in general, did not perform well in regions with overlapping lines. In comparing the two derivative based algorithms, the authors noted that the

Marquardt algorithm was a bit more robust and was less sensitive to the errors in the initial estimates of the parameters. Due to the dependence of the SIMPLEX algorithm on the initial parameter estimates and its general inability to estimate the precision of the refined parameters, it is not considered in the ensuing discussion.

In general, the ability of the derivative-based optimization algorithms to perform their tasks is primarily affected by three parameters:

- The quality of the initial estimates for the parameters;
- The degree of correlation between the parameters;
- The number of parameters undergoing refinement.

If the initial values for the parameters are significantly different from the true values then these algorithms may not be able to determine the adjustments. This is particularly true for the Gauss-Newton algorithm. The Marquardt algorithm performs well even if the starting values differ significantly from the true values. In fact, relatively poor estimates can be given and this algorithm will still perform satisfactorily. Although errors for R_{wp} greater than 50% to 60% may cause the Gauss-Newton algorithm to fail, R_{wp} values of more than 100% to 200% may still not deter the Marquardt algorithm. When the estimates are relatively close, the Gauss-Newton algorithm tends to converge more rapidly.

The effect of parameter correlation on the efficacy of the optimization techniques is two-fold: (1) it will increase the number of iterations required before the optimization algorithm converges, and (2) it will reduce the ability of the algorithm to determine the optimum parameter values. The ability of the algorithms to determine the optimum parameter values also decreases as the number of parameters increases. These algorithms may be insensitive to those parameters that cause very small changes in the overall error when their values are changed. This problem may be further compounded by a computer's finite representation of numbers and the propagation of the associated round-off error in computations. The greater the number of mathematical operations the lesser will be the precision and accuracy of the final results. The best recourse, in light of these facts, is to reduce the number of parameters to be refined at any given time.

<u>The convergence criterion for the optimization algorithms.</u> Convergence criteria specify the conditions that must be met before the optimization algorithm is halted. There are basically two convergence criteria that may be used. The first is based on the size of the parameter adjustments, $\Delta\beta_i$, relative to the magnitude of the respective parameter, β_i. This effectively allows for ending the optimization procedure when a desired precision of the parameters has been achieved. The refinement is halted when all parameters pass the following test:

$$\frac{\Delta\beta_i}{|\beta_i| + \tau} \leq \epsilon , \qquad (38)$$

where τ prevents a division by zero and ϵ effectively specifies the required precision. (The authors have found useful values to be on the order of $\tau = 0.001$ and $\epsilon = 0.001$.)

A more reasonable convergence criterion is based on the estimated standard deviations of the parameters which can be derived based on the errors associated with the measured intensity values. The variance of the ith parameter is estimated from the formula:

$$\sigma_i^2 = \left[M_{ii}^{-1} \Sigma_j \{w(2\theta_j)[I(2\theta_j)^{obs} - I(2\theta_j)^{calc}]^2\} / (N-P) \right]^{1/2}, \qquad (39)$$

where M_{ii}^{-1} is the corresponding diagonal element in the inverted normal-equation matrix.

Convergence is achieved when all the parameter shifts calculated on a given cycle satisfy the condition $\Delta\beta_i < \alpha \times \sigma_i$, where α is a user-selected value and may be on the order of 0.3. This condition is more realistic as the precision of the individual parameters may vary significantly from one to another. This avoids requiring the optimization algorithm to continue until the change in a parameter value is in the sixth significant digit when the optimization algorithm may only be able to estimate the precision of the parameter to three significant digits.

<u>Goodness-of-fit.</u> The calculation of the goodness-of-fit (GOF) parameter includes the number of variables undergoing refinement. Thus, the GOF may help in determining whether a change in the number of parameters, e.g., a change to a PSF having a different number of parameters, or the addition of another line to a peak group, has significantly decreased the residual error. The GOF is calculated from the following:

$$\text{GOF} = \left[\frac{\Sigma [w(2\theta_i) \times (I(2\theta_i)^{obs} - I(2\theta_i)^{calc})^2]}{N - P} \right]^{1/2}. \qquad (40)$$

Choice of profile-shape-function

The profile-shape-function (PSF) is used to "distribute" the integrated line intensity with the goal of matching the shape of the observed diffraction line. Table 4 lists the attributes of the PSFs available in program SHADOW (Howard and Snyder, 1983; Howard, 1989). The number of profiles is relatively large in order to allow for the wide variety of diffraction line shapes encountered in practice. The following attributes need to be considered when choosing the PSFs:

- the fundamental line function (e.g., Gaussian or Lorentzian),
- whether or not the line function is split, and
- the number of lines comprising the PSF, i.e., whether or not it is a single-line or a compound PSF.

<u>Single-line PSFs.</u> The simplest PSFs consist of a single-line profile generated from a single function, e.g., a Lorentzian function, and are symmetric in shape. An asymmetry factor, such as that used by Rietveld (1969) may be applied to generate an asymmetric profile. Split profiles, such as the one illustrated in Figure 4, use two "half-profiles" to generate a single asymmetric line profile. The two halves share a common Bragg angle and peak intensity and their shape-related parameters may be either constrained or allowed to vary independently. The more flexible PSFs employ a larger number of shape-related parameters. Although use of the more flexible PSFs usually results in lower residual errors, the higher degree of parameter correlation associated with the increased number of parameters may limit their usefulness.

Figures 8 and 9 show the results of fitting the low and high S/N patterns using a Lorentzian function as the basis for a single-line PSF. The bottom halves of the figures show the calculated pattern without the introduction of noise and the fit of the refined pattern. The refined pattern is the sum of the six best-fit PSFs and the refined background value. In both figures, the differences between the calculated and refined patterns are so small that they can not be observed at the scale the figures are presented. The full-width at half-maximums, H_k, for all the lines in the calculated pattern were the same. Variations in the refined H_k values are evident in both the low and high S/N data with the deviations being greatest in the profiles fit to the low S/N pattern. The $\alpha_1:\alpha_2$ intensity ratios also vary to a greater extent in the low S/N pattern. Although one would expect to obtain the true calculated profiles since there are no statistical fluctuations in the calculated pattern, the effects of parameter correlation, the approximations made by the optimization algorithm, and the near equivalence of overall refined errors obtained by more then one set of parameters compounded to produce the observed results.

The top halves of Figures 8 and 9 show the calculated patterns with the addition of the statistical fluctuations in the intensities. (The fluctuations in the low S/N pattern are more apparent simply due to the scale of the plots.) The results obtained from fitting the high S/N data more closely match the calculated values than those obtained from the low S/N pattern. A simplified reason for this is that the program can discern between the component lines by the inflection points in the pattern in much the same way we would from a visual inspection of the data. In the low S/N pattern, the statistical fluctuations mask these inflection points. In this case, varying the number of lines may produce refined patterns that fit the calculated pattern almost as well. This poses a problem when the number of lines in a group is unknown. In summary, the greater the S/N ratio, i.e., the greater the intensity of the lines, the more likely the results will match the true values.

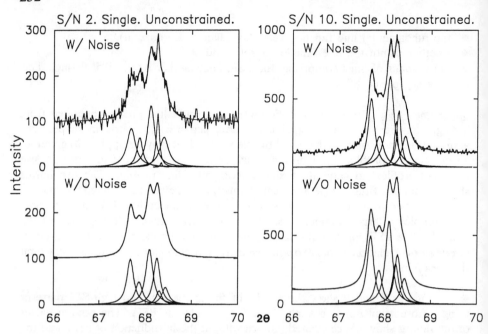

Figure 8 (left). The signal-to-noise (peak to background) ratio for the data in this figure is 2:1. A Lorentzian single-line profile-shape-function was refined for each of the α_1 and α_2 components in the three doublets without requiring their shapes to be the same. In the bottom half of the figure, the profiles were fit to a calculated pattern without the presence of the statistical intensity fluctuations (noise) normally found in patterns. In the top half of the figure, the profiles were fit to a calculated pattern to which intensity fluctuations were added according to the statistics associated with X-ray diffraction. A solid line, representing the sum of the refined profiles and an additive constant to accommodate background, is also shown.

Figure 9 (right). The signal-to-noise ratio for the data in this figure is 10:1. A Lorentzian single-line profile-shape-function was refined for each of the α_1 and α_2 components in the three doublets without requiring their shapes to be the same.

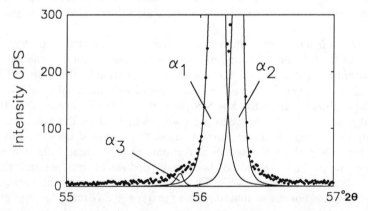

Figure 10. The misnamed Cu $K\alpha_3$ line, which is actually a multiplet of lines, is evident in this plot of refinement results obtained from a Si pattern. The low angle α_3 line has a peak intensity equal to 1.27% of that of the α_1, or 26 CPS.

Correlation may prove to be an insurmountable obstacle even when a simple single-line PSF is used, particularly when both the α_1 and α_2 components must be fit in a group of overlapping lines. It is often advantageous to include a second line in the PSF to account for the α_2 and base its position and intensity on those of the α_1 line (Naidu and Houska, 1982; Howard and Snyder, 1983; Enzo et al., 1988). In this text, a PSF composed of more than one line will be referred to as a compound PSF.

Accommodating satellite lines. In cases where reflections are particularly strong, three lines may be required when Cu radiation is employed; one for each of the Cu $K\alpha_1$ and α_2 lines, and one, commonly referred to as the α_3 line, summing up the contributions of the satellite lines (Edwards and Toman, 1970; Huang and Parrish, 1975). The satellite lines are due to a multiplicity of transitions while the atom is in an ionized state. The intensity of the K-α satellite group is approximately 1% and cannot be neglected in detailed work. The results of refining three single-line split-pearson VII profiles functions against the set of Si reflections at 56.15° are shown in Figure 10. The α_3 line, which appears at the lowest angle, has a refined intensity equal to 1.27% of the α_1 value, or 26 CPS. Unless otherwise stated in the ensuing discussions, only the α_1 and α_2 lines are assumed to be observable.

Compound profile-shape-functions. "Compound" PSFs are designed to reduce the parameter correlation involved in refining separate profiles for the overlapping α_1 and α_2 lines. A compound PSF is the sum of two functions: one for the α_1 and another for the α_2. The position and intensity of the α_2 line is based on the position and intensity of the α_1 line. The simplest compound PSF might consist of, for example, two Lorentzian lines with a common H_k. The most flexible and least-constrained compound PSFs might employ split-line functions with only the position and intensity of α_2 line constrained. Constraints on split-line based compound PSFs may include, for example, setting the values of the shape-related parameters for the low-angle sides of the α_1 and α_2 lines equal and the values of the shape-related parameters for the high-angle sides equal. In this case, an asymmetric PSF is generated with equivalently shaped α_1 and α_2 lines. The lines comprising such a PSF and the aggregate profile are illustrated in Figure 11.

A relative correction for the Lorentz and polarization effects is usually applied to the component lines in a compound PSF. The value Lp of the correction at the angle $2\theta_k$ is determined from:

$$\text{Lp}(2\theta_k) = \frac{1 + C \cos^2(2\theta_k)}{\cos(\theta_k) \sin^2(\theta_k)}, \tag{41}$$

where C is a monochromator correction factor. The value of C is 1.0 if no monochromator is present and 0.8005 for a typical graphite monochromator.

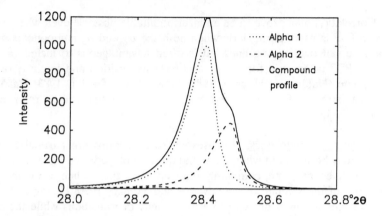

Figure 11. A compound profile (solid line) composed of three split-Pearson VII line functions (broken lines). The positions and intensities of the α_2 and α_3 lines are based on those of the α_1 line.

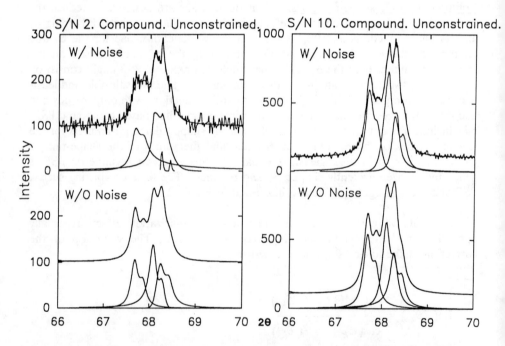

Figure 12 (left). The results of fitting compound profiles based on split-Pearson VII functions to the low S/N pattern. The shape of the compound profiles were allowed to vary independently.

Figure 13 (right). The results of fitting compound profiles based on split-Pearson VII functions to the high S/N pattern. The shape of the compound profiles were allowed to vary independently.

Taking the Lp factor to influence the peak intensities in the same proportion as the integrated intensities, the corrected peak intensity for the α_2 line is:

$$I'_{\alpha 2} = I_{\alpha 2}\, Lp(\,2\theta_{k\text{-}\alpha 1}\,) / Lp(\,2\theta_{k\text{-}\alpha 2}\,) \quad . \tag{42}$$

The α_2 line intensity used in the PSF, $I'_{\alpha 2}$, is thus calculated from $I_{\alpha 2}$, the intensity of the α_2 line based on the intensity of the α_1 line and the relative intensities of the respective wavelengths, and the relative contributions of the LP factors to the α_1 and α_2 lines. The α_3 line is corrected in a similar manner.

The advantage of using a compound PSF is most apparent when refining multiple overlapping lines. In most cases, the shape of the compound PSF will act to limit the number of ways it will "fit" under the group. Figures 12 and 13 illustrate the use of split-Pearson-VII-based compound profiles described at the beginning of this section. The fits to the high S/N pattern without noise were nearly perfect. The corresponding fits to the low S/N pattern show unrealistic line profiles and incorrect intensities. The fits to the low S/N pattern with noise show exaggerations of these distortions but the fit to the high S/N pattern with noise yielded acceptable results. Again, one is more likely to obtain more realistic results when the peaks are more intense. With this low S/N pattern, it is necessary to constrain the shapes of the profiles.

<u>Additional constraints on the profile shapes.</u> In heavily overlapped regions or in groups with a low S/N ratio, the optimization algorithm may be unable to produce a valid set of parameters even when a compound PSF is used. If one is willing to assume that the shapes of the diffraction lines are similar when they are in close proximity, then constraining the shape of all PSFs to be identical will aid in the refinement process. This assumption may not hold true when the lines are from different phases or there is pronounced anisotropic line broadening. In short, keeping the shape of all lines the same allows for the refinement of fewer variables thus lessening the degree of parameter correlation and increasing the likelihood of finding a set of physically realistic parameters.

Figures 14-17 correspond to figures 9-10 and 12-13, but the results shown in the former figures are a result of constraining all lines to have the same shape during refinement. In all cases, the line shapes are now closer to what are expected. For the single-line PSFs, the α_1:α_2 ratios are clearly incorrect for the low S/N pattern and are only approximate for the high S/N pattern. Note that the intensity for a line in Figure 14, a low S/N pattern with noise, has been refined to a near zero value thus effectively eliminating it from the pattern. The results with the compound profiles constrained to have the same shape are significantly better. The line position and intensities are nearly the same in the high S/N pattern refinements regardless of whether the noise is present or not. One should also note that the refined lines tend to be broadened as the intensity of the lines approaches the background level.

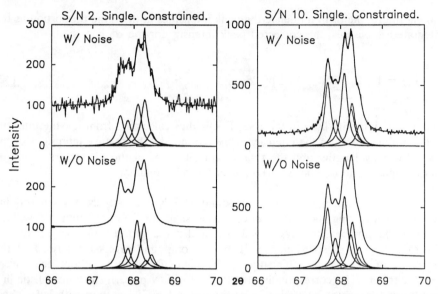

Figure 14 (left). The signal-to-noise ratio for the data in this figure is 2:1. A Lorentzian single-line profile-shape-function was refined for each of the α_1 and α_2 components in the three doublets. The shape of each profile was required to be the same during refinement.

Figure 15 (right). The signal-to-noise ratio for the data in this figure is 10:1. A Lorentzian single-line profile-shape-function was refined for each of the α_1 and α_2 components in the three doublets. The shape of each profile was required to be the same during refinement.

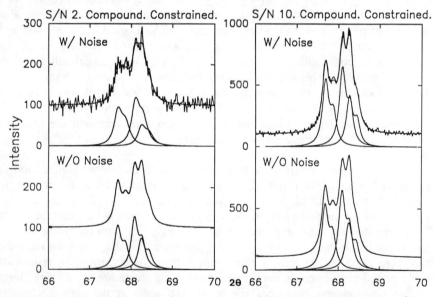

Figure 16 (left). The results of fitting compound profiles based on split-Pearson VII functions to the low S/N pattern. The shape of each profile was required to be the same during refinement.

Figure 17 (right). The results of fitting compound profiles based on split-Pearson VII functions to the high S/N pattern. The shape of each profile was required to be the same during refinement.

The quality of the results obtained from profile fitting is in part determined by the step width and step count time used in the pattern collection process. An analytical evaluation of profile separation and detectibility has been given, for example, by Naidu and Houska (1982) and Van de Velde and Platbrood (1985). In general, one can separate peaks if their centers are approximately greater than one half-width apart, but this will depend on the shape of the profile and the statistical noise present (Van de Velde and Platbrood, 1985). Naidu and Houska (1982) recommended that the step width used in data collection be approximately 1/7 the width of the sharpest line in a group. Step counting time has a most significant effect on the ability to determine accurate and precise profile parameters as has been illustrated in figures 9-17.

Alpha-2 stripping. The Rachinger technique (1948) of α_2 stripping may be employed in an attempt to reduce the interference from the α_2 lines. Figures 18-21 show the results of fitting the test patterns after they had been α_2 stripped. The advantage of using the stripped data is that only single-line profiles are required. However, new errors are introduced into the data since the stripping removes the α_2 line's intensity yet leaves the statistical fluctuations that were associated with the original intensities. This reduces the validity of the weighting scheme used in the determination of the residual error being refined. Examination of the figures show trends similar to those observed with the non-stripped data, i.e., an overall dependence of the accuracy on the S/N ratio and a broadening of the profiles in the low S/N data. It is interesting to note that although the data in Table 3 indicate the results from the stripped data show a larger deviation in peak and integrated areas from the calculated values than those obtained from the non-stripped data, the angular accuracy obtained from the stripped data on the whole is improved.

Digital filtering. Smoothing or, more precisely, digital filtering would seem to offer a means of rescuing buried lines in patterns either having an inherently low S/N ratio or obtained through rapid scanning. Digital filters typically employ a low order polynomial to model the characteristics of the pattern over a small angular range. In practice a zero, second or fourth order polynomial based filter is used over a series of 5 to 25 points in the pattern. In general, as the order of the filter increases, the ability of the filter to traverse sharper maxima without distorting the shape of the peak increases. As the filter width, i.e., the number of points used in the filter, increases, the smoothing effect increases. The optimum filter is based on the line width, sharper lines require a higher order filter and a smaller width. Conversely, broad lines are best approached with a lower order filter and larger width. Some distortion of the line shape usually can not be avoided when filters are applied.

The most frequently used digital filter is that described by Savitzky and Golay (1964). The interested reader may wish to read the comments on the characteristics and application of this algorithm given by Steiner et al. (1972), Betty and Horlick (1977), Madden (1978), and Bromba and Ziegler (1981). Bromba and Ziegler (1979) have presented a variation of the Savitzky and Golay filter that is

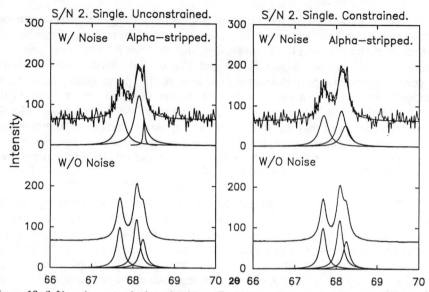

Figure 18 (left). An α_2 stripping algorithm (Rachinger, 1948) was used to remove the α_2 components from the low S/N patterns. A Lorentzian profile-shape-function was refined for each of the α_1 components. The shape of each profile varied independently during refinement.

Figure 19 (right). An α_2 stripping algorithm (Rachinger, 1948) was used to remove the α_2 components from the low S/N patterns. A Lorentzian profile-shape-function was refined for each of the α_1 components. The shape of each profile was required to be the same during refinement.

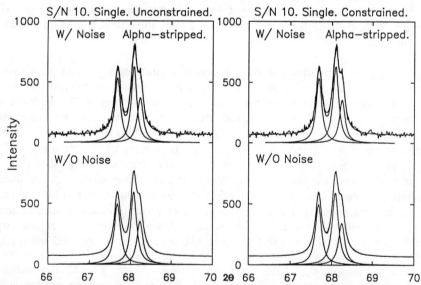

Figure 20 (left). An α_2 stripping algorithm (Rachinger, 1948) was used to remove the α_2 components from the high S/N patterns. A Lorentzian profile-shape-function was refined for each of the α_1 components. The shape of each profile varied independently during refinement.

Figure 21 (right). An α_2 stripping algorithm (Rachinger, 1948) was used to remove the α_2 components from the high S/N patterns. A Lorentzian profile-shape-function was refined for each of the α_1 components. The shape of each profile was required to be the same during refinement.

based on the use of integer arithmetic and thus performs a bit faster and does not propagate computer round-off errors. Effective use of digital filters through the control of filter parameters has been addressed by Enke and Nieman (1976).

Fourier smoothing techniques are employed to a lesser extent in pattern smoothing. The step-scanned data in our patterns can be thought of as a signal that is a function of time. This is conceptually easy to do if one remembers that the data are collected at a fixed scan rate, regardless of whether a synchronous or step-scan data collection algorithm is used. The Fourier techniques effectively transform the step-scanned data from the time domain to the frequency domain where the pattern is represented by a series of frequencies and associated amplitudes. The noise, assumed to be represented by the high frequency components in the spectrum, is reduced by either damping or setting the high frequency amplitude coefficients to zero. Some of the considerations in the use of Fourier filtering can be found in the articles by Hieftje (1972a,b), Aubanel and Oldham (1985), and Cameron and Moffatt (1984).

There are two caveats the reader should keep in mind. One should be aware that the basic assumption behind most filters is that the noise is normally distributed and its magnitude is constant throughout the data. This, of course, is not the case with diffraction data that are collected in the typical fixed-time count mode. The Poisson statistics governing diffraction data dictate that the number of counts at each step, N, determine the standard deviation of the measurement, $\sqrt{N}$. Application of these filters to data collected in fixed-count mode is statistically correct. Secondly, as stated earlier, a digital filter approximates the character of a pattern over a small angular range. Conversely, profile-fitting models the pattern over a significantly larger range and uses a PSF to do so. Smoothing data to be profile fit is redundant and may introduce significant error into the data.

The data shown in Figures 22-25 were filtered using the algorithm described by Bromba and Ziegler (1979). The data in these figures were filtered using either a second or fourth order filter with a width of 5 and 11 points. The results generally indicate that a higher order, small width filter distorts the data to a lesser degree.

<u>Accommodating an amorphous profile in the pattern</u>

Difficulties may arise when one or more diffraction lines are superimposed on an amorphous line. Perhaps the simplest way to account for the presence of an amorphous profile is to add a Gaussian PSF to the set of lines to be refined. This approach may not always work due to the asymmetry of the amorphous profile and/or because the amorphous phase may produce more than one diffuse profile. If one includes a broad Gaussian PSF to accommodate the diffuse profile, the initial angles and intensities of all input lines should be specified with reasonably good accuracy. Failure to provide estimates with sufficient accuracy may result in the optimization algorithm using the amorphous profile to fit a Bragg diffraction line. Conversely, the program may use a sharp diffraction line to account for the

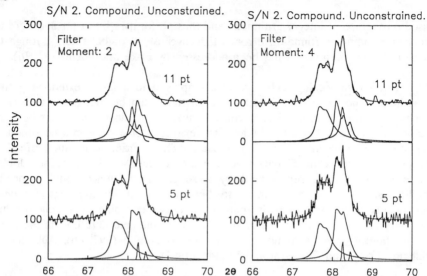

Figure 22 (left). The low S/N pattern is shown in this figure after being smoothed using a 2nd order digital filter. The widths of the filters employed in the upper and lower patterns was 11 and 5, respectively. The results of fitting compound profiles based on split-Pearson VII functions are shown. The shape of each profile varied independently during refinement.

Figure 23 (right). The low S/N pattern is shown in this figure after being smoothed using a 4th order digital filter. The widths of the filters employed in the upper and lower patterns was 11 and 5, respectively. The results of fitting compound profiles based on split-Pearson VII functions are shown. The shape of each profile varied independently during refinement.

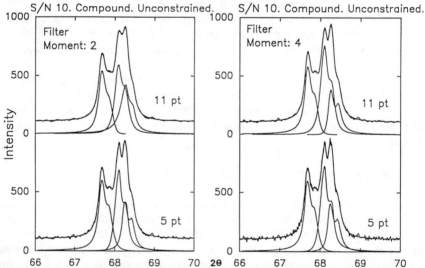

Figure 24 (left). The high S/N pattern is shown in this figure after being smoothed using a 2nd order digital filter. The widths of the filters employed in the upper and lower patterns was 11 and 5, respectively. The results of fitting compound profiles based on split-Pearson VII functions are shown. The shape of each profile varied independently during refinement.

Figure 25 (right). The high S/N pattern is shown in this figure after being smoothed using a 4th order digital filter. The widths of the filters employed in the upper and lower patterns was 11 and 5, respectively. The results of fitting compound profiles based on split-Pearson VII functions are shown. The shape of each profile varied independently during refinement.

amorphous profile. The amorphous profile's position, intensity, and H_k may be refined with the other pattern parameters. Ideally, one should base the amorphous profile on a known radial distribution function (Larson and von Dreele, 1987). This approach not only fixes the asymmetry but also accounts for multiple diffuse profiles.

Convolution-based profiles

The principles discussed here relate to the determination of the intrinsic diffraction profile produced by a powder diffractometer and its use in evaluating the specimen-related mechanisms causing line broadening. The basic principles have been discussed in the Introduction under 'Direct Convolution Products'. The details of the convolution technique used in the program SHADOW (Howard and Snyder, 1983; Howard, 1989) are described below as an example.

To determine the instrument profiles, patterns are recorded from a set of specimens having a minimum of defects, i.e., the crystallite size is sufficiently large to preclude line broadening yet small enough to give the best particle statistics, and there is no strain-related broadening present. A profile-shape-function is fit to the diffraction lines and the values of its adjustable parameters are evaluated as a function of angle. The following sections detail the use of these data in determining the instrument profile calibration curves.

W - the factor determining the number of lines in the instrument PSF. The wavelength spectrum W for an incident-beam monochromatized or synchrotron diffractometer may be described by a single-line function, perhaps a simple Lorentzian function. Two functions may be required to reproduce the instrument profile (W*G) by convolution; one for W and one for G. In practice, since the primary concern is the description of the instrument profile (W*G), it is advantageous to model the instrument profile with a single function:

$$P = (W*G) \quad , \tag{43}$$

where P is a function with enough flexibility to describe (W*G) over the entire angular range of the instrument.

The spectrum for most diffraction work, e.g., from a Cu X-ray target, consists primarily of two characteristic lines. If the wavelength distribution functions for each component are superimposed, then the following is true:

$$(W*G) = G * (W_1 + W_2) \quad . \tag{44}$$

In practice, the instrument profiles are modeled by the simple addition of three PSFs:

$$(W*G) = P_1 + P_2 + P_3 \quad . \tag{45}$$

In the program SHADOW, a compound PSF is used to model $(W*G)$. The PSF consists of one split-Pearson VII function for each of the wavelengths; the shape of the α_2 is constrained to follow that of the α_1. The relative intensities of the component lines were corrected for Lorentz and polarization effects.

Development of the instrument profile calibration curves. The values of the full-width at half-maximum, H_k, and exponent parameters, m, obtained by fitting the compound profile to the lines of defect-free standards are used to determine the coefficients of two polynomial "profile calibration curves". These calibration curves are then used to generate the values of the H_k and exponent parameters for the $(W*G)$ PSF at any angle.

The polynomial expression used in program SHADOW to describe the angular dependence of the H_k values was originally derived for a neutron powder diffractometer (Cagliotti et al., 1958):

$$H_k = U \tan^2\theta_k + V \tan\theta_k + W \quad . \tag{46}$$

Two sets of U, V, and W coefficients are determined; one for the low-angle side of the profiles and one set for the high-angle sides. This expression has been routinely used in X-ray powder diffraction as it adequately describes the broadening of the diffraction profiles.

A similar polynomial expression which experimentally describes the angular dependence of the exponents is used without theoretical justification:

$$m = a (2\theta_k)^2 + b (2\theta_k) + c \quad . \tag{47}$$

Again, two sets of a, b, and c coefficients must be determined. An example of these calibration curves obtained from fitting a Si specimen is shown in Figure 26 (Yau and Howard, 1989).

The method of profile generation. The component lines in the $(W*G)$ compound profile are generated by SHADOW in the following manner. The algorithm begins at the peak of a line and generates the profile towards the low-angle side until a minimum limit of 0.01 counts per second above background is reached. Then,

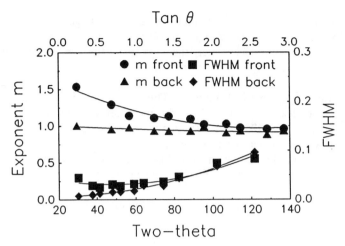

Figure 26. A plot of the exponent m as a function of 2θ and the full-width at half-maximum (FWHM) values as a function of $\tan\theta$ for the low- and high-angle sides of a split-Pearson VII profile. The coefficients of the second-order polynomial curves shown here can be used to generate the instrument profiles used in convolution based profile-shape-functions (Howard and Snyder, 1989).

starting at one point over from the high-angle side of the peak, the upper half of the profile is generated until the same minimum limit is reached. This method is used rather than generating the profiles out to a fixed number of H_ks from $2\theta_k$. The H_k limit may compromise both the execution speed and the precision of an optimization algorithm. Those lines that are wider than the n×H_k window are not utilizing all the available information. In contrast, those lines that are narrower than their window are not generating any additional information and cause excess computing time to be consumed. Further complications arise when the sample consists of phases with unequally broadened lines.

The α_1 line position is used to calculate the values of H_k and m for the component lines from the profile-calibration-curve coefficients. The position and intensity of the α_2 line are calculated based on those of the α_1 line. The peak intensity for the α_1 line is set to 100.0 and all component lines are generated until the tails fall below the limit of 0.01 counts. The profile is then normalized to reflect unit peak intensity.

<u>Generation of the specimen profile.</u> In program SHADOW, either a single Lorentzian or a Gaussian function may be used to model the contributions of S, or both may be used simultaneously. In all cases, profiles normalized for unit peak intensity are generated until the tails drop below the lower limit of 0.0002 for the Lorentzian function and 0.0005 for the Gaussian function. These limits are sufficient to generate more than 99% of the integrated line area.

The distance from the peak position to the limit of the tails will vary since the integral-breadth of S is related to the degree of specimen defects. S for an ideal specimen is described by a single point regardless of the specimen PSF. By contrast, the integral-breadth of S may be relatively large for a specimen with defects. The tails of a Gaussian function will fall to the limits at a relatively short distance from the center. However, the tails for a Lorentzian function with the same integral-breadth may extend to impractical limits.

Numerical convolution of (W*G) and S. The (W*G) and S profiles in SHADOW are generated at discreet values of 2θ; the incremental difference between these values is referred to as $\Delta 2\theta$. It is also convenient to refer to the value of (W*G) as (W*G)(i), where the index i runs over all points in the profile. The values for S are S(j) and, since the specimen profile is symmetric, j runs from -l to l where l is the number of points from the peak to the limits of the tails.

A point in the convoluted profile is obtained using the following numerical integration:

$$(W*G*S)(i) = \sum_{j=-l}^{+l} [\, (W*G)(i+j) \times S(j) \,] \quad . \tag{48}$$

Figures 27-29 demonstrate the relationship between the crystallite size and the integral breadth of the specimen profile. The specimen profiles in the figures are representative of the specimen contribution to broadening resulting from crystallites having a size of 70 nm, 143 nm and 286 nm, respectively. Convoluting these functions with the instrument profile, shown in the center of the graphs, yields the profiles shown on the right. The integrated line intensity remains the same while the peak broadens and the peak intensity decreases. The (W*G*S) profile on the right can be regarded as a "smeared" representation of the (W*G) profile. The degree of smearing depends on the integral-breadth of S.

If the specimen profile is based on either the Lorentzian or Gaussian functions individually, the above integration will generate the required convolution product. If both profile functions are required in the convolution product, the Lorentzian function is first convoluted with the (W*G) function, then the Gaussian function is convoluted with this product to yield the final profile.

Maintaining the definition of the convolute profile. It would be easiest to generate the values of (W*G) and S with the same 2θ increments as that of the pattern, however, difficulties are encountered when the H_k of S is on the order of $\Delta 2\theta$ or less. (Conversion between the integral-breadth of a normalized profile and the H_k of that profile involves only a constant. The following considerations are best illustrated when considering the H_k of the S function.) In practice, a $\Delta 2\theta$ of 0.01° is only sufficient to describe the specimen profile for those cases where broadening is clearly visible. Convolution cannot be performed when $\Delta 2\theta$ is increased to a value of 0.05° or more. Hence, the step-width of (W*G) and S cannot be constrained to be that of the pattern.

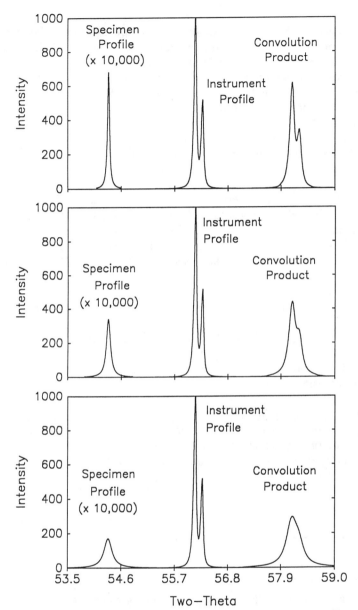

Figure 27 (top). The line shape (right) resulting from the convolution of a profile representing instrument (center) and a profile representing the specimen contributions to the line shape (left). The integrated area of the convolution product remains constant whereas the peak intensity and shape will vary with the specimen's crystallite size and micro-strain. The specimen profile represents broadening corresponding to a crystallite size of 285.8-nm.

Figure 28 (middle). The line shape (right) resulting from the convolution of a profile representing instrument (center) and a profile representing the specimen contributions to the line shape (left). The specimen profile represents broadening corresponding to a crystallite size of 142.9-nm.

Figure 29 (bottom). The line shape (right) resulting from the convolution of a profile representing instrument (center) and a profile representing the specimen contributions to the line shape (left). The specimen profile represents broadening corresponding to a crystallite size of 70-nm.

To enhance the precision of the convolution algorithm, the $\Delta 2\theta$ steps at which the S and (W*G) profiles are generated are a fraction of the step-width of the pattern. This allows for using larger angular increments for pattern collection and decreases the lower limit for the specimen profile H_k. In program SHADOW, a maximum of 3100 points is allocated for the (W*G) profile. Two further constraints are placed on the system:

- the 3100 points must represent a range of at least 10° 2θ, and
- the $\Delta 2\theta$ of the pattern must be an even multiple of the $\Delta 2\theta$ steps used to generate the (W*G) and S profiles.

For example, if the step-width of the pattern is 0.06° then the angular increment for (W*G) and S would be (0.06°/18) = 0.0033°. In this example there are 18 points in the (W*G) and S profiles between each point in the observed pattern. This gives an angular range of 3100 × 0.0033° = 10.33° 2θ. A pattern step scanned at 0.01° $\Delta 2\theta$ would also have the values of (W*G) and S calculated at (0.01/3) = 0.0033° $\Delta 2\theta$.

The parameters refined when using a convolute PSF include:

- the angle $2\theta_k$ for the α_1 line in the (W*G) profile (all other parameter values are determined using the profile calibration curves),
- the integral-breadth of the specimen profile, and
- the intensity of the convolute profile.

<u>Specimen broadening as a function of angle: particle-size and strain constraints.</u> The angular dependence of the specimen profile's integral-breadth β can be constrained according to relations derived for crystallite size and strain effects. The Scherrer (1918) equation is used to model the broadening due to small crystallite size:

$$\beta_\tau = \frac{\text{RTOD } \lambda}{\tau \cos(\theta_k)}, \tag{49}$$

and the strain broadening is constrained according to (Wilson, 1940):

$$\beta_\epsilon = 4 \, \epsilon \, \text{RTOD} \, \tan(\theta_k), \tag{50}$$

where: τ = the crystallite thickness, or X-ray crystallite size,
        ϵ = the crystallite strain, and
        RTOD = the constant converting units of radians to degrees.

When both strain and crystallite-size broadening are present and both contributing specimen profiles are Lorentzian, then the integral-breadth of the S profile is a linear addition of the two components:

$$\beta_c = \beta_\tau + \beta_\epsilon \ . \tag{51}$$

When both contribution are modeled by a Gaussian function, the integral-breadth of the S profile can be calculated from:

$$\beta_G = \sqrt{\{(\beta_\tau)^2 + (\beta_\epsilon)^2\}} \ . \tag{52}$$

If, for example, the crystallite-size broadening contributions are modeled with a Lorentzian function having integral-breadth β_c, and the strain contributions by a Gaussian function having integral breadth β_g, the integral-breadth of the specimen profile, which is the convolution product of the two functions is:

$$\beta = \beta_g \frac{\exp[-(\beta_c/\beta_g)^2 \pi]}{1 - \mathrm{erf}[(1/\pi)^{1/2}(\beta_c/\beta_g)]} \ . \tag{53}$$

A simple approximation relating the respective integral-breadths is:

$$\frac{\beta_c}{\beta} = 1 - \left\{\frac{\beta_g}{\beta}\right\}^2 , \tag{54}$$

The resultant convolution product, normalized for unit integrated area, is:

$$I(2\theta) = \beta_g^{-1} \mathrm{Re}\left\{\Omega\left[\frac{\sqrt{\pi}}{\beta}|2\theta i - 2\theta k| + i\frac{\beta_c^2}{\beta_g^2 \pi}\right]\right\} , \tag{55}$$

which is, of course, a Voigt function. Since the shape of the convolute profile is determined by β_c and β_g, the values of β_c and β_g can, in principle, be determined from profile analysis of a single line. However, the angular dependence of the size (β_c) and strain (β_g) contributions should be used to minimize misinterpretation of any analyses; typically two orders of a reflection should be used. A more detailed discussion regarding the Voigt function is given above.

One has two options after choosing a convolution-based PSF. The first allows for constraining the values of each specimen function's integral-breadth by the relation expected for broadening as a result of small crystallite size, the second for constraining the specimen function's integral-breadth values by the relation expected for broadening as a result of strain. The options are not mutually exclusive and, hence, three relations are manifest:

(1) No crystallite size or strain constraints are used. If no constraints are used then the integral-breadth for each specimen function is refined individually. In this case, the integral-breadths values of either the specimen profile or the component integral-breadths of the Voigt profile may be plotted to reflect the anticipated line broadening mechanisms. For example, size-related broadening may be modeled by plotting β versus $\lambda/\cos\theta_k$, and strain as β versus $4\epsilon\tan\theta_k$. The slopes of the resulting lines would yield $1/\tau$ and ϵ, respectively. The most insightful method would be to graph the refined data in the form of a Williamson-Hall (1953) style plot. The significance of this plot can be demonstrated from the following relations:

$$\beta = \beta_c + \beta_g = \text{RTOD} [\lambda/\cos\theta_k + 4 \epsilon \tan\theta_k] , \qquad (56)$$

$$\beta \cos\theta_k / \text{RTOD} = \lambda/\tau + 4\epsilon\sin\theta_k . \qquad (57)$$

Plotting $\beta\cos\theta_k/\text{RTOD}$ as a function of $4\epsilon\sin\theta_k$ would result in a line with a slope of ϵ and a y-intercept of λ/τ. Examining the graph would make the relative contribution immediately recognizable; lines with little or no slope would indicate minor contributions from strain-related broadening, whereas an intercept value approaching zero indicates large crystallite sizes and little contributions to line broadening.

(2) Either crystallite size or strain constraints are used. The appropriate relation will determine the angular dependence of specimen PSFs integral-breadth. Either the particle size, τ, or the particle strain, ϵ, will be refined.

(3) Both particle size and strain constraints are used. The integral-breadth of the specimen function will be determined by the appropriate additivity laws for the specimen profiles involved. Both the crystallite size and strain parameters will be refined.

Miscellaneous considerations

Background. Accommodation of the background is necessary and should not be neglected. How background is handled is dependent upon the angular range of the

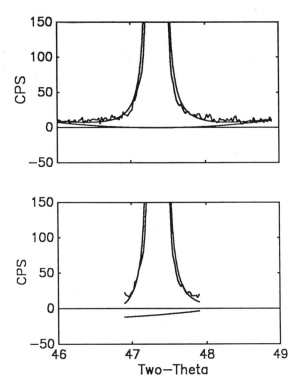

Figure 30. This plot illustrates possible problems encountered when refining background simultaneously with the profile parameters. In the top half of the figure, the use of a 2nd order polynomial to accommodate background resulted in a background having an unrealistic curvature. The bottom half shows an unrealistically low background resulting from not including a large enough sampling of the background in the region of the tails.

region under investigation. In many cases, a low-order polynomial can be refined along with the profiles. Consider, for example, a first-order polynomial or line. When the angular range is short, the slope and intercept of the background line may fluctuate with relative ease. If the intensity of the background erroneously increases, the intensities of small peaks may undergo severe reduction. Increasing the angular range of refinement will increase the amount of pattern information available to the optimization algorithm thereby reducing the uncertainty in the refined background. The degree of curvature that the calculated background can have increases with an increase in the order of the polynomial. In wide refinement regions, a high-order polynomial has the ability to conform to complex background shapes without becoming unstable. However, a high-order polynomial used in a short angular range might become flexible enough to describe the peaks as well as the background. Examples of these problems are shown in Figure 30.

<u>Adding lines after the initial refinement.</u> The ability to add lines after the refinement of a set of line parameters is advantageous in the case of overlapping strong and weak lines. The strong line should be refined in the initial round. The addition of the second line in the second round will force the optimization algorithm

to account for the intensity remaining after refinement of the first line with the second line. Initially adding both lines may result in the algorithm using both lines to describe the strong line. This problem is less likely to manifest itself when a compound PSF is used since the shape of a compound PSF reduces the likelihood of both lines fitting under the strong line.

<u>Discrepancies indicated by the difference patterns.</u> Plotting the differences between an observed and refined set of data may show a number of discrepancies. These discrepancies may arise from a number of sources:

- The optimization algorithm may not be able to find the best value for a parameter when the change in the value of that parameter has little effect on the overall residual error. This is particularly true when refining a segment containing a large number of parameters and/or points. The result may be a relatively poor fit of one or more of the profiles.

- A distorted line profile may be generated when refining overlapping lines. This line profile, although not physically acceptable, is generated because its shape will lower the overall residual error. (This is, of course, the sole goal of the optimization algorithm!) When split profiles are involved, this distortion may manifest itself by a profile having an extremely wide or narrow full-width at half-maximum on one side. Distorted symmetric lines may have a very wide full-width at half-maximum and a relatively low intensity thus spreading the profile under several lines.

- A major source of differences may arise from the counting statistics. The standard deviation, σ, of the number of counts received in a time interval, N, is equal to $\sqrt{N}$. The relative error in the intensity measurement is then determined from:

$$R_{rel} = \frac{\sigma}{N}. \tag{58}$$

Although the absolute errors for the points of highest intensity are larger than for those points with lower intensity, the relative errors are lower. The regions in the segment where the α_2 component have been subtracted will have the original absolute deviations. The related intensity fluctuations will be noticeable when compared to the neighboring background intensities. The magnitude of these fluctuations will depend on the original intensities of the points; low intensity lines will show fluctuations approaching that of the background.

- Any misfit between the calculated PSF and the observed diffraction profile, particularly with high intensity lines, will almost always be evident.

- Anisotropic line broadening may be present. This may be particularly difficult to handle in groups of overlapping lines where parameter correlation forces a reduction in variable parameters.

- More than one phase with differing degrees of line broadening may be present.

- One or more lines may be unaccounted for in the peak group.

SUMMARY

Line positions, peak intensities, and widths taken from the traditional strip-chart recorder tracings are sufficient for a number of analyses. However, many applications require accurate and precise values for these parameters in order to obtain meaningful results. When these parameters are required for lines found in overlapping groups, or an analysis based on the shape of the diffraction line is to be performed, it becomes highly advantageous to collect and analyze the diffraction data with the aid of a computer. The numerical modeling of a diffraction line requires the use of a mathematical function, a profile-shape-function (PSF), that will accurately fit the measured line. The shape, flexibility, and parametric representation of the PSF are of primary concern in profile fitting. The diffraction line shape is primarily dependent on the configuration of the diffractometer and the source of radiation, e.g., a neutron source, a synchrotron, or a traditional laboratory X-ray tube. A flexible PSF is necessary in order to accommodate the change of line shape with a change of specimen condition. The parameters associated with a PSF should represent physical characteristics of the specimen and/or the instrument.

The investigation of a variety of PSFs employed on data collected with different radiation sources and in different instrumental geometries has been reported in the literature. In general, the most widely used PSFs are the pseudo-Voigt or Voigt function and the Pearson VII function since their shape can vary between that of a Lorentzian and a Gaussian function. The Lorentzian and Gaussian functions are of fundamental importance since they reflect the intrinsic contributions to the diffraction line shape from the radiation wavelength distribution, the instrumental geometry, and the specimen. In many cases, the angular dependence of the line broadening can be used to assess the contributions from the various sources.

Ideally, one must be able to separate the instrumental and radiation source contributions from those of the specimen. Indeed, the traditional Warren-

Averbach (1950) type of analysis accomplishes this separation by mathematically deconvoluting a line measured from a standard specimen from the corresponding line measured from a "defective" specimen. Unless the instrumental contributions are very small, e.g., the data are measured using finely tuned synchrotron X-rays, or the line broadening is significantly large in relation to the instrumental contributions, an accurate and precise evaluation of the specimen condition will not be possible. In the latter case, only when such separation is performed will the true specimen character be revealed. Standards are currently being commercially prepared for the purpose of measuring the instrument contributions to the line shape. Data derived from the measurement of these standards can be used with newly emerging techniques which convolute a specimen related function with synthesized instrumental line profiles. These techniques will shed new light on the defect nature of many materials.

Synchrotron radiation is providing a source of tunable X-rays that can be finely collimated while still maintaining significant intensities. The use of profile fitting with synchrotron powder diffraction data can provide information about changes in crystal systems from minute line splittings, the presence of trace phases, and the physical characteristics of the specimen since there is almost no instrumental contributions to the measured line widths. Whole-pattern fitting (WPF), where the entire pattern is fit but there is no reference to a structural model as with the Rietveld (1969) method, will play a significant role in powder diffraction. The Rietveld method has provided a means of performing crystal structure refinement using powder data but cannot be used in the *determination* of a crystal structure. WPF has the potential for structure determination in small-cell problems. The high resolution obtainable from these instruments, coupled with the integrated intensity determination via profile fitting, should provide intensities that can be employed in the solution of crystal structures using direct methods. An *a priori* knowledge of the line shape, in addition to intensity and lattice-parameter constraints will be necessary in order to circumvent the problem of correlations between parameters undergoing refinement.

Profile-fitting techniques give the materials engineer or mineralogist a powerful tool for routine as well as state-of-the-art materials characterization. Software packages are available either through instrument manufacturers or by outside sources which are friendly enough to be learned and employed almost immediately and robust and flexible enough to tackle difficult problems. While we continue to push the advances of science, one should keep in mind that there will probably always be a bit of art in the application of these techniques.

REFERENCES

Alexander, L.E. (1954) The synthesis of X-ray spectrometer line profiles with application to crystallite size measurements. J. Appl. Phys. 25, 155-161.

Ahtee, M., Unonius, L., Nurmela, M. and Suortti, P. (1984) A Voigtian as profile shape function in Rietveld refinement. J. Appl. Cryst. 17, 352-357.

Armstrong, B.H. (1967) Spectrum line profiles: The Voigt function. J. Quant. Spectrosc. Radiat. Transfer 7, 61-88.

Asthana, B.P. and Kiefer, W. (1982) Deconvolution of the Lorentzian linewidth and determination of fraction Lorentzian character from the observed profile of a Raman line by a comparison technique. Appl. Spectrosc. 36, 250-57.

Benedetti, A., Fagherazzi, A., Enzo, S., and Battagliarin, M. (1988) A profile-fitting procedure for analysis of broadened X-ray diffraction peaks. II. Application and discussion of the methodology. J. Appl. Cryst. 21, 543-549.

Betty, K.R. and Horlick, G. (1977) Frequency response plots for Savitzky-Golay filter functions. Anal. Chem. 49, 351-352.

Beu, K.E. (1967) The Precise and Accurate Determination of Lattice Parameters. Handbook of X-rays, ed. Emmett F. Kaelble. McGraw-Hill Book Co.

Bish, D.L. and Howard, S.A. (1988) Quantitative phase analysis using the Rietveld method. J. Appl. Cryst. 21, 86-91.

Bromba, M.U.A. and Ziegler, H. (1979) Efficient computation of polynomial smoothing digital filter. Anal. Chem. 51, 1760-1762.

Bromba, M.U.A. and Ziegler, H. (1981) Application hints for Savitzky-Golay digital smoothing filters. Anal. Chem. 50, 1583-1586.

Brown, A., and Edmonds, J.W. (1980) The fitting of powder diffraction profiles to an analytical expression and the influence of line broadening factors. Adv. X-ray Anal. 23, 361.

Caglioti, G., Paoletti, A., and Ricci, F.P. (1958) Choice of collimators for a crystal spectrometer for neutron diffraction. Nucl. Inst. and Meth. 3, 223-226.

Chipera, S.J. and Bish, D.L. (1988) Pitfalls in profile refinement of clay mineral diffraction patterns. Presented as a poster session, 25th Annual Mtg. Clay Miner. Soc., Grand Rapids, MI.

Copland, L.E. and Bragg, R.H. (1958) Quantitative X-ray diffraction analysis. Anal. Chem. 30, 196-201.

Delhez, R., de Keijser, Th.H. and Mittemeijer, E.J. (1982) Determination of crystallite size and lattice distortions through X-ray diffraction line profile analysis (Recipes, Methods, Comments). Fresenins Zeit. Analyt. Chemie 312, 1-16.

de Keijser, Th.H., Langford, J.I. and Mittemeijer, E.J. (1982) Use of the Voigt function in a single-line method for the analysis of X-ray diffraction line broadening. J. Appl. Cryst. 15, 308-14.

de Vreede, J.P.M., Mehrotra, S.C., Tal, A. and Dijkerman, H.A. (1982) Frequency modulation detection and its application to absorption line shape parameters. Appl. Spectrosc. 36, 227-235.

Edwards, H.J. and Toman, K. (1970) Correction for the satellite group in the variance method of profile analysis. J. Appl. Cryst. 3, 157-164.

Enzo, S., Fagherazzi, G., Benedetti, A. and Polizzi, S. (1988) A profile-fitting procedure for analysis of broadened X-ray diffraction peaks. I. Methodology. J. Appl. Cryst. 21, 536-542.

Fiala, J. (1976) Optimization of powder diffraction identification. J. Appl. Cryst. 9, 429-432.

Frevel, L.K. (1965) Computational aids for identifying crystalline phases by powder diffraction. Anal. Chem. 37, 471-482.

Frevel, L.K. and Adams, C.E. (1968) Quantitative comparison of powder diffraction patterns. Anal. Chem. 40, 1739-1741.

Gonzalez, C.R. (1987) General theory of the effect of granularity in powder X-ray diffraction. Acta Cryst. A43, 769-774.

Hastings, J.B., Thomlinson, W. and Cox, D.E. (1984) Synchrotron X-ray powder diffraction. J. Appl. Cryst. 17, 85-95.

Hermann, H. and Ermrich, M. (1987) Microabsorption of X-ray intensity in randomly packed powder specimens. Acta Cryst. A43, 401-405.

Howard, S.A. (1989) SHADOW: A system for X-ray powder diffraction pattern analysis. Adv. X-ray Anal. 37, in press.

Howard, S.A. and Snyder, R.L. (1983) An evaluation of some profile models and the optimization procedures used in profile fitting. Adv. in X-ray Anal. 26, 73-81.

Howard, S.A. and Snyder, R.L. (1985) Simultaneous crystallite size, strain and structure analysis from X-ray powder diffraction pattern. Materials Science Research; Symposium on Advances in Materials Research 19, 57-71.

Howard, S.A., Yau, J.K. and Anderson, H.U. (1989) Structural characteristics of $Sr_{1-x}La_xTiO_{3\pm\delta}$ as a function of oxygen partial pressure at 1400°C. J. Appl. Phys. 65, 1492-1498.

Howard, S.A. and Snyder, R.L. (1989) The use of direct convolution products in profile and pattern fitting algorithms I: Development of the algorithms. J. Appl. Cryst. 22, 238-243.

Hubbard, C.R. (1983) Certification of Si powder diffraction standard reference material 640a. J. Appl. Cryst. 16, 285-288.

Huang, T.C. and Parrish, W. (1975) Accurate and rapid reduction of experimental X-ray data. Appl. Phys. Letters 27, 123-4.

Jansen, E., Schafer, W. and Will, G. (1988) Profile fitting and the two-stage method in neutron powder diffractometry for structure and texture analysis. J. Appl. Cryst. 21, 228-239.

Jenkins, R., (1984) Effects of diffractometer alignment and aberrations on peak positions and intensities. Adv. X-ray Anal. 26, 25-33.

Johnson, G.G., Jr. and Vand, V. (1967) A computerized powder diffraction identification system, Ind. Eng. Chem. 59, 19-31

Jones, F.W. (1938) The measurement of particle size by the X-ray method. Proc. Roy. Soc. A 166, 16-43.

Khattak, C.P. and Cox, D.E. (1977) Profile analysis of X-ray powder diffractometer data: Structural refinement of $La_{0.75}Sr_{0.25}CrO_3$. J. Appl. Cryst. 10, 405-11.

Kimmel, G. and Schreiner, W.N. (1988) Accuracy in temperature factor determination in powder diffractometry. Aust. J. Phys. 41, 719-728.

Kohlbeck, F. and Horl, E. M. (1976) Indexing program for powder patterns especially suitable for triclinic, monoclinic and orthorhombic lattice. J. Appl. Cryst. 9, 28-33.

Klug, H.P. and Alexander, L.E. (1974) X-ray Diffraction Procedures, 2nd ed., John Wiley and Sons, New York.

Larson, A.C. and von Dreele, R.B. (1987) GSAS - Generalized Crystal Structure Analysis System. Los Alamos National Lab. Rpt. (LA-UR-86-748).

Langford, J.I. (1978) A rapid method for analyzing the breadths of diffraction and spectral lines using the Voigt function. J. Appl. Cryst. 11, 10-14.

Lehmann, M.S., Christensen, A. Norlund, Fjellvag, H., Feidenhans'l, R. and Nielsen, M. (1987) Structure determination by use of pattern decomposition and the Rietveld method on synchrotron X-ray and neutron powder data; the structures of $Al_2Y_4O_9$ and I_2O_4. J. Appl. Cryst. 20, 123-129.

Madden, H.H. (1978) Comments on the Savitzky-Golay convolution method for least-squares fit smoothing and differentiation of digital data. Anal. Chem. 50, 1383-1386.

Marquardt, D.W. (1963) An algorithm for least-squares estimation of nonlinear parameters. J. Soc. Indust. Appl. Math. 11, 431-441.

Malmros, G. and Thomas, J.O. (1977) Least-squares structure refinement based on profile analysis of powder film intensity data. J. Appl. Cryst. 10, 7-11.

Nandi, R.K. and Sen Gupta, S.P. (1978) The analysis of X-ray diffraction profiles from imperfect solids by an application of convolution relations. J. Appl. Cryst. 11, 6-9.

Naidu, S.V.N. and Houska, C.R. (1982) Profile separation in complex powder patterns. J. Appl. Cryst. 15, 190-198.

Nelder, J.A. and Mead, R. (1965) A simplex method for function minimization, Comput. J. 7, 308-313.

Parrish, W. and Huang, T.C., and Ayers, G.L. (1976) Profile fitting: A powerful method of computer X-ray and instrumentation analysis. IBM Research Rpt. RJ1761.

Parrish, W. and Huang, T.C. (1980) Accuracy of the profile fitting method for X-ray polycrystalline diffraction. In: Accuracy in Powder Diffraction, Nat'l Bur. Stds Special Pub. 567, 95-110.

Pawley, G.S. (1981) Unit-cell refinement from powder diffraction scans. J. Appl. Cryst. 14, 357-361.

Rachinger, W.A. (1948) A correction for the $\alpha_{1,2}$ doublet in the measurement of widths of X-ray diffraction lines. J. Sci. Instrum. 25, 254-5.

Rietveld, H.M. (1969) A profile refinement method for nuclear and magnetic structures. J. Appl. Cryst. 2, 65-71.

Rotella, F.J., Jorgenson, J.D., Biefeld, R.M. and Morosin, B., (1982) Location of deuterium sites in the defect pyrochlore $DTaWO_6$ from neutron powder diffraction data. Acta Cryst. B38, 1697-1703.

Sabine, T.M. (1977) Powder neutron diffraction-refinement of the total pattern. J. Appl. Cryst. 10, 277-80.

Sakata, M. and Cooper, M.J. (1979) An analysis of the Rietveld profile refinement method. J. Appl. Cryst. 12, 554-563.

Savitzky, A. and Golay, J.E. (1964) Smoothing and differentiation of data by simplified least squares procedures. Anal. Chem. 36, 1627-1638.

Scherrer, P. (1918) Gött. Nachr 2, 98.

Schoening, F.R.L. (1965) Strain and particle size values from X-ray line breadths. Acta. Cryst. 18, 975-6.

Schreiner, W.N., Surdukowski, C. and Jenkins, R. (1982a) A probability based scoring technique for phase identification in X-ray powder diffraction. J. Appl. Cryst. 15, 524-530.

Schreiner, W.N., Surdukowski, C. and Jenkins, R. (1982b) An approach to the isotypical and solid solution problems in multiphase X-ray analysis. J. Appl. Cryst. 15, 605-610.

Snyder, R.L. (1983) Accuracy in angle and intensity measurements in X-ray powder diffraction. Adv. X-ray Anal. 26, 1-10.

Snyder, R.L., Johnson, Q.C., Kahara, E., Smith, G.S. and Nichols, M.C. (1978) An analysis of the powder diffraction file. Lawrence Livermore Laboratory (UCRL-52505), 61 pp.

Steiner, J., Termonia, Y. and Deltour, J. (1972) Comments on smoothing and differentiation of data by simplified least-square procedure. Anal. Chem. 44, 1906-1909.

Sundius, T. (1973) Computer fitting of Voigt profiles to Raman lines. J. Raman Spectroscopy 1, 471-88.

Suortti, P., Ahtee, M. and Unonius, L. (1979) Voigt function fit of X-ray and neutron powder diffraction profiles. J. Appl. Cryst. 12, 365-69.

Taupin, D. (1973) Automatic peak determination in X-ray powder patterns. J. Appl. Cryst. 6, 266-73.

Thompson, P., Cox, D.E. and Hastings, J.B. (1987) Rietveld refinement of Debye-Sherrer synchrotron X-ray data from Al_2O_3. J. Appl. Cryst. 20, 79-83.

Toraya, H. (1986) Whole-powder-pattern fitting without reference to a structural model: Application to X-ray powder diffractometer data. J. Appl. Cryst. 19, 440-447.

Van de Velde, H. and Platbrood, G. (1985) Detectability of two overlapping Pearson VII peaks in simulated X-ray diffraction patterns. J. Appl. Cryst. 18, 114-119.

Visser, J.W. (1969) A fully automatic program for finding the unit cell from powder data. J. Appl. Cryst. 2, 89-95.

Warren, B.E., and Averbach, B.L. (1950) The effect of cold-work distortion on X-ray patterns. J. Appl. Phys. 21, 595.

Will, G., Bellotto, M., Parrish, W. and Hart, M. (1988) Crystal structures of quartz and magnesium germanate by profile analysis of synchrotron-radiation high resolution powder data. J. Appl. Cryst. 21, 182-191.

Will, G., Masciocchi, N., Parrish, W. and Hart, M. (1987) Refinement of simple crystal structures from synchrotron radiation powder diffraction data. J. Appl. Cryst. 20, 394-401.

Will, G., Parrish, W. and Huang, T.C. (1983) Crystal-structure refinement by profile fitting and least-squares analysis of powder diffractometer data. J. Appl. Cryst. 16, 611-622.

Williamson, G.K. and Hall, W.H. (1953) X-ray line broadening from filed aluminium and wolfram. Acta Metall. 1, 22-31.

Wilson, A.J.C. (1949) X-ray Optics. Methuen, London. p. 5.

Yau, J.K. and Howard, S.A. (1989) The use of direct convolution products in profile and pattern fitting algorithms II: Application of the algorithms. J. Appl. Cryst. 22, 244-251.

Young, R.A., Mackie, P.E. and Von Dreele, R.B. (1977) Application of the pattern-fitting structure refinement to X-ray powder diffractometer patterns. J. Appl. Cryst. 10, 262-69.

Young, R.A. and Wiles, D.B. (1982) Profile shape functions in Rietveld refinements. J. Appl. Cryst. 15, 430-8.

9. RIETVELD REFINEMENT OF CRYSTAL STRUCTURES USING POWDER X-RAY DIFFRACTION DATA

J.E. Post & D.L. Bish

INTRODUCTION

Most of our detailed understanding of the structures of crystalline materials has resulted from single-crystal X-ray and neutron diffraction experiments. Unfortunately, for many substances it has not been possible to grow, or in the case of minerals, find crystals that are suitable for single-crystal diffraction studies. It is not simply coincidence that many important minerals for which structures have not been refined typically are finely crystalline and/or poorly ordered, for example clay minerals, manganese and iron oxides, zeolites, etc. In many of these cases where good single-crystals cannot be found, powder X-ray or neutron diffraction experiments provide an alternative to single-crystal methods for structural studies. In this chapter we will restrict our discussion to using the Rietveld method for the derivation of crystal structure information from powder X-ray diffraction data (from conventional X-ray sources), with emphasis on applications to geological materials. Detailed discussions of neutron and synchrotron X-ray powder diffraction methods and applications are the subjects of separate chapters in this volume.

Powder X-ray diffraction (XRD) has been an important standard tool for many decades for identification and characterization of minerals and other crystalline materials. Until relatively recently, however, conventional wisdom held that powder XRD data were unsuitable for serious crystal structure studies, primarily because of problems of peak overlap and the difficulty of measuring accurate Bragg intensities. In the past decade, however, two developments have kindled an interest in structural analysis using powder XRD data. Firstly, the introduction of computer-automated diffractometers has made it possible to collect routinely digitized data using a step-scan procedure. Secondly, the application of the Rietveld (Rietveld, 1967, 1969) refinement method to powder XRD data, which at least in part circumvents the peak overlap problem, has yielded successful refinements for an increasing number of crystal structures.

Two methods generally are used for extracting structural information from powder diffraction data. One approach is to measure the integrated intensities of individual Bragg reflections, convert them to structure factors, and use them as you would single-crystal data to solve or refine structures. Will et al. (1983) summarized this so-called profile-fitting procedure and reported refinements for Si, quartz and corundum. This technique has been used to refine a variety of structures, including: boehmite (Christoph et al., 1979), germanite (Tettenhorst and Corbato, 1984), the zeolite theta-1 (Highcock et al., 1985), $ZrNaH(PO_4)_2$ (Rudolf and Clearfield, 1985), huntite (Dollase and Reeder, 1986), and Rb-pollucite (Torres-Martinez and West, 1986). The profile-fitting procedure works well for relatively simple, high-symmetry structures that yield diffraction patterns with minimal peak overlap. For complex structures with low symmetry, however, patterns might consist of hundreds of Bragg reflections with such severe peak overlap that it would be difficult if not impossible to decompose the pattern into its constituent Bragg reflections (see for example Fig. 1). The profile-fitting method is perhaps best suited for use with synchrotron powder XRD data

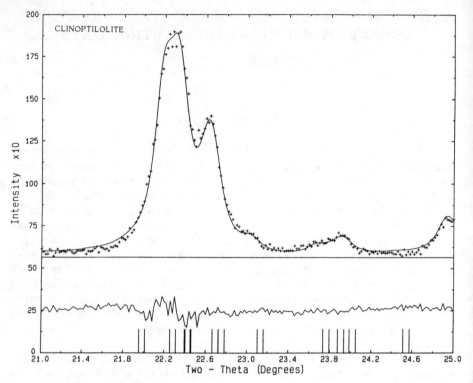

Figure 1. Portion of observed (crosses) and calculated powder X-ray diffraction pattern for clinoptilolite. The background is indicated by the horizontal line, and the vertical lines mark the positions of the Bragg reflections ($K\alpha_1$ and $K\alpha_2$). The plot near the bottom of the figure is the difference between the observed and calculated patterns.

where the increased resolution, compared to conventional diffractometer data, improves the chances of successfully decomposing complex patterns (e.g. Will et al., 1987, 1988). A more detailed discussion of profile fitting procedures is given in Chapter 8.

The Rietveld method, in contrast to the profile-fitting procedure, does not use integrated peak intensities but employs the entire powder diffraction pattern, thereby overcoming the problem of peak overlap and allowing the maximum amount of information to be extracted from the pattern (Rietveld, 1967, 1969). In the Rietveld method, each data point (2θ step) is an observation, and during the refinement procedure, structural parameters, background coefficients, and profile parameters are varied in a least-squares procedure until the calculated powder profile, based on the structure model, best matches the observed pattern. This method was first applied to powder neutron diffraction data but was later adapted for use with X-ray data (Malmros and Thomas, 1977; Young et al., 1977). Powder XRD data have been used only relatively recently in Rietveld refinements because whereas peak shapes arising from constant-wavelength powder neutron diffraction experiments are close to Gaussian, those from powder XRD are considerably more complex and therefore more difficult to model. This disadvantage is, however, countered by the easy accessibility and smaller sample requirements for XRD as opposed to neutron diffraction experiments. These factors along with improvements in functions to describe powder XRD peak shapes have spawned a rapidly growing number of publications reporting Rietveld structure refinements using powder XRD data, e.g., quartz and fluorapatite (Young et al., 1977), $La_{0.75}Sr_{0.25}CrO_3$ (Khattak and Cox, 1977), nacrite (Toraya et al., 1980), muscovite (Sato et al., 1981), $(Ni,Mg)_3(PO_4)_2$ solid solutions (Nord and Stefanidis, 1983), synthetic hercynite (Hill, 1984), olivine (Hill and Madsen, 1984), farringtonite (Nord, 1984), synthetic natrotantite (Ercit et al., 1985), the zeolites TPA-ZSM-5 (Baerlocher, 1984), ZK-14 (Cartlidge et al., 1984), and gobbinsite (McCusker et al., 1985), kaolinite (Suitch and Young, 1983; Bish and Von Dreele, 1989), synthetic pargasitic amphiboles (Raudsepp et al., 1987b), synthetic eckermannites and nyboite (Raudsepp et al., 1987a), Na-gallosilicate sodalite (McCusker et al., 1986), todorokite (Post and Bish, 1988), akaganeite (Post, 1988), and coronadite (Post and Bish, 1989). A limitation of the Rietveld method is that one must start with a model that is a reasonable approximation of the actual structure, and it is therefore primarily a structure refinement, as opposed to structure solution, technique. In some cases, however, as discussed below, partial structure solutions are possible.

In addition to providing crystal structure information for materials that cannot be studied by single-crystal techniques, powder XRD and the Rietveld method offer several other attractive features. Powder XRD data collection is rapid compared to most single-crystal experiments, allowing many Rietveld refinements to be completed in the time it takes to complete a single-crystal study. Thus Rietveld refinements are well suited for phase transformation studies, heating or cooling experiments, pressure studies, etc. Twinning is not a problem for powder diffraction studies, and in fact, might serve to help randomize the crystallite orientations in the powder sample. Rietveld refinements can yield very precise and, with care, accurate unit-cell parameters, as well as quantitative analyses of mixtures (Bish and Howard, 1988; Bish and Post, 1988). On the negative side, in addition to being mainly a refinement technique, atomic structural parameters (especially temperature factors) are, in general, less accurately determined than by comparable single-crystal studies. Also, sample-related problems such as structure disorder and preferred orientation of crystallites can affect the accuracies of the Bragg peak intensities, resulting in a poor or incorrect refined structure.

RIETVELD METHOD IN PRACTICE

The basic requirements for any Rietveld refinement are: (1) accurate powder diffraction intensity data measured in intervals of 2θ (i.e. step-scan), (2) a starting model that is reasonably close to the actual crystal structure of the material of interest, and (3) a model that accurately describes shapes, widths and any systematic errors in the positions of the Bragg peaks in the powder pattern.

Data collection

In the great majority of studies reporting Rietveld refinements using powder XRD data, data were collected using automated powder diffractometers, primarily because accurate digitized step-scan intensity information can be collected quickly and easily. Successful studies have also been completed, however, using Guinier-Hägg (Malmros and Thomas, 1977; Nord and Stefanidis, 1983), Debye-Scherrer (Thompson and Wood, 1983), and Gandolfi (L. Finger, unpublished results) film data. Film patterns were converted to digitized data using scanning microdensitometers. In some cases the film methods offer advantages over diffractometer experiments because data can be collected from smaller amounts of sample and preferred orientation effects are minimized. On the other hand absorption problems are more severe, and it is difficult to measure accurate intensity data from films.

Without a doubt, sample preparation is one of the most critical stages of any structure study based on powder XRD data. In particular, most minerals exhibit some cleavage and therefore are prone to yield powder samples having nonrandomly oriented crystallites. Consequently, observed peak intensities will be inaccurate to some unknown degree, resulting in an incorrect refined structure. In many cases, grinding samples to very small particle sizes and using proper sample mounting techniques can greatly reduce preferred orientation problems. Even for samples that do not show a pronounced cleavage, it is essential that they be ground to particle sizes generally less than about 10 µm, depending on their absorption characteristics. Note that this is much finer than material obtained by passing through either a 325 mesh (45 µm) or 400 mesh (38 µm) sieve. Klug and Alexander (1974) found, for example, that accurate and reproducible peak intensities were obtained for quartz only with samples consisting of particles in the <15 µm size fraction. On the other hand, they observed an 18.2 mean percent deviation in peak intensities for 10 quartz samples from the 15 to 50 µm size fraction.

Several methods are described in the literature for preparing random XRD powder mounts, for example side- or back-loading (Klug and Alexander, 1974), dilution with a second phase, dusting the sample onto glass fiber-filters (Davis, 1986), and spray drying (Smith et al., 1979). These methods are discussed in detail in Chapter 4. Unfortunately, none of these preparation techniques will guarantee a random sample and it is therefore important to compare data from several sample mounts to assess the magnitude of any preferred orientation effects. Our experience has shown that sieving powder onto a Whatman glass microfibre filter (934-AH) is an effective method for preparing a random sample. One drawback, however, is that for weakly absorbing materials it might not be possible to prepare a sample layer on the filter substrate that is infinitely thick to X-rays, thereby contributing background (from the filter) and transparency effects to the resulting powder pattern that must be accounted for during the Rietveld refinement. These effects, if not properly modeled in the refinement can result in significant errors, particularily in temperature factors. Whatever the sample preparation technique, the mounted sample should be effectively infinitely thick (depends on radiation used and absorption properties of the sample) and be large enough to contain fully the X-ray beam at the lowest diffraction angle of interest.

A survey of published Rietveld refinements reveals that a wide range of step intervals and counting times has been used for data collections, and in most cases no reasons are given for selecting those values. Obviously, increasing the number of steps (N) (by decreasing the step width) and the counting time (T) will improve the precision of the data but with the trade-off of a longer experiment. It should also be considered that even though the number of observations in a Rietveld experiment (N) can be increased by using a smaller step interval, it is the Bragg intensities that are fundamental to any structure analysis, not the step intensities (Hill and Madsen, 1987). Hill and Madsen (1984, 1986) evaluated the effects of using different combinations of N and T on the values of refined atomic parameters and associated estimated standard deviations. They demonstrated that, despite the improved precision as N and T are increased, at some point the counting variance becomes negligible compared to other sources of errors (e.g. errors in structure or peak shape models). Additional counts will not yield more structure information but rather will lead to inappropriate weights (usually the inverse of the counting error) and underestimated parameter deviations. Hill and Madsen (1987) demonstrated that in the case of corundum the refined values of structure parameters did not differ significantly for data collected using increasingly larger step intervals until the interval exceeded the minimum full-width at half maximum for peaks in the pattern. They also reported a Rietveld refinement for olivine, using data collected with a 0.04° step width and a counting time of 0.05 s/step, that yielded structure parameter values that are not significantly different from those determined by single-crystal diffraction experiments. Obviously, long counting times are not a requirement for successful refinements. Hill and Madsen (1984, 1986) recommended using counting times that accumulate no more than a few thousand counts for the strongest peaks, and step intervals that are one-fifth to one-half of the minimum full-width at half maximum of resolved peaks in the pattern. They also suggested that if experiment time is limited, it is more efficient to improve precision by decreasing the step interval than increasing the counting time.

Another factor to consider when formulating a data collection strategy is that peak intensities fall off with increasing 2θ, primarily due to the decrease in atomic scattering factors (Klug and Alexander, 1974), and therefore longer counting times might be used for high versus low-angle data. Bish and Von Dreele (1989), for example, when collecting powder XRD data for kaolinite used step count times that were a factor of five to ten times greater for high versus low-angle portions of the pattern. Finally, data should be collected to the highest 2θ angle possible. High-angle data are especially important for successful refinement of temperature factors. Diffraction patterns with severely broadened reflections at intermediate angles (80° to 100°) probably will not contain much useful information at high angles, but even severely overlapped reflections at high angles still are useful and should be included.

Starting model

Because the Rietveld method is primarily a refinement technique, it is necessary to start with a model that is a reasonable approximation of the actual crystal structure. Commonly, a starting model, or at least partial model, is conceived by analogy to similar structures, for example other members of a solid solution series or isostructural materials whose crystal structures are known (McCusker et al., 1985; Post and Bish, 1989). Alternately, a model might be developed using distance-least-squares (Meier and Villiger, 1969), electrostatic energy minimization (e.g. Busing, 1981; Post and Burnham, 1986), or other computer modeling procedures. In some cases, high-resolution transmission electron microscope images can provide sufficient information from which to assemble a structure model (Post and Bish, 1988).

Computer programs

Most researchers today are using extensively modified or totally new versions of Rietveld's (1967, 1969) original computer program. Several of the most commonly used Rietveld refinement programs and packages are listed in Table 1. DBW (Wiles and Young,

1981) has to date probably been the most widely distributed Rietveld package. GSAS (Larson and Von Dreele, 1988) currently seems to offer the greatest flexibility, for example single-crystal and powder X-ray and neutron diffraction data can be refined simultaneously or independently. The package also provides the capability to calculate easily Patterson and Fourier maps and bond distances and angles. Both GSAS and Baerlocher's (1982) program include provisions for applying constraints (e.g. on bond lengths and angles) to parameters during the refinements. Unfortunately, GSAS is portable only to the VMS operating system on VAX computers.

Table 1. Rietveld refinement computer programs and packages

Rietveld Program	Rietveld (1969)
Rietveld Program	Hewat (1973)
PFLS	Toraya and Marumo (1980)
DBW	Wiles and Young (1981)
X-ray Rietveld System	Baerlocher (1982)
LHPM1	Hill and Howard (1986)
GSAS	Larson and Von Dreele (1988)

The diffraction profile

During a Rietveld refinement, structure parameters (atom positions, temperature and occupancy factors), scale factor, unit-cell parameters and background coefficients, along with profile parameters describing peak widths and shape, are varied in a least-squares procedure until the calculated powder pattern, based on the structure model, best matches the observed pattern. The quantity that is minimized by the least-squares procedure is:

$$R = \sum_i w_i(Y_{io} - Y_{ic})^2 , \qquad (1)$$

where Y_{io} is the observed intensity and Y_{ic} the calculated intensity at step i, and w_i is the weight assigned to each step intensity and is typically equal to the inverse of the counting error (generally equal to the square root of the observed intensity) at step i. The minimization is performed over all of the data points in the diffraction pattern (or only over the data points in Bragg peaks if the background is not refined).

The observed intensity at each step in a diffraction pattern consists of a contribution from Bragg reflections at that step plus background. The background can arise from several factors, including fluorescence from the sample, detector noise, thermal-diffuse scattering from the sample, disordered or amorphous phases in the sample, incoherent scattering, and scatter of X-rays from air, diffractometer slits and sample holder (including substrate for thin samples). It is essential in any Rietveld refinement to describe the background accurately. Probably the most straightforward approach is to select several points in the pattern that are away from Bragg peaks and then model the background by linear interpolation between these points. Obviously, interpolation can only be used with relatively simple patterns that have several, well-spaced intervals where background intensities can be measured between Bragg peaks. For complex diffraction patterns, and in recent years for most Rietveld studies, it is normal practice to include background coefficients as variables in the refinement. In the computer program DBW (Wiles and Young, 1981), for example, the background intensity at step i is calculated by the function:

$$Y_{ib}(c) = B_0 + B_1 TT_i + B_2 TT_i^2 + B_3 TT_i^3 + B_4 TT_i^4 + B_5 TT_i^5 , \qquad (2)$$

where $TT_i = 2\theta_i - 90.0$, and the B coefficients are determined by the least squares during the Rietveld refinement. Alternatively, the GSAS (Larson and Von Dreele, 1988) refinement package provides a cosine Fourier series with up to 12 refineable coefficients:

$$Y_{ib}(c) = B_1 + \sum_{j=2}^{12} B_j \cos(2\theta_{j-1}) \ . \qquad (3)$$

If the peak shape is well known, the background can be fitted even for complex patterns. If the pattern is poorly resolved, however, the background parameters tend to correlate with other parameters, particularily temperature factors, which can result in systematic underestimates of the standard deviations of these other parameters if the background is not included in the refinement (Prince, 1981; Albinati and Willis, 1982).

The Bragg peaks in a powder diffraction pattern are the integrated intensities as distributed by a peak-shape function. The intensities are determined by the atom positions and other structure parameters but can be affected by other factors such as absorption, extinction, and preferred orientation of particles in the sample. The peak shape in an X-ray powder diffraction pattern is a complex convolution of several sample and instrumental effects and therefore can vary significantly depending on the data collection geometry and type of sample. The pure diffraction maxima have natural profiles determined largely by crystallite-size distribution, crystal structure distortions, and the spectral distribution in the incident radiation (Klug and Alexander, 1974). Klug and Alexander (1974) identified six functions arising from a typical X-ray diffractometer that modify the pure Bragg diffraction peaks: (1) geometry of the X-ray source, (2) varying displacements of different portions of a flat sample surface from the focusing circle (flat-specimen error), (3) axial divergence of the X-ray beam, (4) specimen transparency, (5) effects of receiving slits, and (6) misalignment of the diffractometer.

During a Rietveld refinement, the calculated intensity, Y_{ic}, at point i in the diffraction pattern is determined according to the following equation:

$$Y_{ic} = Y_{ib} + \sum_{k=k1}^{k2} G_{ik} I_k \ , \qquad (4)$$

where Y_{ib} is the background intensity, G_{ik} is a normalized peak profile function, I_k is the Bragg intensity, and k1....k2 are the reflections contributing intensity to point i (Hill and Madsen, 1987). As a consequence of the many combined sample and instrumental effects listed above, the shapes of the Bragg peaks in powder XRD patterns deviate significantly from Gaussian, and therefore a more complex function is required to describe accurately their profiles. Young and Wiles (1982) concluded, after comparing results for a large number of Gaussian, Lorentzian, and combined Gaussian and Lorentzian profile functions, that Gaussian was consistently the worst performer and that the combined Gaussian and Lorentzian, such as the pseudo-Voigt and Pearson VII, were the best. They recommended the pseudo-Voigt because it is relatively simple to calculate and because the refineable constant eta can be interpreted as degree of Lorentzian versus Gaussian character.

The pure Voigt, a convolution of Gaussian and Lorentzian functions, represents the mathematically correct way in which various phenomena combine with one another to form the final peak shape (David, 1986). Sample-related effects, e.g. particle-size broadening, can be well described by a Lorentzian function and instrumental contributions by a Gaussian (David and Matthewman, 1985). The relative complexity of the Voigt function, however, has delayed its use, and to date most Rietveld programs have used the pseudo-Voigt because of its simplicity and functional similarity to the Voigt (Wertheim et al., 1974). David (1986) presented a method of parameterizing the pseudo-Voigt to make it essentially equivalent to the Voigt that allows refinement of separate half-width functions for the Lorentzian and Gaussian components. Consequently, angular dependencies of Lorentzian and Gaussian contributions can be independently modeled, making it possible to extract particle-size information (Keijser et al., 1983; David and Matthewman, 1985; Larson and Von Dreele, 1988). Madsen and Hill (1988) using a Voigt profile function, were able to determine as part of a Rietveld refinement crystal size and strain properties for Si and CaF_2 powders that compare very well with results from scanning electron microscope and neutron powder diffraction studies. A more detailed discussion of profile shape functions is presented in Chapter 8.

Another approach for modeling powder XRD profiles is use of a so-called learned peak-shape function (Baerlocher, 1986). Here a nonanalytical function is used that contains symmetric and asymmetric parts that are determined in numerical form ("learned") from a single resolved peak in the pattern (Hepp, 1981). It has the advantage that it is easily calculated, can account for peak asymmetry, and can describe virtually any peak-shape (Baerlocher, 1986). Unfortunately, the minimum requirement for this method, that there is a completely resolved reflection, is not met for many diffraction patterns.

Several workers have noted that the Lorentzian character of the peak shapes in XRD patterns increases with diffraction angle (Hill, 1984; Hill and Howard, 1985; Toraya, 1986). Hill (1984) reported that the fit between the observed and calculated diffraction profiles is improved if the pseudo-Voigt coefficient is allowed to vary as a function of 2θ. He also observed that for his refinement of hercynite neglecting the angular dependence of peak shape resulted in overestimation of the temperature factors but atomic parameters were not significantly affected. Many of the newer Rietveld codes (e.g. DBW3.2S and GSAS) allow for an angular dependence of the peak shape.

The Cu $K\alpha_{12}$ doublet provides an additional complexity in modeling the peak shapes from most powder XRD experiments. The problem can be avoided by using strictly monochromatic radiation as in Guiner-Hägg or synchrotron experiments or by using Kß radiation (Khattak and Cox, 1977). It should be noted, however, that the Cu Kß peak is also technically a doublet of $Kß_1$ and $Kß_3$ (International Tables for X-ray Crystallography, 1974). Also, the much lower intrinsic intensity of Cu Kß versus $K\alpha$ radiation greatly limits its usefullness for many diffraction experiments. Most modern Rietveld refinement programs calculate both $K\alpha_1$ and $K\alpha_2$ profiles, assuming a constant intensity ratio between the two (typically $K\alpha_1/K\alpha_2 = 2.0$, assuming a well-aligned diffracted-beam monochromater). We have not obtained improved Rietveld refinement results using Cu Kß instead of Cu $K\alpha$ radiation.

The widths of the Bragg maxima in powder XRD patterns are typically modeled in a Rietveld refinement as a function that is quadratic in $\tan\theta$ and gives the full-width at one-half of the maximum peak height (FWHM) as a function of the diffraction angle (Cagliotti et al., 1958):

$$H^2 = U\tan^2\theta + V\tan\theta + W , \qquad (5)$$

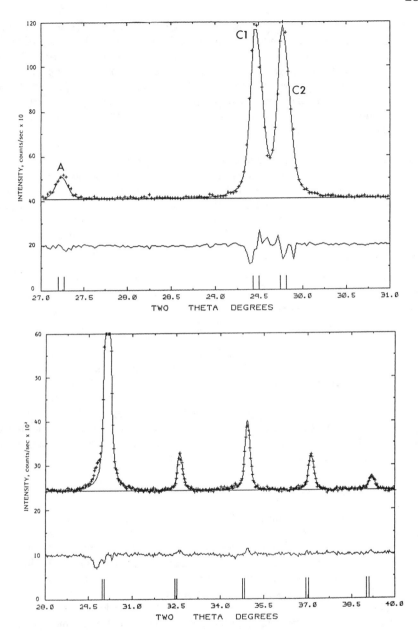

Figure 2 (top). Portions of observed (crosses) and calculated powder X-ray diffraction patterns for a mixture containing two calcite phases (C1 and C2) and aragonite (A). The background is indicated by the horizontal line, and the vertical lines mark the positions of the Bragg reflections (Kα_1 and Kα_2). The plot near the bottom of the figure is the difference between the observed and calculated patterns.

Figure 3 (bottom). Portion of observed and calculated powder X-ray diffraction patterns for calcite. The plot near the bottom of the figure is the difference between the observed and calculated patterns and in this case reveals the presence of an additional calcite phase (low-angle side of peak near 30°2θ) that was not accounted for in the Rietveld refinement.

where H is the FWHM and U, V, and W are refineable parameters. The values of U, V, and W for a given sample depend on the instrumental configuration and choice of profile shape function. This peak-width expression, however, does not readily accommodate broadening that is a function of crystallographic direction, such as that caused by sample crystallites with greatly anisotropic shapes (platelets or fibers) or by structural disorder. The Rietveld package GSAS (Larson and Von Dreele, 1988) uses a modification of the pseudo-Voigt profile function described by Thompson et al. (1987) to accomodate anisotropic Lorentzian strain and crystallite-size broadening.

Most Rietveld refinement computer codes perform the least-squares fitting over a 2θ region extending on either side of each Bragg peak, defined as being equivalent to a certain number of peak-widths or a certain fraction of the total peak intensity. Toraya (1985) and Ahtee et al. (1984) recommended extending the definition range of a Voigtian profile function to ten to twenty times the FWHM. Toraya (1985) observed that truncation to a narrower range results in an increase in the refined background and a concomitant increase in the temperature factors, and also can affect values of atomic positional parameters.

Most of the commonly used profile shape functions, including the Pearson VII and pseudo-Voigt, are symmetrical about the nominal Bragg peak positions. A variety of instrumental and sample effects, for example axial divergence of the X-ray beam and structure disorder in the sample, however, give rise to pronounced asymmetry in the observed peak shape, especially at low diffraction angles. Consequently, most Rietveld refinement computer programs include as a variable a semiempirical asymmetry correction term (Klug and Alexander, 1974). This factor corrects primarily asymmetry at low angles caused by axial divergence of the X-ray beam and does not affect the integrated intensity but will shift the apparent peak position. The use of both incident- and diffracted-beam Soller slits during data collection significantly reduces asymmetry on the low-angle side of peaks (and improves resolution). Also, because asymmetry is largely associated with low-angle peaks, in some cases results might be improved if only the higher-angle data are used in the refinement.

Unit-cell parameter refinement

One of the important strengths of the Rietveld method is the capability of refining precise and accurate unit-cell parameters, even for materials having complex diffraction patterns with severe peak overlap. Because the Rietveld technique does not require decomposition of patterns into component Bragg peaks (or measurement of individual peak positions) it is perhaps the only and certainly the quickest method for refining accurate unit-cell parameters from complex diffraction patterns. For example, in Figure 1 what appears to be two clinoptilolite reflections is in fact a composite of six Bragg peaks. It would be virtually impossible, even using some type of peak-fitting routine, to index and measure accurate positions for each of these peaks, as required for conventional unit-cell parameter least-squares refinements. A second advantage of the Rietveld approach is that for multiphase samples the contribution of each component to the diffraction pattern can be modeled and summed, and unit-cell parameters can be determined for each phase (Fig. 2). Additionally, comparison of the observed and calculated diffraction patterns at the completion of the refinement can reveal the presence of previously undetected minor phases (Fig. 3).

Typically, in conventional unit-cell parameter determinations, systematic errors in peak positions, such as those resulting from zeropoint shift, sample displacement, and sample transparency, are compensated for, at least in part, by using an internal standard. A major drawback to this approach is the contamination of the sample by the internal standard, which is not a desirable prospect if, for example, one is working with a small quantity of a rare mineral. Also for complex patterns, it can be challenging to identify an internal standard that yields sufficient well-resolved peaks to permit an accurate correction. Most of the Rietveld codes used today include as refineable parameters correction factors for systematic errors arising from peak asymmetry, sample displacement, and zero-shift. If these correction terms are incorporated into the refinement, it is typically possible to obtain unit-cell parameters that are at least comparable to, and in many cases more accurate and precise than those derived from conventional methods using internal standards. In Table 2, we list unit-cell parameters determined by Rietveld refinements using a correction term for sample displacement (which also accounts for most zeropoint error in the case of thick samples) for several standard materials and compare them to the generally accepted values.

Another potential advantage of using the Rietveld method to determine unit-cell parameters is that it is commonly possible with little additional effort to refine simultaneously other structural parameters, and/or obtain quantitative analyses of mixtures. For example, refining the occupancy factor for the Ca site in calcite serves as an independent check on the use of determinative curves to estimate mole percent Mg, and if more than one calcite phase is present, the amount and composition of each can be determined. In Table 3, we list unit-cell parameters and corresponding refined occupancy factors for water sites in analcime for a range of temperatures. Refinement of atom positions along with the unit-cell parameters might, in some cases, provide insights into the nature of phase transformations observed in heating or pressure experiments. In the case of analcime, structure refinements revealed that the intensity changes occurring on heating are primarily due to loss of water.

If evaluating unit-cell parameters is the primary goal, then shorter counting times can be used to collect the data than are required for detailed Rietveld refinements of crystal structures. Data collection times comparable to those used for conventional unit-cell parameter refinements (e.g. 1°/min, depending on size and purity of sample) are normally suitable. Similarily, no special sample preparation techniques are required. We have obtained consistently good results from smear as well as packed sample mounts. As is the case for any Rietveld refinement, it is necessary to have a structure model for the phase(s) of interest. Fortunately, single-crystal structure refinements have been published for most minerals and can be used for this purpose. In order to obtain accurate unit-cell parameters it is important to include in the refinement all of the phases that are in the sample. During the refinement, only scale factor, background coefficients, systematic error correction factors, and peak-shape parameters are routinely varied along with the lattice parameters. As mentioned above, however, it might be informative in some cases to include selected atomic parameters in the refinement.

Several authors have noted that the estimated standard deviations reported for Rietveld derived unit-cell parameters usually are too small, considering the observed accuracies (Sakata and Cooper, 1979; Hill and Madsen, 1984, 1987, etc.), and our results in Table 2 generally confirm that observation. During a Rietveld refinement, the peak positions are evaluated very precisely because each data point that defines the peak profile, and not just

Table 2. Unit-cell parameters from Rietveld refinements (Å)

	a	c	Reference
silicon(640B)	5.430940(35)		NBS certified
(1)	5.43065(1)		This study
(2)	5.43062(1)		This study
(3)	5.43062(1)		This study
(4)	5.43066(1)		This study
(5)	5.43069(1)		This study
corundum	4.7589(1)	12.9911(3)	This study
	4.75855(2)	12.9906(1)	Thompson et al. (1987)
hematite	5.0352(1)	13.7469(2)	This study
	5.0356(1)	13.7489(7)	Morris et al. (1981)
quartz	4.9110(1)	5.4021(4)	This study
	4.91239(4)	5.40385(7)	Will et al. (1988)

Notes: 1) Unit-cell parameters refined using Rietveld program DBW3.2. Refinements included correction factors for sample displacement errors.

2) Silicon samples 1-5 were displaced from the focal plane during data collection by 0, 1, 2, 4, and 8 mil, respectively. For the sample displaced by 8 mil, without refinement of the sample displacement correction, a = 5.43968(3).

Table 3. Rietveld parameters for analcime vs. temperature

T (°K)	a (Å)	c (Å)	Water occupancy
294	13.6901(5)	13.685(2)	0.665(4)
323	13.6903(7)	13.686(3)	0.655(5)
373	13.6880(6)	13.683(2)	0.622(4)
423	13.6819(5)	13.690(2)	0.558(5)
473	13.6711(3)	13.686(1)	0.469(4)
523	13.6618(6)	13.675(2)	0.373(5)
573	13.6390(4)	13.657(1)	0.302(4)

Note: 1) XRD data were collected at temperature under vacuum.

2) The analcime sample is from the Faroe Islands.

the peak top, for example, is considered a measurement of the peak position. Furthermore, the peak positions are relatively insensitive to errors in the peak shape function or structure model (Hill and Madsen, 1987). As is the case for any unit-cell parameter refinement, however, the error estimates are indications of precision only and commonly do not accurately account for the effects of systematic errors in the data. One way of obtaining reliable estimated deviations is to refine unit-cell parameters for an internal standard simultaneously with those of the phase of interest. The magnitudes of the actual deviations for the refined cell parameters of the standard from the accepted values provide estimates of the appropriateness of the errors for the phase of interest.

A sample Rietveld refinement

Let us assume that we have high-quality powder XRD step-scan data from a sample whose crystal structure we want to refine using the Rietveld method. Furthermore, we have a starting atomic structure model derived by one of the methods described above. If our sample is multiphasic, then we must include structure models for each phase. It is important to assign reasonable beginning values for background, peak width, and unit-cell parameters, and these can be approximated from the diffraction pattern or from results for similar phases. Initially, only one background and one peak-width parameter is defined, and they are constants over 2θ. In the early stages of the refinement, the atomic parameters are normally held fixed to their starting values. During the first least-squares cycles, typically only the scale factor and background coefficients are adjusted, and then gradually in successive cycles other parameters are included. Throughout the refinement it is essential to examine plots of the observed versus calculated diffraction patterns in order to detect problems, such as improperly fit background and peak-shape irregularities. Also, the plot of the difference between the observed and calculated patterns will reveal extra phases that have not been accounted for. During the next cycles of refinement, unit-cell parameters are allowed to vary along with a specimen displacement or zeropoint error correction term. If the refinement is converging smoothly, as indicated by the weighted profile residual (R_{wp}) or goodness-of-fit (GofF) factor (both will be discussed below), and the plots show that the calculated background and unit-cell parameters are accurately fitting the observed data, then the first peak-width parameter can be refined. Next, a second and perhaps third peak-width parameter are added to the refinement, followed by peak-shape (e.g. eta, for the psuedo-voigt profile function) and asymmetry parameters.

The emphasis to this point in the refinement has been the accurate modeling of the background and peak shape; the only structure-related variables that have been refined are the scale factor and unit-cell parameters for each phase. Generally the first atomic parameters added to the refinement are the positions and occupancy factors for any heavy atoms and an overall temperature factor. Gradually, the positions and occupancy factors for the remaining atoms in the structure are also refined. Finally, if the refinement looks good, individual atomic temperature factors are included. As mentioned above, temperature factors tend to accomodate model deficiencies for background and absorption, and commonly exhibit large discrepancies relative to values determined by single-crystal diffraction experiments (Thompson and Wood, 1983; Nord and Stefanidis, 1983; Baerlocher, 1986). Consequently, in Rietveld refinements using powder XRD data, individual atomic temperature factors are routinely fixed to reasonable values (e.g. from single-crystal refinements for similar compounds) and an overall temperature factor is refined. We have found, however, that if data are collected over a large angular range (up to about 160° 2θ), then it is commonly possible to obtain reasonable refined individual temperature factors or, alternatively, temperature factors for groups of like atoms, e.g., all O atoms, all tetrahedral cations, all octahedral cations, etc.

Compared with refinements using single-crystal diffraction data, Rietveld refinements generally converge more slowly and are more prone to settle into false minima, and typically it is not possible to refine all parameters together from the start (Baerlocher, 1986). The major reason for this is that many fewer Bragg reflections are available for Rietveld refinements than can be measured in a comparable single-crystal experiment. Whereas the Bragg reflection-to-parameter ratios are typically ten or twenty to one for single-crystal studies, values of three to five, or less, to one are not uncommon in Rietveld refinements. Consequently, it is important to have a good initial structure model and proceed with the Rietveld refinements slowly and carefully. The application of soft geometric constraints (Waser, 1963) during the refinement, i.e., constraining bond distances and angles to values measured for related compounds, in many cases hastens convergence to the correct structure (Baerlocher, 1986). Ideally, the constraints are gradually given less weight as the the refinement proceeds and are eventually removed completely.

The quantities used to measure the progress of a Rietveld refinement and the agreement between the observations and the calculated model are summarized in Table 4. The R_{wp} and GofF are the most meaningful in that the numerator in each of these expressions is the quantity being minimized by the least-squares during the refinement. The R_{exp} is based on counting statistics only and is an estimation of the minimum R_{wp} obtainable for a given problem. For a perfect refinement with correctly weighted data, the final R_{wp} would equal R_{exp} and GofF would be 1.0. Scott (1983) cautioned that if the number of observations (data points) is much larger than the number of Bragg peaks, as is typically the case for Rietveld refinements, the GofF index might be relatively insensitive to errors in the atomic structure model. R_B is a measure of the agreement between the "observed" and calculated Bragg intensities. In a Rietveld refinement, the observed Bragg intensities for overlapping reflections are determined by assuming that they are in the same proportion as their calculated counterparts. Consequently, the observed intensities are heavily biased by the structure model, and the R_B tends to be overly optimistic (Baerlocher, 1986). Even so, the R_B is a valuable indicator because it depends more heavily on the fit of the crystal structure parameters than do the other agreement indices (Hill and Madsen, 1987). Our experience suggests that an R_B larger than approximately 0.10 indicates significant structural inaccuracies and/or systematic errors in the data, such as preferred orientation of the crystallites in the powder. On the other hand, a low value of R_B does not indicate to the same degree of confidence as for a single-crystal refinement that the structure is correct. For example, we have observed errors in atom positions of as much as 0.3 Å for some Rietveld refinements yielding R_B values of 0.08-0.09. Obviously, other criteria must be used along with the agreement indices discussed above to assess accurately the quality of a refined structure. Probably the most important test of a structure is whether occupancy factors and bond distances and angles make reasonable chemical sense. Tetrahedral Si-O bond lengths, for example, of 1.8 or 1.4 Å suggest serious problems with the refined structure, regardless of the R_B value.

An additional important check of any Rietveld refinement is the comparison of plots of the final observed and calculated powder patterns (e.g. Figs. 2, 3 and 4). In the case of a good refinement, the plot of the difference between the two patterns should be essentially flat. Features in the difference plot might reveal systematic errors in the data, the presence of previously undetected phases, inaccuracies in the peak-shape model, or problems with background fit.

Estimated standard deviations

There has been much discussion in the literature in recent years about the reliability of estimated standard deviations (esd's) reported for Rietveld refinements. Prince (1981) concluded that the error estimates for Rietveld refinements are correct if counting statistics are the sole source of error. Baharie and Pawley (1983) pointed out, however, that in most Rietveld analyses, long counting times ensure that errors arising from counting variance are insignificant and that the sole cause of deviations in final structure parameters are systematic errors, such as preferred orientation of sample crystallites, inadequate peak-shape or atomic structure models, poor background fit, and the presence of unaccounted for phases in the sample. Pawley (1980) went so far as to assert that no published Rietveld study is limited by counting statistics. Additionally, Sakata and Cooper (1979) suggested that calculations of esd's for Rietveld refinements are fundamentally invalid because they ignore serial

Table 4. Agreement indices for Rietveld refinements

$R_p = \Sigma |Y_{io} - Y_{ic}|/\Sigma Y_{io}$

$R_{wp} = [\Sigma w_i(Y_{io} - Y_{ic})^2/\Sigma w_i Y_{io}^2]^{0.5}$

$R_B = \Sigma |I_{ko} - I_{kc}|/\Sigma I_{ko}$

$R_{exp} = [(N - P)/\Sigma w_i Y_{io}^2]^{0.5}$

$GofF = \Sigma w_i(Y_{io} - Y_{ic})^2/(N - P) = (R_{wp}/R_{exp})^2$

$d = \sum_{i=2}^{N} ((Y_{io} - Y_{ic})/\sigma_i - (Y_{io-1} - Y_{ic-1})/\sigma_{i-1})^2 / \sum_{i=1}^{N} ((Y_{io} - Y_{ic})/\sigma_i)^2$

Notes: Y_{io} and Y_{ic} are observed and calculated intensities, respectively, at point i; $\sigma^2 = Y_{io}$; w_i is the weight assigned each step intensity; I_{ko} and I_{kc} are the observed and calculated intensities, respectively, for Bragg reflection k; N is the number of data points in the pattern and P is the number of parameters refined.

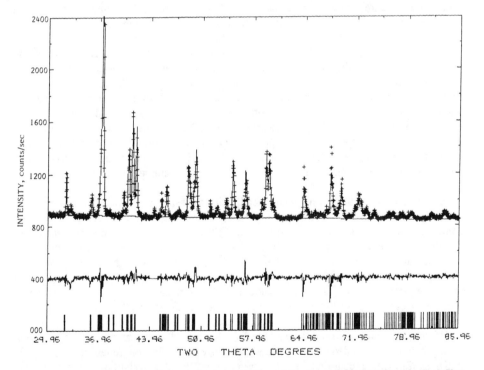

Figure 4. Final observed (crosses) and calculated powder X-ray diffraction patterns for hollandite (fiber-filter mounted sample). The background is indicated by the horizontal line, and the vertical lines mark the positions of the Bragg reflections ($K\alpha_1$ and $K\alpha_2$). The plot near the bottom of the figure is the difference between the observed and calculated patterns.

correlation among neighboring residuals resulting from systematic differences between observed and calculated peak shapes and/or Bragg intensities. Recently, however, Prince (1985) has shown that the Rietveld method is mathematically identical to properly formulated integrated intensity methods, and provided there are no unaccounted for systematic errors, that the calculated estimated errors are correct. Furthermore, he points out that if serial correlation is present, then no esd is a valid measurement of uncertainty. Serial correlation indicates that the model is inadequate, and biases in the refined parameters depend on the unknown correlation between the refined and missing parameters.

By increasing the counting time and the number of step intervals, the esd's for a Rietveld refinement can be made arbitrarily small, even if the model provides a poor fit to the data (Scott, 1983). Of course, esd's are measures of precision, and in cases with significant model errors they provide little information about accuracy. Only if the model is completely correct, implying that any systematic errors in the data must be appropriately described, are least-squares error estimates reliable indicators of accuracy (Sakata and Cooper, 1979; Prince, 1981). Cooper (1982) concluded that published Rietveld results indicate that in general the model does not fit adequately on the basis of statistical criteria, and so therefore the estimated errors are unreliable (in most cases underestimated).

Several approaches have been suggested for obtaining estimated errors for Rietveld refinements that are consistent with observed levels of accuracy. The brute-force approach is to repeat the experiment several times and then calculate the standard deviations of the replicate results (Hill and Madsen, 1987). Unfortunately, this can be a very time-consuming and perhaps costly exercise. Pawley (1980) found that multiplying the final esd's by the square root of the number of observations in the average full-width-at-half-maximum for resolved peaks (values for the factor are typically in the range of 2.0 to 3.0) yields error estimates consistent with observed accuracy. The drawback to this approach is that it is empirical with no statistical basis. Scott (1983) proposed converting Rietveld esd's to equivalent integrated intensity deviations on the grounds that these better reflect the effects of model errors. He reported that this procedure generally increases the estimated errors by a factor of three to five. He also conceded, however, that the statistical basis of this conversion is dubious and that it only applies to structure parameters.

Hill and Madsen (1987) contended that adjusting the estimated standard deviations after the fact is less desireable than modifying the experiment itself to ensure that errors from counting statistics are at least comparable to those from other sources in the refinement. As mentioned above, this can be accomplished by selecting smaller counting times and fewer step intervals for the data collection, thereby limiting the precision of the intensity measurements (Hill and Madsen, 1984, 1986). Adjusting step interval or collection time so as to make the statistical uncertainty comparable to the systematic error does not, however, remove the systematic errors, it merely "sweeps them under the rug".

Hill and Madsen (1986) and Hill and Flack (1987) have demonstrated that the application of the Durbin-Watson d statistic provides a means of assessing the reliability of Rietveld esd's, by furnishing quantitative information about serial correlation in the residuals. Furthermore, the d statistic is a sensitive measure of the progress of a refinement and remains discriminating when other agreement indices fail (Hill and Madsen, 1987). For example, Hill and Flack (1987) showed that conventional agreement indices do not distinguish between refinements using data collected from a sample of anglesite at different step widths, ranging between 0.05° and 0.35°, even though the esd's change by a factor of 2.6. On the other hand, the d statistic does change with step width and in the case of anglesite indicates that the calculated deviations are valid for refinements only when the step interval is greater than 0.24°. The Durbin-Watson d statistic is calculated in the most recently released version of DBW (Wiles and Young, 1981).

The conclusion drawn from the above discussion seems to be that if there are no unaccounted for systematic errors in the data or model, then the esd's calculated by Rietveld refinements are probably about right. Unfortunately, this situation is probably the exception rather than the rule, and therefore, for most Rietveld refinements the calculated errors are probably not true indications of accuracy. Even so, if the estimated errors are considered in the context of the above discussion and along with other criteria that evaluate the quality of the model, then the uncertainty about significance of the esd's should not detract seriously from the merits of the Rietveld method. Scott (1983) concluded that parameter estimates are unbiased and almost certainly more accurate than results from powder integrated intensity refinements because the Rietveld technique uses information in the peak shapes. It is after all the values of the structure parameters that are the important results of most Rietveld studies, and in many cases the structure cannot be refined by any other method.

EXAMPLES OF RIETVELD REFINEMENTS

Refinements using powder neutron diffraction data comprise the great majority of Rietveld studies reported to date, and it has been well demonstrated that this method normally yields results comparable to those from single-crystal diffraction studies. During the past decade or so, however, the number of published Rietveld refinements using powder X-ray diffraction data has increased dramatically, and it is not unreasonable to project that X-ray refinements will eventually outnumber those using neutrons. This surge in interest in X-ray Rietveld refinements stems largely from improvements in model X-ray profile functions and the simplicity of collecting X-ray, versus neutron, data. Unfortunately, as discussed above, experiments using powder XRD data are more prone to certain systematic errors, such as preferred orientation of sample crystallites, and X-ray peak profiles are more complex to model than their neutron counterparts. Consequently, although very good Rietveld refinements are possible with XRD data, the potential for error is probably greater than for neutron refinements. Given the interest in, and some of the problems involved with structure refinements using powder XRD data, it is perhaps suprising that more studies have not been published that compare Rietveld results to those from refinements using single-crystal X-ray diffraction data, especially for complex structures. In the following section we summarize several such comparisons along with results of some Rietveld refinements of mineral structures that illustrate some of the strengths as well as weaknesses of the technique.

Some of the earliest reported Rietveld refinements using powder X-ray data are by Young et al. (1977) for quartz and synthetic fluorapatite. Despite using only Gaussian and simple Lorentzian peak profile functions, all of the atomic positional parameters except one agree within two esd's of the corresponding values from single-crystal studies. Thompson and Wood (1983) used Debye-Scherrer data to refine a structure for quartz in which all atom positions are within 0.005 Å of single-crystal values, but the temperature factors are too large. In the structure they refined for AlPO$_4$ (berlinite), all of the positional coordinates are within two esd's of those determined by a refinement using single-crystal data, but again temperature factors show less good agreement. Single-crystal and Rietveld-refined structures for the olivine-related Ni$_3$(PO$_4$)$_2$ (Nord and Stefanidis, 1983) compare within about five esd's, but the Rietveld structure exhibits negative temperature factors and errors in the Ni-O bond lengths of as much as 0.06 Å. Hill and Madsen (1984) reported good agreement between single-crystal and Rietveld positional parameters for olivine, corundum, and plattnerite. On the other hand, the Rietveld structure reported by Sato et al. (1981) for muscovite shows differences in T$_1$-O and Al-O bond lengths, compared with published distances from single-crystal studies, of as much as 0.13 Å and 0.27 Å, respectively. They attributed the discrepancies to errors in the Bragg intensities caused by preferred orientation of the platy crystallites in the sample.

Zeolites

Some of the more robust Rietveld structure refinements using XRD data have been reported for zeolites, and although typically there are no single crystal results with which to compare, in some cases the validity of the refined structures has been confirmed by Rietveld refinements using powder neutron data or by comparison to similar structures. Cartlidge and Meier (1984) used powder XRD data to perform Rietveld refinements of Na-exchanged zeolite ZK-14 at three different temperatures in order to monitor structural changes leading to transformation to a sodalite-type phase. They used a starting structure model derived from a DLS calculation and refined as many as 57 structural parameters. Soft constraints were applied during the least-squares refinement in the forms of specified T-O distances and O-T-O angles. The constraints were successfully removed during the final cycles of refinement only for the room temperature structure. The bond distances determined for the refined structures appear reasonable, but the thermal parameters exhibit a range of values and in general are unusually large. Similarily, McCusker et al. (1985) determined a structure for the zeolite mineral gobbinsite, refining 64 structural parameters using 50 soft constraints on bond distances and angles. When the constraints were removed, however, unacceptable interatomic distances resulted, with no improvement in the R-values, possibly due to pseudosymmetry and/or to the fact that XRD data alone could not support the refinement of 64 structural parameters. They concluded that in any case the refined model must be a correct approximation to the structure since it fits both the X-ray data and the geometric requirements. This brings up the question of just how sensitive are the X-ray data to the correct structure. McCusker et al. (1985) cited attempts to refine gobbinsite in incorrect space groups using soft constraints which yielded reasonable bond distances and angles, but fits to the X-ray data were unsatisfactory. This is consistent with our experiences that it is not possible to achieve a satisfactory Rietveld refinement, as judged by all criteria, if there are gross errors in the model structure.

Kaolinite

Although kaolinite is one of the more common clay minerals in the Earth's crust, it does not occur in well-ordered crystals large enough for single-crystal studies and therefore its structure has never been accurately determined. The average kaolinites exhibit two-dimensional diffraction effects in their powder diffraction patterns, and thus it has proven very difficult to measure accurate Bragg intensities. An occurrence of kaolinite in geodes in Iowa has provided samples of kaolinite exhibiting little detectable two-dimensional diffraction effects, making these samples suitable for obtaining data for Rietveld refinement.

Suitch and Young (1983) (powder X-ray diffraction data) and Adams (1983) (powder neutron diffraction data) presented the results of the first attempts to use the Rietveld method to refine the kaolinite structure. Suitch and Young (1983) assumed a lower symmetry than used by others and they obtained a final structure with unrealistic bond lengths. In an attempt to improve the quality of the final structure model, one of us (DLB) obtained powder diffraction data for the well-crystallized kaolinite and attempted a Rietveld refinement using the Suitch and Young (1983) final structure as a starting model. Initially, the Rietveld program DBW3.2 (Wiles and Young, 1981) was used for refinement, and the resultant structure was very similar to that obtained by Suitch and Young (1983). Bond distances were not improved over those of Suitch and Young (1983). The conclusion at the time was that an inadequate preferred orientation correction was the primary reason for the poor results. Subsequently, Young and Hewat (1988) reported the results of a Rietveld refinement of the kaolinite structure using neutron powder diffraction data. Again, the refinement yielded a structure with unrealistic bond lengths.

Using the Rietveld package GSAS (Larson and Von Dreele, 1988), Bish and Von Dreele (1989) resumed refining the kaolinite structure using XRD data and the Suitch and Young (1983) result as their starting model. GSAS differs significantly from DBW3.2 in a variety of ways, including the use of the March function preferred orientation correction (Dollase, 1986) and the ability to apply soft constraints to distances and angles during refinement. In spite of attempts to influence the refinement, including varying the preferred orientation correction from 1.0 (no correction) to as low as 0.2 and dramatically raising the weights of the soft constraints, the refinement again converged to a structure similar to the result of Suitch and Young (1983). When the preferred orientation correction was manually reduced to 0.2, it smoothly and rapidly returned to 0.99, the value that was ultimately obtained with the final refinements.

At this point in refinement, Bish and Von Dreele (1989) performed a DLS refinement of the kaolinite structure, using observed lattice parameters and bond distances from similar structures. Using this DLS result as the starting model, the refinement converged quickly to a crystal-chemically reasonable structure, with an Rwp of 0.123 versus over 0.18 for refinements with the other starting model. This experience highlights one of the major pitfalls that can be encountered in Rietveld refinement of complex structures, that of false minima. In the refinement of the kaolinite structure, the false minimum was so distinct and deep that at least three groups encountered it. The kaolinite result also emphasizes the importance of assessing the chemical reasonability and accuracy (not precision) of the resulting structure.

Although the refinement of kaolinite presents good examples of the problems inherent in performing Rietveld refinement, it also provides an example of the quantity and quality of the structural information that can be obtained for complex samples. As noted above, the lattice parameters obtainable with Rietveld refinement are potentially of very high precision. In spite of (or because of) the fact that their diffraction pattern contained in excess of 3700 reflections, Bish and Von Dreele (1989) obtained precision of lattice parameters of better than one part in 40,000. The precision on atomic positional parameters for kaolinite is between 6 and 18 in the fourth decimal place, yielding precisions in bond lengths between 5 and 6 in the third decimal place. As discussed above, however, it is difficult to assess the reliability of the calculated esd's for Rietveld refinements, and it is likely that those determined for kaolinite are underestimated.

The refinement of Bish and Von Dreele (1989) also illustrates the types of additional information that can be obtained from a diffraction pattern using the Rietveld refinement procedure. The Rietveld program used by Bish and Von Dreele (1989), GSAS (Larson and Von Dreele, 1988), allows the calculation of anisotropic particle size and strain from broadening of individual peaks. The profile parameters for kaolinite indicated a coherent domain size of >960 Å in the a-b plane and a domain size of >5500 Å perpendicular to the a-b plane, opposite of what would be expected for a platy, layer-structure mineral. These are useful and unusual data and would be very difficult to obtain using conventional methods, given the highly overlapped nature of the kaolinite diffraction pattern. The rod-shaped domains may result from nucleation of domains within a single crystallite in which the vacant octahedral site in the dioctahedral mineral is in one of three possible structural positions 120° apart. The profile parameters indicated <0.1% strain in the a-b plane (i.e. <0.1% fluctuation in the a and b lattice parameters) and <0.4% strain along [001], consistent with the weaker interlayer bonding along [001] than within the individual layers.

When Bish and Von Dreele (1989) first conducted the Rietveld refinement of kaolinite, it was obvious from the plot of observed and calculated diffraction data that an impurity phase existed in the sample. Close inspection revealed that the impurity was dickite, a closely related kaolin mineral. Explicit inclusion of dickite in the refinement accounted for all extra intensity in the observed diffraction pattern and allowed determination of lattice parameters for, and the amount of, dickite in the sample (~4.6%).

It is interesting to note the two opposing points of view taken regarding the kaolinite results obtained by Young and coworkers. Young (1988) used his kaolinite refinement results as an example of the considerable structural detail that can be obtained with Rietveld refinement, using kaolinite as an example of "pressing the limits of Rietveld refinement." Thompson et al. (1989), however, took a different point of view, and considered that the application of Rietveld methods to the kaolinite structure was "beyond the limits of Rietveld refinement." It appears that the opinion offered by Young (1988) is probably more correct, as long as careful attention is paid to sample preparation, data collection, and the soundness of the results.

Hollandite

For many of the tetravalent manganese oxide minerals, crystals suitable for single-crystal diffraction experiments have not been found, and therefore powder diffraction techniques offer the best hope for understanding their detailed structures and crystal chemistries. In order to gauge better the reliability and explore the limitations of Rietveld refinements for manganese oxides, we performed a series of structure refinements for hollandite [(Ba,Pb)Mn$_8$O$_{16}$] from Stuor Njuoskes, Sweden (USNM #127118) and compared the results to a structure refined using data collected from a single-crystal that came from the same sample (Post et al., 1982). We were particularly interested in effects of preferred orientation on the Rietveld results. Scanning and transmission electron microscope images indicate that the hollandite crystallites, like those of many of the manganese oxide minerals, are laths or needles elongated parallel to b.

The first hollandite sample was hand-ground under acetone with a mortar and pestle to <325 mesh and packed into a cavity mount. In attempts to minimize preferred orientation of the sample crystallites, subsequent samples were prepared by: (1) grinding in a Brinkmann Microrapid Mill under acetone to particle sizes <10 μm, loosely packing into a cavity mount, and roughening the sample surface with a razor blade, and (2) dusting the milled sample onto a glass-fiber filter. The starting atomic model was the structure determined by Post et al. (1982), and the refinements were performed using the computer program DBW3.2 (Wiles and Young, 1981), as modified by S. Howard (personal communication). The backgrounds were fit using a third-order polynomial, and the observed peak shapes were approximated with a pseudo-Voigt profile function limited to ten full-widths on each side of the peak position. Individual isotropic temperature factors were held fixed to the values determined by the single-crystal study and overall temperature factors were refined. Final refinement parameters are listed in Table 5, and atomic parameters and selected bond distances resulting from the three Rietveld refinements are compared in Tables 6 and 7, along with the comparable values reported from the single-crystal study by Post et al. (1982). Final observed, calculated, and difference plots for the filter-mount sample are shown in Figure 4.

The Mn positions refined for each of the three sample preparations compare well with single-crystal results, but there are significant differences in some of the O atom positions, especially O1 and O4 in refinements 1 and 2 (Table 6). Comparison of the experimental hollandite patterns with one calculated using the single-crystal parameters (Post et al., 1982) clearly reveals evidence of preferred orientation, especially in the hand-ground sample (Fig. 5). The intensities of the (hkl) reflections are low in the observed pattern, relative to the calculated, and the observed (h0l) intensities are relatively high, which is consistent with a nonrandom orientation of the lath-like crystallites. The agreements between the single-crystal atom positions and the Rietveld values improve progressively for the packed milled and the milled and fiber-filter mounted samples, in support of the conclusion that the discrepancies in the O atom parameters are largely the result of incorrect intensities arising from a nonrandomly oriented powder sample. Furthermore, the R_{wp} improved from 0.177 for the unmilled to 0.132 for the filter mounted sample, and the R_B dropped from 0.139 to 0.078 (Table 5). It is interesting that for a Rietveld refinement using the hand-ground sample and atomic parameters fixed to the values of the single-crystal hollandite structure, the R_B is 0.209. Subsequent refinement of the atomic parameters resulted in a lower R_B (0.139) and a structure that, although is incorrect, gives a better fit to the observed profile and Bragg intensities. Obviously then, for cases in which powder data suffer from significant preferred orientation effects, the R_B is not a sensitive indicator of the quality of the refinement. By comparison, a Rietveld refinement using the fixed single-crystal structure and the diffraction data from the fiber-filter mounted sample yielded a R_B of 0.079, which did not change significantly when atomic parameters were refined.

Mn-O bond distances even from the Rietveld refinement using the filter-mounted sample differ from the single-crystal values by as much as 0.06 Å (Mn-O distances are in error by as much as 0.30 Å for sample 1), which is at least three times the esd's for the Rietveld results

Table 5. Final Rietveld refinement parameters for hollandite

	1	2	3
a (Å)	10.010(1)	10.009(1)	10.013(1)
b (Å)	2.8796(1)	2.8796(2)	2.8801(2)
c (Å)	9.729(1)	9.730(1)	9.733(1)
β (°)	90.970(3)	90.974(4)	90.970(4)
Pseudo-Voigt Coefficient	0.94(3)	0.77(3)	0.89(3)
Parameters	31	31	31
Bragg Reflections	111	163	141
2θ Range (°)	30 - 80	30 - 100	30 - 90
Time/step (s)	10	40	18
R_p	0.138	0.095	0.102
R_{wp}	0.177	0.127	0.132
R_{exp}	0.095	0.056	0.089
R_B	0.139	0.099	0.078

Samples: 1) hand-ground, <325 mesh; 2) milled, packed mount 3) milled, glass-fiber filter mount

Table 6. Atom parameters for hollandite   Samples: as in Table 5.

		Single Crystal*	1	2	3
Mn1	x	0.85180(3)	0.854(1)	0.8527(4)	0.8500(5)
	z	0.33266(4)	0.332(1)	0.3330(5)	0.3322(5)
Mn2	x	0.33670(3)	0.339(1)	0.3371(4)	0.3365(5)
	z	0.15345(3)	0.155(1)	0.1586(5)	0.1534(5)
O1	x	0.6583(2)	0.647(2)	0.667(2)	0.656(2)
	z	0.3022(2)	0.266(2)	0.294(2)	0.295(2)
O2	x	0.6552(2)	0.655(2)	0.647(2)	0.652(2)
	z	0.0414(2)	0.035(2)	0.037(2)	0.043(2)
O3	x	0.2940(2)	0.289(2)	0.292(1)	0.293(2)
	z	0.3502(2)	0.367(2)	0.354(2)	0.349(2)
O4	x	0.0415(2)	0.076(2)	0.050(2)	0.035(2)
	z	0.3222(2)	0.334(2)	0.330(2)	0.326(2)
Ba	occ	0.74	0.82(1)	0.68(1)	0.74(1)
Pb	y	0.20(1)	0.29(1)	0.22(1)	0.22(1)
	occ	0.30(1)	0.50(1)	0.46(1)	0.37(1)

* - Post et al., 1982

Table 7. Hollandite Mn-O bond lengths (Å)

	Single-Crystal[a]	Rietveld[b]
Mn1 -O1	1.958(2)	1.97(2)
-O1(x2)	1.949(2)	1.90(2)
-O2(x2)	1.892(2)	1.88(2)
-O4	1.907(2)	1.85(2)
<Mn1-O>	1.925	1.90
Mn2 -O2	1.899(2)	1.92(2)
-O3	1.969(2)	1.96(2)
-O3(x2)	1.946(2)	1.94(2)
-O4(x2)	1.899(2)	1.94(2)
<Mn2-O>	1.926	1.94

a) Post et al., 1982   b) milled, glass-fiber filter mount

Table 8. Model expressions for preferred orientation effects

1) $I_{corr} = I_{obs}\exp(-G\phi^2)$ Rietveld (1969)
2) $I_{corr} = I_{obs}\exp[G(\pi/2 - \phi)^2]$ Will et al. (1983)
3) $I_{corr} = I_{obs}(G^2\cos^2\phi + \sin^2\phi/G)^{-3/2}$ Dollase (1986)

Note: G is a variable in the Rietveld refinement, and φ is the acute angle between the preferred orientation plane and the diffracting plane.

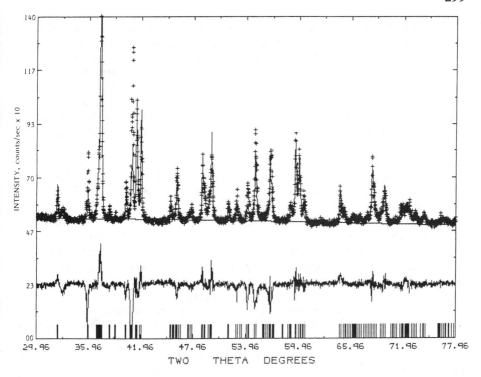

Figure 5. Observed (crosses) powder X-ray diffraction pattern for the hand-ground hollandite sample compared to a pattern calculated using the single-crystal structure of Post et al. (1982). The background is indicated by the horizontal line, and the vertical lines mark the positions of the Bragg reflections ($K\alpha_1$ and $K\alpha_2$). The plot near the bottom of the figure shows the significant differences between the observed and calculated patterns, caused by preferred orientation of the crystallites in the sample.

(Table 7). The average Mn-O distances for individual octahedra for sample 3, however, agree quite well with the comparable single-crystal values.

The results of the hollandite refinements underscore the fact that preferred orientation of sample particles is probably the most difficult problem to overcome when using powder XRD data to refine crystal structures. The theory of powder diffraction assumes a random distribution of crystallites of equal size. In practice this is seldom the case, and the consequent errors in relative Bragg intensities are the principal limitation on the accuracy of structures refined from powder data (Thompson, et al., 1987; Will et al., 1988). In general, preferred orientation problems are much more severe for powder X-ray than neutron diffraction experiments. The large sample volume and cylindrical sample holders used in typical neutron experiments generally ensure a nearly random sample, and consequently effects of preferred orientation can largely be ignored. The packed flat sample mounts and much smaller sample sizes used in most powder XRD experiments almost guarantee a nonrandom particle orientation. The problem is exacerbated for minerals which commonly exhibit good cleavages in one or more directions, resulting in anisotropically shaped particles.

In spite of the potentially severe limitations posed by nonrandom samples on X-ray Rietveld refinements, no guaranteed methods for dealing with the problem have been developed. Consequently, in many cases preferred orientation effects are overlooked in refinements or deliberately ignored (Baerlocher, 1986). Several algorithms have been proposed and tested for modeling during the Rietveld refinement intensity errors arising from samples with nonrandomly oriented crystallites (Rietveld, 1969; Will et al., 1983; Dollase, 1986; Thompson et al., 1987) (Table 8). In general, all of these expressions model an intensity correction factor as a function of the angle between the diffraction vector for a given reflection and the unique crystallographic direction of preferred orientation. None of these functions has proven generally effective for samples with high degrees of preferred orientation but might yield some improvement in cases where the problem is relatively slight. They seem to be most effective for the case of platey particles. Will et al. (1988) compared results using several functions to model preferred orientation effects and concluded that whereas there were significant improvements in all of the cases, the March expression (Dollase, 1986) performed best overall. They were not successful, however, in using the March function to model two preferred orientation directions. Repeating the hollandite refinements described above incorporating a March function to correct for preferred orientation effects did not yield significantly improved results. In the absence of a generally effective software correction procedure, strategies must be adopted for minimizing preferred orientation problems during the sample preparation stage. Alternatively, in some cases it might be possible to carry out structure analyses substituting synthetic analogues for mineral samples. Commonly, synthetic materials can be prepared to be sufficiently finely crystalline as to greatly reduce preferred orientation problems. Recently, Ahtee et al. (1989) introduced a revised Rietveld code that accounts for effects of preferred orientation by expanding the orientation distribution in spherical harmonics.

STRUCTURE SOLUTION USING RIETVELD METHODS

Because the Rietveld method is primarily a refinement technique, for unknown structures it must be preceded by a structure determination involving indexing and measuring intensities of individual Bragg reflections. This is not always a straightforward process because of peak overlap and consequently is most suited for high-resolution data, such as that from Guinier-Hägg or synchrotron experiments. Once a diffraction pattern has been decomposed into the component Bragg reflections, standard Patterson and direct methods procedures can be used to solve the basic crystal structure (e.g. Berg and Werner, 1977; Rudolf and Clearfield, 1985; Attfield et al., 1986; Lehmann et al., 1987; McCusker, 1988).

If at least a portion of a structure is known, then in some cases Rietveld refinements can be combined with Fourier syntheses to locate missing atoms (e.g. Rudolf and Clearfield, 1985; McCusker et al., 1985; Post and Bish, 1988; Shiokawa et al., 1989). During the initial cycles of Rietveld refinement using the partial structure model, the scale factor and background, unit-cell, and profile parameters are refined. The observed Bragg intensities as assigned by the Rietveld program are corrected for multiplicity and Lorentz-polarization effects and converted into structure factors, and a difference Fourier map is calculated using the observed structure factors and partial structure model. New atoms appearing on the difference map are added to the model and the procedure is repeated until the structure is complete, at which time a Rietveld refinement of the entire structure is performed.

When interpreting Fourier maps based on Rietveld observed intensities, it is important to remember that the results are biased by the model structure, typically to a greater degree than in single-crystal experiments. This is because, as mentioned above, the observed Bragg intensities determined by the Rietveld program for overlapping peaks are assumed to have the same ratios as the corresponding calculated intensities. Consequently, the amount of information contained in the resulting Fourier map is typically less (and electron-density maxima are lower) than on comparable maps resulting from single-crystal data. In some cases, it might be necessary to calculate several successive maps in order to locate all of the atoms in a structure.

The combination of Rietveld refinement and Fourier analysis has proven to be an effective technique for locating cavity and tunnel atoms and molecules in framework structures such as zeolites and certain manganese oxide phases. McCusker et al. (1985) assumed a gismondine-type framework as a starting model for a Rietveld refinement of the structure of the new zeolite mineral gobbinsite, and they used Fourier difference maps to locate the Na and K cation and five water molecule sites in the cavities. Similarily we calculated Fourier difference maps for analcime and Cs-exchanged clinoptilolite using models for the Si,Al-O frameworks determined from single-crystal studies (Mazzi and Galli, 1978; Koyama and Takéuchi, 1977) and Rietveld-derived observed structure factors. The difference map for analcime shows 2 to 5 $e^-/Å^3$ peaks for the Na and water sites which agree very well with the corresponding positions from the single-crystal refinement (Table 9). There are no other peaks on the difference map larger than 0.5 $e^-/Å^3$. A layer from the difference map calculated for Cs-bearing clinoptilolite is compared in Figure 6 with a comparable map constructed by Koyama and Takéuchi (1977) from single-crystal diffraction data for a (Ca, Na, K)-bearing clinoptilolite. The similarity of the two maps supports the conclusion that the Rietveld-based map is showing real structural information.

Todorokite

The crystal structure of the manganese oxide mineral todorokite [$(Na,Ca,K,Ba,Sr)_{0.3-0.7}(Mn,Mg,Al)_6O_{12}$3.2-4.5$H_2O$] has been the subject of interest and speculation for many years. Much of this attention stems from its role as a major manganese phase in ocean manganese nodules and studies showing that it exhibits dehydration and cation-exchange behavior similar to many zeolites (Bish and Post, 1984, 1989). Unfortunately, todorokite typically occurs as poorly crystalline masses, and to date no crystals have been found that are suitable for single-crystal diffraction studies. High-resolution transmission electron microscope images indicate that the todorokite structure is constructed of triple chains of edge-sharing Mn-O octahedra that corner-share to form large tunnels (Chukhrov et al., 1978; Turner and Buseck, 1981) (Fig. 7). Post and Bish (1988) used this tunnel framework as the starting model for their Rietveld refinement of the todorokite structure, and they were able to locate the major sites of the water molecules and cations in the tunnels from Fourier difference maps (Fig. 8). The similarity of the these maps for the two different todorokite samples indicates that they are showing real structural information.

The refinement for todorokite was complicated by the presence of significant amounts of chain-width disorder in the samples, giving rise to anisotropically broadened reflections. Structural defects such as chain-width disorder or stacking faults can be viewed as reducing

Table 9. Rietveld, difference-Fourier results for analcime

		Rietveld-Fourier[a]	Single Crystal[b]
Na1	x	0.25	1/4
	y	0.124	0.1254(4)
	z	0.0	0
	ΔF peak/occ	5.8 e$^-$/Å^3	0.61-0.84
Na2	x	0.0	0
	y	0.25	1/4
	z	0.125	1/8
	ΔF peak/occ	4.2 e$^-$/Å^3	0.17-0.80
H$_2$O	x	0.127	0.120(1)
	y	0.143	0.131(1)
	z	0.150	1/8
	ΔF peak/occ	2.3 e$^-$/Å^3	1.0

a. atom positions from Fourier difference map
b. Mazzi and Galli (1978)

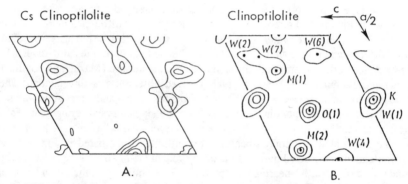

Figure 6. Difference Fourier maps showing cavity cation/water sites calculated for (a) Cs-bearing clinoptilolite, and (b) for a (Ca, Na, K)-bearing clinoptilolite (from Koyama and Takéuchi, 1977) using Rietveld-derived Bragg intensities and the tetrahedral framework of Koyama and Takéuchi (1977). Contours are at: (a) every 1 e$^-$/Å^3 and (b) 1, 5, and 10 e$^-$/Å^3.

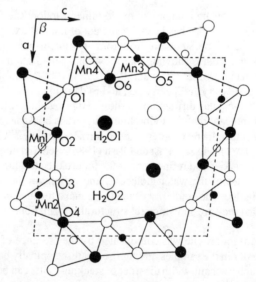

Figure 7. Projection of the Rietveld-refined todorokite structure down b. The filled circles represent atoms at y = 0, and the open circles, those at y = 0.5.

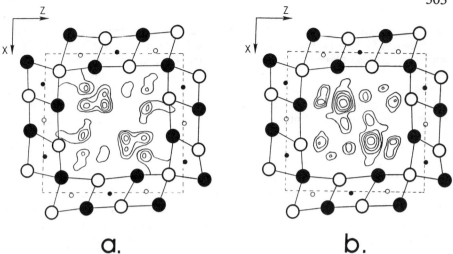

Figure 8. Difference Fourier maps at y = 0, superimposed onto projections of the todorokite framework looking down b, for samples from (a) South Africa, and (b) Cuba. The maps were calculated using the DLS framework and Rietveld-derived observed Bragg intensities. The filled circles represent atoms at y = 0, and the open circles, those at y = 0.5. The large circles indicate oxygen atoms and the small circles Mn. The contour interval is 0.4 e$^-$/Å^3. The unit cell is indicated by the dotted lines.

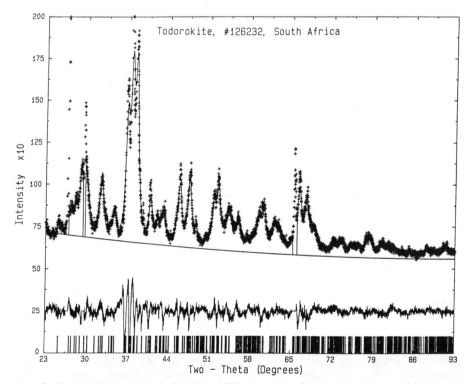

Figure 9. Observed (crosses) and calculated X-ray diffraction patterns for todorokite. The smooth line beneath the patterns is the calculated background. The plot near the bottom of the figure is the difference between the observed and calculated patterns, and the vertical lines mark the positions of the Bragg reflections ($K\alpha_1$ and $K\alpha_2$).

the effective crystallite dimension in the stacking direction, and therefore the reflections corresponding to that direction are broadened (Baerlocher, 1986). Similar types of peak broadening are also observed for samples with anisotropically shaped crystallites, e.g., platelets or fibers. Obviously, a single peak-width function is not adequate for patterns exhibiting such peak broadening effects (see, for example, Fig. 9). Greaves (1985) introduced a modification to the Caglioti et al. (1958) peak-width function that at least partially corrects for anisotropic broadening:

$$H = [U\tan^2\theta + V\tan\theta + W]^{0.5} + X\cos\phi/\cos\theta \ , \qquad (6)$$

where ϕ is the angle between the scattering vector and the broadening direction, and X is a refineable parameter. He also found that neglecting anisotropic peak broadening results in larger esd's but seems to have little impact on structural parameters. As mentioned above, the GSAS Rietveld package (Larson and Von Dreele, 1988) is able to model anisotropic broadening using a modified pseudo-Voigt profile function (Thompson et al., 1987).

In refinements for todorokite and several other manganese oxide minerals, we noted with some puzzlement that the mixing parameter for the pseudo-Voigt profile function refined to physically unreasonable values larger than 1.0. Normally, the mixing coefficient has values between 0.0 and 1.0, corresponding to the pure Gaussian and Lorentzian limits, respectively. Recently, Young and Sakthivel (1988) used computer simulations to demonstrate that "super-Lorentzian" profiles might occur if each reflection were actually the composite of two distinct profiles, one narrow and one broad. For example, a material might have two distinct distributions, with different means, of crystallite sizes or strain (or in the case of the manganese oxides, structure disorder), resulting in the superposition of a broad and narrow profile at the position of each Bragg reflection in the diffraction pattern. According to Young and Sakthivel (1988), probably the most serious consequence of "super-Lorentzian" profiles is that they can introduce errors into the refined background.

SUMMARY

The surge of interest in recent years in using powder XRD data for crystal structure analysis has been triggered primarily by the development of the Rietveld refinement method and the increased availability of computer-automated diffractometers. The major attraction of this approach is for studying structures of minerals and other materials that do not typically form crystals suitable for single-crystal diffraction experiments. Additionally, Rietveld refinements have the capability of providing accurate and precise unit-cell parameters, even for very complex diffraction patterns, and quantitative analyses of mixtures. Because powder data can typically be collected in a fraction of the time needed for a comparable single-crystal diffraction experiment, Rietveld methods can be used to advantage when several refinements are mandated, such as for heating or phase transformation studies.

The Rietveld method has benefited from many significant improvements and additions in recent years, the most notable probably coming in the area of peak profile functions. Numerous studies have demonstrated that with good data it is routinely possible to obtain refined atomic parameters that with the exception of temperature factors agree reasonably well with comparable single-crystal values. Unfortunately, most X-ray powder diffraction data suffer the effects of one or more systematic errors which are only approximately modeled during the refinement or simply ignored. Potentially the most serious error, especially for most minerals, is nonrandom orientation of the sample crystallites giving rise to systematically incorrect relative Bragg intensities. This can result in significant inaccuracies in the refined structure. To date no general software corrections have been developed that adequately deal with severe preferred orientation problems. Systematic errors if not properly accounted for in the model also contribute to unreliable esd's (generally they are underestimated), thus making it difficult to assess the accuracy of the final structure. It is therefore important to use criteria other than esd's and agreement indices when evaluating the validity of a refined structure, such as difference plots and chemical reasonability of the occupancy factors, bond distances, and angles.

In conclusion, Rietveld refinement using powder X-ray data can be a valuable tool for studying materials that are inaccessible to single-crystal diffraction methods, but at least for now the method can not in general yield results as consistently accurate and precise as those from refinements based on single-crystal diffraction data. Based on the progress of the past several years, however, it seems probable that we can anticipate significant advancements in various aspects of Rietveld refinements, particularily in the profile shape functions and modeling of systematic errors, that will lead to improved and more consistent results.

REFERENCES

Adams, J.M. (1983) Hydrogen atom positions in kaolinite by neutron profile refinement. Clays Clay Minerals, 31, 352-356.
Ahtee, M., Nurmela, M. and Suortti, P. (1989) Correction for preferred orientation in Rietveld refinement. J. Appl. Cryst., 22, 261-268.
Ahtee, M., Unonius, L., Nurmela, M. and Suortti, P. (1984) A Voigtian as profile shape function in Rietveld refinement. J. Appl. Cryst., 17, 352-357.
Albinati, A. and Willis, B.T.M. (1982) The Rietveld method in neutron and X-ray powder diffraction. J. Appl. Cryst., 15, 361-374.
Attfield, J.P., Sleight, A.W. and Cheetham, A.K. (1986) Structure determination of α-CrPO$_4$ from powder synchrotron X-ray data. Nature, 322, 630-632.
Baerlocher, Ch. (1982) "The X-ray Rietveld system". Institute of Crystallography and Petrography, ETH, Zurich.
Baerlocher, Ch. (1984) The possibilities and the limitations of the powder method in zeolite structure analysis; the refinement of the silica end-member of TPA-ZSM-5. Proceedings of the 6th International Zeolite Conference, Reno, Ed. D. Olson and A. Bisio, Butterworth, Surrey.
Baerlocher, Ch. (1986) Zeolite structure refinements using powder data. Zeolites, 6, 325-333.
Baharie, E. and Pawley, G.S. (1983) Counting statistics and powder diffraction scan refinements. J. Appl. Cryst., 16, 404-406.
Berg, J.E. and Werner, P.E. (1977) On the use of Guinier-Hägg film data for structure analysis. Z. Krist., 145, 310-320.
Bish, D.L. and Howard, S.A. (1988) Quantitative phase analysis using the Rietveld method. J. Appl. Cryst., 21, 86-91.
Bish, D.L. and Post, J.E. (1984) Thermal behaviour of todorokite and romanechite. Geol. Soc. Amer. Abstr. Programs, 16, 625.
Bish, D.L. and Post, J.E. (1988) Quantitative analysis of geological materials using X-ray powder diffraction data and the Rietveld refinement method. Geol. Soc. Amer. Abstr. Programs, 20, A223.
Bish, D.L. and Post, J.E. (1989) Thermal behavior of complex, tunnel-structure manganese oxides. Amer. Mineral., 74, 177-186.
Bish, D.L. and Von Dreele, R.B. (1989) Rietveld refinement of non-hydrogen atomic positions in kaolinite. Clays Clay Minerals, 37, 289-296.

Busing, W.R. (1981) WMIN, a computer program to model molecules and crystals in terms of potential energy functions. U.S. National Technical Information Service, ORNL-5747.
Cagliotti, G., Paoletti, A. and Ricci, F.P. (1958) Choice of collimators for a crystal spectrometer for neutron diffraction. Nucl. Inst., 3, 223-228.
Cartlidge, S., Keller, E.B. and Meier, W.M. (1984) Role of potassium in the thermal stability of CHA- and EAB-type zeolites. Zeolites, 4, 226-230.
Cartlidge, S. and Meier, W.M. (1984) Solid state transformations of synthetic CHA- and EAB-type zeolites in the sodium form. Zeolites, 4, 218-225.
Christoph, G.G., Corbato, C.E., Hofmann, D.A. and Tettenhorst, R.T. (1979) The crystal structure of boehmite. Clays Clay Minerals, 27, 81-86.
Chukhrov, F.V., Gorshkov, A.L., Sivtsov, A.V. and Berezovskaya, V.V. (1978) Structural varieties of todorokite. Izvestia Akademia Nauk, SSSR, Series Geology, 12, 86-95.
Cooper, M.J. (1982) The analysis of powder diffraction data. Acta Cryst., A38, 264-269.
David, W.I.F. (1986) Powder diffraction peak shapes. Parameterization of the pseudo-Voigt as a Voigt function. J. Appl. Cryst., 19, 63-64.
David, W.I.F. and Matthewman, J.C. (1985) Profile refinement of powder diffraction patterns using the Voigt function. J. Appl. Cryst., 18, 461-466.
Davis, B.L. (1986) A tubular aerosol suspension chamber for the preparation of powder samples for X-ray diffraction analysis. Powder Diffraction, 1, 240-243.
Dollase, W.A. (1986) Correction of intensities for preferred orientation in powder diffractometry: application of the March model. J. Appl. Cryst., 19, 267-272.
Dollase, W.A. and Reeder, R. J. (1986) Crystal structure refinement of huntite, $CaMg_3(CO_3)_4$, with X-ray powder data. Amer. Mineral., 71, 163-166.
Ercit, T.S., Hawthorne, F.C. and Cerny, P. (1985) The crystal structure of synthetic natrotantite. Bull. Mineral., 108, 541-549.
Greaves, C. (1985) Rietveld analysis of powder neutron diffraction data displaying anisotropic crystallite size broadening. J. Appl. Cryst., 18, 48-50.
Hepp, A. (1981) Beiträge zur Strukturverfeinerung mit Röntgenpulverdaten. PhD Thesis, University of Zürich, Zürich.
Hewat, A.W. (1973) UK Atomic Energy Authority Research Group Report No. RRL73/897. Harwell, Oxon, England:UKAEA.
Highcock, R.M., Smith, G.W. and Wood, D. (1985) Structure of the new zeolite theta-1 determined from X-ray powder data. Acta Cryst., C41, 1391-1394.
Hill, R.J. (1984) X-ray powder diffraction profile refinement of synthetic hercynite. Amer. Mineral., 69, 937-942.
Hill, R.J. and Flack, H.D. (1987) The use of the Durbin-Watson d statistic in Rietveld analysis. J. Appl. Cryst., 20, 356-361.
Hill, R.J. and Howard, C.J. (1985) Peak shape variation in fixed-wavelength neutron powder diffraction and its effect on structure parameters obtained by Rietveld analysis. J. Appl. Cryst., 18, 173-180.
Hill, R.J. and Howard, C.J. (1986) A computer program for Rietveld analysis of fixed wavelength X-ray and neutron powder diffraction patterns. Rep. No. M112, Australian Atomic Energy Commission (now ANSTO), Lucas Heights Research Laboratories, Menai, NSW, Australia.
Hill, R.J. and Madsen, I.C. (1984) The effect of profile step counting time on the determination of crystal structure parameters by X-ray Rietveld analysis. J. Appl. Cryst., 17, 297-306.
Hill, R.J. and Madsen, I.C. (1986) The effect of profile step width on the determination of crystal structure parameters and estimated standard deviations by X-ray Rietveld analysis. J. Appl. Cryst., 19, 10-18.
Hill, R.J. and Madsen, I.C. (1987) Data collection strategies for constant wavelength Rietveld analysis. Powder Diffraction, 2, 146-162.
International Tables for X-ray Crystallography (1974). Vol. IV. Kynoch Press, Birmingham.
Keijser, Th. H. De, Mittemeijer, E.J. and Rozendaal, H.C.F. (1983) The determination of crystallite-size and lattice-strain parameters in conjunction with the profile-refinement method for the determination of crystal structures. J. Appl. Cryst., 16, 309-316.
Khattak, C.P. and Cox, D.E. (1977) Profile analysis of X-ray powder diffractometer data: structural refinement of $La_{.75}Sr_{.25}CrO_3$. J. Appl. Cryst., 10, 405-411.
Klug, H.P. and Alexander, L.E. (1974) "X-ray Diffraction Procedures for Polycrystalline and Amorphous Materials". John Wiley and Sons, New York.
Koyama, K. and Takéuchi, Y. (1977) Clinoptilolite: the distribution of potassium atoms and its role in thermal stability. Z. Krist., 145, 216-239.

Larson, A.C. and Von Dreele, R.B. (1988) GSAS Generalized structure analysis system. Laur 86-748. Los Alamos National Laboratory, Los Alamos, NM 87545.

Lehmann, M.S., Christensen, A.N., Fjellvåg, Feidenhans'l, R. and Nielsen, M. (1987) Structure determination by use of pattern decomposition and the Rietveld method on synchrotron X-ray and neutron powder data; the structures of $Al_2Y_4O_9$ and I_2O_4. J. Appl. Cryst., 20, 123-129.

Madsen, I.C. and Hill, R.J. (1988) Effect of divergence and receiving slit dimensions on peak profile parameters in Rietveld analysis of X-ray diffractometer data. J. Appl. Cryst., 21, 398-405.

Malmros, G. and Thomas, J.O. (1977) Least-squares structure refinement based on profile analysis of powder film intensity data measured on an automatic micro densitometer. J. Appl. Cryst., 10, 7-11.

Mazzi, F. and Galli, E. (1978) Is each analcime different? Amer. Mineral., 63, 448-460.

McCusker, L.B. (1988) The ab initio structure determination of sigma-2 (a new clathrasil phase) from synchrotron powder diffraction data. J. Appl. Cryst., 21, 305-310.

McCusker, L.B., Baerlocher, Ch. and Nawaz, R. (1985) Rietveld refinement of the crystal structure of the new zeolite mineral gobbinsite. Z. Krist., 171, 281-289.

McCusker, L.B., Meier, W.M., Suzuki, K. and Shin, S. (1986) The crystal structure of a sodium gallosilicate sodalite. Zeolites, 6, 388- .

Meier, W.M. and Villiger, H. (1969) Die methode der Abstandsverfeinerung zur bestimmung der Atomkoordinaten idealisierter Gerüststrukturen. Z. Krist., 129, 411-423.

Morris, M.C., McMurdie, H.F., Evans, E.H., Paretzkin, B., Parker, H.S., Panagiotopoulos, N.C. and Hubbard, C.R. (1981) Standard X-ray diffraction powder patterns. NBS Monograph 25, Section 18, NBS, Washington, D.C.

Nord, A.G. (1984) Use of the Rietveld technique for estimating cation distributions. J. Appl. Cryst., 17, 55-60.

Nord, A.G. and Stefanidis, T. (1983) Crystallographic studies of some olivine-related $(Ni,Mg)_3(PO_4)_2$ solid solutions. Phys. Chem. Minerals, 10, 10-15.

Pawley, G.S. (1980) EDINP, the Edinburg powder profile refinement program. J. Appl. Cryst., 13, 630-633.

Post, J.E. (1988) Rietveld refinement of the akaganeite structure using powder X-ray diffraction data. Geol. Soc. Amer. Abstr. Programs, 20, A102.

Post, J.E. and Bish, D.L. (1988) Rietveld refinement of the todorokite structure. Amer. Mineral., 73, 861-869.

Post, J.E. and Bish, D.L. (1989) Rietveld refinement of the coronadite structure. Amer. Mineral., In Press.

Post, J.E. and Burnham, C.W. (1986) Ionic modeling of mineral structures in the electron gas approximation: TiO_2 polymorphs, quartz, forsterite, diopside. Amer. Mineral., 71, 142-150.

Post, J.E., Von Dreele, R.B. and Buseck, P.R. (1982) Symmetry and cation displacements in hollandites: structure refinements of hollandite, cryptomelane and priderite. Acta Cryst., B38, 1056-1065.

Prince, E. (1981) Comparison of profile and integrated-intensity methods in powder refinement. J. Appl. Cryst., 14, 157-159.

Prince, E. (1983) The effect of finite detector slit height on peak positions and shapes in powder diffraction. J. Appl. Cryst., 16, 508-511.

Prince, E. (1985) Precision and accuracy in structure refinement by the Rietveld method. In A.J.C. Wilson (Ed.), "Structure and Statistics in Crystallography", Adenine Press.

Raudsepp, M., Turnock, A.C. and Hawthorne, F.C. (1987a) Characterization of cation ordering in synthetic scandium-fluor-eckermannite, indium-fluor-eckermannite, and scandium-fluor-nyboite by Rietveld structure refinement. Amer. Mineral., 72, 959-964.

Raudsepp, M., Turnock, A.C., Hawthorne, F.C., Sherriff, B.L. and Hartman, J.S. (1987b) Characterization of synthetic pargasitic amphiboles $(NaCa_2Mg_4M^{3+}Si_6Al_2O_{22}(OH,F)_2; M^{3+} = Al,Cr,Ga,Sc,In)$ by infrared spectroscopy, Rietveld structure refinement, and ^{27}Al, ^{29}Si, and ^{19}F MAS NMR spectroscopy. Amer. Mineral., 72, 580-593.

Rietveld, H.M. (1967) Line profiles of neutron powder-diffraction peaks for structure refinement. Acta Cryst., 22, 151-152.

Rietveld, H.M. (1969) A profile refinement method for nuclear and magnetic structures. J. Appl. Cryst., 2, 65-71.

Rudolf, P. and Clearfield, A. (1985) The solution of unknown crystal structures from X-ray powder diffraction data. Technique and an example, $ZrNaH(PO_4)_2$. Acta Cryst., B41, 418-425.

Sakata, M. and Cooper, M.J. (1979) An analysis of the Rietveld profile refinement method. J. Appl. Cryst., 12, 554-563.

Sato, M., Kodama, N. and Matsuda, S. (1981) An application of pattern fitting structure refinement to muscovite $KAl_2(Si_3Al)O_{10}(OH)_2$. Mineral. J. (Japan), 10, 222-232.

Scott, H.G. (1983) The estimation of standard deviations in powder diffraction Rietveld refinements. J. Appl. Cryst., 16, 159-163.

Shiokawa, K., Ito, M. and Itabashi, K (1989) Crystal structure of synthetic mordenites. Zeolites, 9, 170-176.
Smith, S.T., Snyder, R.L. and Brownell, W.E. (1979) Minimization of preferred orientation in powders by spray drying. Advances in X-ray Analysis, 22, 77-88.
Suitch, P.R. and Young, R.A. (1983) Atom positions in highly ordered kaolinite. Clays Clay Minerals, 31, 357-366.
Tettenhorst, R.T. and Corbato, C.E. (1984) Crystal structure of germanite, $Cu_{26}Ge_4Fe_4S_{32}$, determined by powder X-ray diffraction. Amer. Mineral., 69, 943-947.
Thompson, J.G., Withers, R.L. and Fitzgerald, J.D. (1989) Beyond the limits of Rietveld refinement? Society of Crystallographers, Australia, 16th Meeting (abstract).
Thompson, P., Cox, D.E. and Hastings, J.B. (1987) Rietveld refinement of Debye-Scherrer synchrotron X-ray data from Al_2O_3. J. Appl. Cryst., 20, 79-83.
Thompson, P. and Wood, I.G. (1983) X-ray Rietveld refinement using Debye-Scherrer geometry. J. Appl. Cryst., 16, 458-472.
Toraya, H. (1985) The effects of profile-function truncation in X-ray powder-pattern fitting. J. Appl. Cryst., 18, 351-358.
Toraya, H. (1986) Whole-pattern fitting without reference to a structural model: application to X-ray powder diffractometer data. J. Appl. Cryst., 19, 440-447.
Toraya, H., Iwai, S. and Marumo, F. (1980) The structural investigation of a kaolin mineral by X-ray powder pattern-fitting. Mineral. J. (Japan), 10, 168-180.
Toraya, H. and Marumo, F. (1980) Report of the Research Laboratory of Engineering Materials, Tokyo Inst. Technology, Vol. 5, 55-64.
Torres-Martinez, L.M. and West, A.R. (1986) New family of silicate phases with the pollucite structure. Z. Krist., 175, 1-7.
Turner, S. and Buseck, P.R. (1981) Todorokites: a new family of naturally occurring manganese oxides. Science, 212, 1024-1027.
Waser, J. (1963) Least-squares refinement with subsidiary conditions. Acta Cryst., 16, 1091-1094.
Wertheim, G.K., Butler, M.A., West, K.W. and Buchanan, D.N.E. (1974) Determination of the Gaussian and Lorentzian content of experimental line shapes. Rev. of Scientific Instruments, 45, 1369-1371.
Wiles, D.B. and Young, R.A. (1981) A new computer program for Rietveld analysis of X-ray powder diffraction patterns. J. Appl. Cryst., 14, 149-151.
Will, G., Bellotto, M., Parrish, W. and Hart, M. (1988) Crystal structures of quartz and magnesium germanate by profile analysis of synchrotron-radiation high-resolution powder data. J. Appl. Cryst., 21, 182-191.
Will, G., Masciocchi, N., Parrish, W. and Hart, M. (1987) Refinement of simple crystal structures from synchrotron radiation powder diffraction data. J. Appl. Cryst., 20, 394-401.
Will, G., Parrish, W. and Huang, T.C. (1983) Crystal-structure refinement by profile fitting and least-squares analysis of powder diffractometer data. J. Appl. Cryst., 16, 611-622.
Young, R.A. (1988) Pressing the limits of Rietveld refinement. Australian J. Phys., 41, 297-310.
Young, R.A. and Hewat, A.W. (1988) Verification of the triclinic crystal structure of kaolinite. Clays Clay Minerals, 36, 225-232.
Young, R.A., Mackie, P.E. and Von Dreele, R.B. (1977) Application of the pattern-fitting structure-refinement method to X-ray powder diffractometer patterns. J. Appl. Cryst., 10, 262-269.
Young, R.A. and Sakthivel, A. (1988) Bimodal distributions of profile-broadening effects in Rietveld refinement. J. Appl. Cryst., 21, 416-425.
Young, R.A. and Wiles, D.B. (1982) Profile shape functions in Rietveld refinements. J. Appl. Cryst., 15, 430-438.

10. SYNCHROTRON POWDER DIFFRACTION

L.W. Finger

INTRODUCTION

As recently as 15 or 20 years ago, synchrotron radiation was an aggravating fact of life for high-energy electron or positron accelerators and storage rings. It contributed to the cost of ring operation because of the electricity needed to replace the energy lost through the radiation process, and it led to heating of the walls of the ring, with the associated degassing and degradation of vacuum. The ions released from the walls interacted with the stored beam, reducing the lifetime and the efficiency of operation. After the first measurement of synchrotron radiation in the X-ray region (Bathow et al., 1966), however, progress in application of these sources to studies of the structure of matter was rapid. Synchrotron radiation facilities, which included the Stanford Synchrotron Radiation Laboratory (SSRL) on the Stanford Positron-Electron Accelerating Ring (SPEAR) storage ring, the Deutsches Electronen Synchrotron (DESY) at Hamburg, a ring named VEPP-3 at Novosibirsk, and the Cornell High Energy Synchrotron Source (CHESS), were operated in conjunction with high-energy physics experiments, or in the so-called "parasitic mode." The demand for access to these facilities, and the natural friction between the investigators interested in high-energy particles and those interested in intense beams of radiation, resulted in the construction of second-generation machines that are dedicated to production of synchrotron radiation and optimized to that end. Included in this category are the National Synchrotron Light Source (NSLS) at Brookhaven National Laboratory and Tsukuba, Japan (Photon Factory). In addition, the operating parameters for older machines have been adjusted to enhance their production of synchrotron radiation. The demands of the scientific community are still not satisfied. Third-generation machines with even higher performance are being designed as the Advanced Photon Source (APS) at Argonne National Laboratory, the European Synchrotron Radiation Facility (ESRF) at Grenoble, France, and a similar machine to be built near Osaka, Japan. Undoubtedly, these machines will be fully subscribed, and an even newer generation of facilities will be built, scientific budgets permitting.

Proper choice of experiment and equipment is critical if efficient operation is to be obtained at a synchrotron beam line. Because of competition for beam line access, the available time is typically short. Furthermore, a synchrotron experiment may be the investigator's first experience with "big" science, i.e., a shared facility in a remote location. This chapter will describe the properties of the beams, how the powder diffraction experiment is affected by the source characteristics, present and future facilities, and some representative experiments that have been conducted.

SOURCE CHARACTERISTICS

A synchrotron is a particle accelerator consisting of a high-vacuum beam

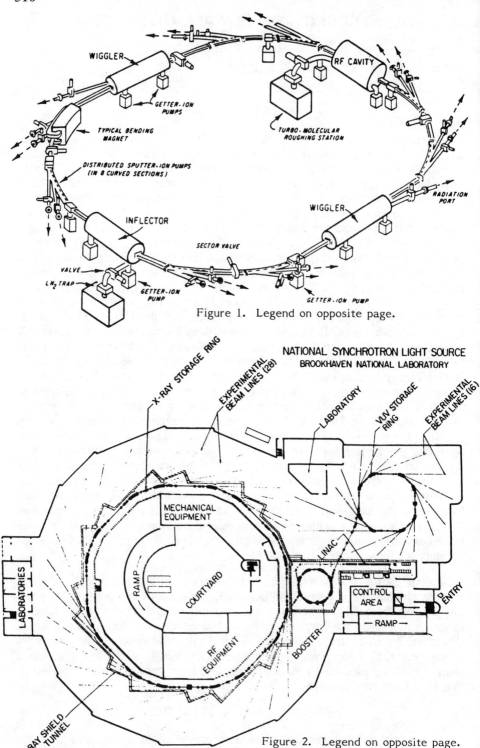

Figure 1. Legend on opposite page.

Figure 2. Legend on opposite page.

tube that is contained in a roughly circular pattern. Particles are injected at relatively low energy and accelerated to high energy through induction in radio frequency cavities. The particles are directed around the ring by the use of magnet assemblies that consist of a dipole with strength on the order of one tesla (10 kgauss) to bend the beam, and associated quadrupoles and sextupoles. These higher order field components are used to focus the electron beams and to compensate for imperfections in the dipole fields. In a synchrotron, the injection and acceleration processes usually have a relatively high repetition rate (60 Hz) as the particles are not stored but are caused to collide with other beams or with fixed targets.

Because charged particles emit photons whenever they are accelerated in an electric or magnetic field, synchrotrons produce radiation; however, these photons are difficult to use for diffraction experiments. The varying energy of the particles during the acceleration cycle and the variations in current, particle position, and beam size from one cycle to the next reduce the usefulness of such devices. Accordingly, most synchrotron radiation is produced in storage rings; devices that are optimized for long-term containment of particle beams. A storage ring has the same type of components that are found in a synchrotron (Fig. 1). The vacuum in a storage ring is more critical, however, as collisions between the residual gas molecules and the particles will determine the lifetime of the beam. At operating conditions, a high-energy storage ring must have a vacuum on the order of 10^{-9} torr. With no current in the device, the pressure characteristically decreases by two orders of magnitude. The layout of the two storage rings that constitute the NSLS is shown in Figure 2. The electrons originate in a linear accelerator (LINAC), are accelerated to an energy of 0.002 GeV, and injected to the booster synchrotron in which the energy is raised to 0.75 GeV. At this point the electrons are transferred to the vacuum ultraviolet (VUV) storage ring or into the X-ray ring. Once the desired current is established in the latter device, the electron energy is ramped to 2.5 GeV.

A synchrotron X-ray beam is very intense, highly polarized, has an intensity distribution with energy or wavelength unlike any conventional X-ray source, and

---

Figure 1. Schematic of an intermediate-energy (0.75 GeV) electron storage ring designed as a source of synchrotron radiation. Tangential beam ports for bending magnet and wiggler radiation sources (see text) are shown. Energy lost by electrons in the form of synchrotron radiation is replenished in the radio-frequency (rf) cavity. The storage ring must operate at high vacuum with no oil contamination; therefore, turbo-molecular pumps, which are used for "rough" pumping, have liquid nitrogen traps. Final vacuum is achieved through the use of getter-ion and sputter-ion pumps (from Winick, 1980b).

Figure 2. Plan of the National Synchrotron Light Source, Brookhaven National Laboratory. Electrons produced in the linear accelerator (LINAC) are injected, at 2 MeV, into the booster synchrotron where the energy is raised to 750 MeV. The electrons are then diverted into one of the two storage rings. The smaller vacuum-ultraviolet (VUV) ring, which operates at 750 MeV, produces radiation in the soft X-ray or ultraviolet region (10–2500 Å). The larger X-ray storage ring is filled at 750 MeV but is ramped to 2.5 GeV for operation. Diagram courtesy of the NSLS staff.

has an intrinsically small divergence perpendicular to the plane of the orbit of the particles in the storage ring. During operation there is a decay in the primary beam intensity as electrons are lost from the ring, thus any quantitative measurements must have provision for normalizing to a standard ring current. The photons also have a time structure that results from the interaction between the electrons and the radio-frequency (rf) equipment used to replace the energy lost through radiation.

In the following discussion, any reference to electrons also applies to positrons. Even though most storage rings in the first two generations use electrons (the Photon Factory is one exception), future rings will use positrons, despite the additional complexity of the sources. In operation, the walls of the ring are bathed in very intense photon fluxes, resulting in photodesorption of impurities. These heavy ions, which are predominately positively charged, are attracted to the electrons and cause rapid current decay. A positron beam repells these ions, with much longer beam life as a consequence. Synchrotron radiation is also possible for proton-storage rings (Winick, 1980a), however, the ring energy, which is proportional to the mass of the particle, would have to be 3 orders of magnitude higher than that for electrons, for comparable spectral distributions.

Beam divergence/resolution

When a charged particle is accelerated by an electric or magnetic field, it emits radiation in a dipole pattern for low velocity particles, as shown in Figure 3. As the particle approaches the speed of light, the details of the emission process, as viewed from the laboratory reference frame, appear to change and the radiation seems to be concentrated in the forward direction of travel for the particle. The transformation from the rest frame to the laboratory system changes the effective angle between a photon and the direction of motion from θ to θ'. For the worst case of $\theta = 90°$, $\theta' \approx mc^2/E$ (Winick, 1980a), where m is the electron mass, c is the speed of light, and E is the energy of the ring. Thus for a storage ring with closed orbits in the horizontal plane, the radiation is emitted tangentially with a vertical divergence given by $\theta_V \approx mc^2/E \approx 5 \times 10^{-4}/E$, where E is given in GeV. This relationship is valid near the critical energy, ϵ_c, which is the value at the midpoint of the power output and is proportional to E^2. Some representative values of ϵ_c, in keV, for bending magnet sources are 4.1 at the Photon Factory, 4.7 at SSRL, 5.0 at NSLS, 11.5 at CHESS, and 19.6 at APS. For readers more used to thinking in terms of wavelength, $\lambda = 12.398/\epsilon$, where λ is in Å and ϵ in keV. At the NSLS with an operating energy of 2.5 GeV, the vertical divergence is approximately 0.2 milliradian leading to a vertical beam thickness on the order of 0.5 cm at a working distance of 25 m. This intrinsic collimation of synchrotron beams in the vertical direction is the factor that permits very high resolution, while maintaining reasonable counting rates. In addition, detector and analyzer systems can be designed with very high signal to noise performance.

The relativistic transformation described above also implies that the radiation will be highly polarized. In Figure 4, the horizontal and vertical components of the polarization are graphed in terms of angle in units of θ_V away from the horizontal plane. For $\epsilon \approx \epsilon_c$, the radiation is highly polarized in the horizontal

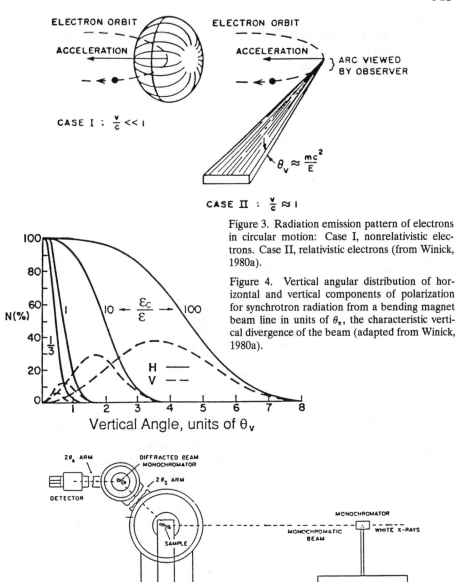

Figure 3. Radiation emission pattern of electrons in circular motion: Case I, nonrelativistic electrons. Case II, relativistic electrons (from Winick, 1980a).

Figure 4. Vertical angular distribution of horizontal and vertical components of polarization for synchrotron radiation from a bending magnet beam line in units of θ_v, the characteristic vertical divergence of the beam (adapted from Winick, 1980a).

Figure 5. Schematic diagram of the triple-axis powder diffractometer at beam-line X7A of the National Synchrotron Light Source (from Cox et al., 1986).

direction. As shown by Reynolds (Chapter 1, this volume), the polarization factor for the component parallel to the Bragg planes is unity; whereas that factor for the perpendicular direction is proportional to $\cos^2 2\theta$, i.e., less than one. To maximize the scattered intensity, instruments are designed with vertical scattering geometry for the sample. One exception is X-ray fluorescence in which diffraction lines could swamp the signal. Accordingly, a horizontal angle of 90° is chosen so that the polarization factor for the horizontal component vanishes (Gordon and Jones, 1985). Figure 4 also shows the manner in which the vertical divergence changes as the energy deviates from ϵ_c.

In any experimental system that has a high intrinsic resolution, it is always possible to trade intensity for resolution, if desired. An instrument optimized for high-resolution powder data was installed at the NSLS beam line X13A (Cox et al., 1986) and has since been moved to X7A at the NSLS. In the original design, this triple-axis spectrometer (Fig. 5) was normally operated with a Ge(111) monochromator, scattering in the horizontal plane, and a Ge(220) or Ge(440) analyzer crystal. For the latter choice, the instrument resolution expressed as $\Delta d/d$ is approximately 3×10^{-4} for $\lambda = 1.5$ Å at the focusing minimum near a 2θ of 60°. The choice of a perfect Ge crystal as the analyzer eliminates the errors in peak position due to specimen transparency or flat-plate geometry due to the stringent acceptance conditions for the analyzer. A recent modification (D.E. Cox, personal communication, 1989) has replaced the Ge crystal with a channel-cut, double-crystal Si(111) monochromator scattering in the vertical plane. Due to a reduction of the energy bandwidth of the primary beam at the sample, the instrument resolution has been substantially increased at a modest loss in intensity.

Parrish and Hart (1985) and Parrish et al. (1986) described an instrument (Fig. 6) with lower resolution than the NSLS machine, but with substantially higher counting rates. The counter aperture can be defined by a set of horizontal parallel slits, a simple receiving slit, or a crystal analyzer. With receiving slits in place, the instrument resolution at minimum width is determined by the angular opening of the slits, ranging from 0.07 to 0.45°. The corresponding range in $\Delta d/d$ is $1-7 \times 10^{-3}$. Configurations with receiving slits are susceptible, however, to asymmetric broadening and peak shifts due to X-ray transparency of the specimen, and to peak shifts due to height errors of the sample.

A hypothetical powder diffractometer for a synchrotron source was described by Sabine (1987), whose design requirements are for an instrument capable of collection of high-resolution intensity data in times comparable to the rate of solid-state reactions. He concluded that a triple-axis spectrometer (with analyzer crystal) is required to obtain the desired resolution, that a reasonable balance between resolution and intensity requires mosaic spread of the monochromator and analyzer crystals to be on the same order as the vertical divergence of the incident beam, and that a Bragg angle of 60° is required for the monochromator.

Photon output

Synchrotrons are very intense sources of ionizing radiation. As a result, safety considerations at the beam lines are very important. Each experimental

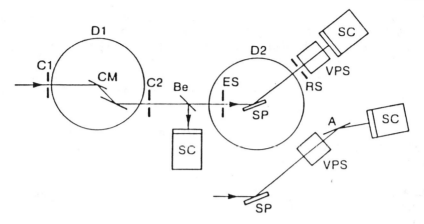

Figure 6. Schematic of powder diffractometer used at SSRL. C1 primary beam collimator, CM channel monochromator on diffractometer D1, C2 radiation shield, Be beryllium foil monitor, ES entrance slit, SP specimen on powder diffractometer D2, RS receiving slit, VPS vertical parallel slits, SC scintillation counter. Crystal analyzer alternative at lower right (from Parrish et al., 1986).

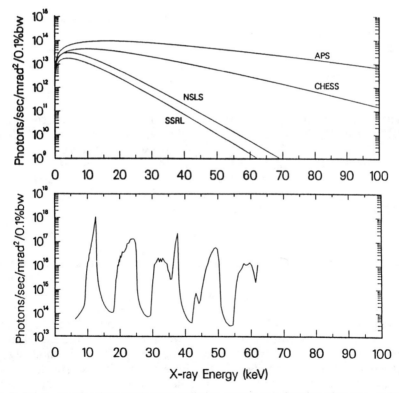

Figure 7. Brightness of various synchrotron radiation sources as a function of energy. a) Selected bending magnet beam lines. The shape of wiggler lines would be similar. (b) Proposed undulator insertion device at APS with a deflection parameter of 1.1 (from Rivers, 1988).

station is interlocked to prevent operating personnel from access to the radiation areas while the beam is on. All adjustments that cannot be made by remote control require that the shutter be closed and the interlock be broken before the investigator enters the radiation enclosure. For further protection of personnel, most faults in the interlock system cause the current in the ring to be dumped. This scheme aborts production of radiation much more quickly than a safety shutter could be closed.

As described by Rivers (1988), there are a number of quantities that describe the photon output of synchrotron sources and these measures will depend upon both the beam line and the experiment. Because the radiation emitted from a bending magnet has a broad energy spectrum (see next section) and it occurs in a horizontal fan, it is necessary to specify a window in the coordinates of both energy and horizontal opening. The usual dimensions are for an energy bandwidth of 0.1%, and for an opening of one milliradian. For an experiment with a relatively large sample such as a flat-plate powder mount, an important quantity is obtained by integrating the entire energy bandwidth over the vertical dimensions of the beam to obtain the *flux* in units of photons per second per milliradian per 0.1% bandwidth. It is unlikely, however, that an experiment could use the entire flux as practical considerations limit the beam height to 2 mm or less.

If small samples, such as found in a very high-pressure, diamond-anvil cell, are under investigation and the beam size is limited by pinhole or slit optics, only a small portion of the vertical divergence is important. Consequently, the flux/vertical opening, or *brightness*, is the important quantity. Most plots of photon output from synchrotron sources (Winick, 1980a; Rivers, 1988; Fig. 7) present this quantity.

For beam lines that employ crystal optics or grazing-angle mirrors for horizontal focusing, the source size as well as the perfection of the optical elements, determine the area of the beam at the focal point. The pertinent quantity is the brightness/source area, or *brilliance*. The second generation machines at the NSLS and the Photon Factory are characterized by smaller source sizes and higher brilliance than earlier models. This trend will be continued with the third generation of machines now under design and construction.

Spectral distribution

The brightness, $B(\epsilon)$, of a bending magnet source (Winick, 1980a) is proportional to $E^2 I H_2(\epsilon/\epsilon_c)$, where E is the electron energy, I is the current in the ring, ϵ is the photon energy, ϵ_c is the critical energy defined above, and H_2 is a modified Bessel function of order 2. The calculated brightness as a function of energy for a number of sources (Rivers, 1988) is reproduced in Figure 7a. The smooth variation with photon energy, which arises from the Bessel function, implies that a user has a wide choice of operating energy in monochromatic experiments. In fact, absorption in the Be windows that isolate the ultra-high ring vacuum from the experiment, and in air in the beam path, coupled with the reduced numbers of photons above 25 keV for rings with $E \sim 2\text{-}3$ GeV, limit the effective working range from 0.5–2.0 Å. Alternatively, the spectrum is well suited for energy-dispersive diffraction, a technique to be described later. The spectral distribution of a monochromatic beam is determined by the characteristics of the monochromator crystal(s). Unlike for a conventional source, there is no intrinsic doublet.

As experience has been gained in the use of synchrotron radiation, it has been realized that bending magnets are not the best sources of synchrotron radiation. A straight section of the ring can be modified by installing magnets to cause the beam to be bent a little, followed by an equivalent reverse bend. With proper choice of the field and the spacing between opposite magnetic poles, the electron beam can emerge undeflected; however, radiation traveling in the forward direction is generated with each deflection. It is possible, therefore, to generate considerably brighter sources by the use of such "insertion" devices (Fig. 1). The characteristics of the radiation produced depend upon the value of a magnetic deflection parameter, K, defined as $K = 0.934\lambda_u B_0$, where λ_u is the period of the deflection magnets, and B_0 is the magnitude of the deflection field (Rivers, 1988). For $K > 10$, which implies a few high-strength magnets with a long period, the insertion device is called a *wiggler* and has a brightness function that is essentially the same as for bending magnets. Note, however, that the field strengths for bending magnets are restricted by the requirement that the electrons traverse a closed orbit. Wigglers, on the other hand, can have magnetic fields that yield a critical energy higher or lower than a bending magnet on the same ring. Two wiggler designs have been proposed for APS with ϵ_c of 9.8 and 32.6 keV, respectively, as opposed to the bending magnet value of 19.6 keV. The wiggler recently installed at beam line X17 of NSLS has ϵ_c equal to 20.8 keV. The resulting brightness is very similar to the APS bending magnet line (Fig. 7a). Unlike a bending magnet beam line, wiggler lines can be arranged to generate beams that are polarized in the vertical direction, or even circularly polarized (Spencer and Winick, 1980).

If the insertion device has a small value of K (~ 1) due to many poles with short period and relatively weak magnetic fields, the insertion device is called an *undulator*, because it undulates, rather than wiggles, the electrons. The brightness spectrum (Fig. 7b) is no longer uniform but has peaks and valleys. In addition, it is highly collimated in both directions. By changing the magnetic field through adjustment of the gap, the energy of the peak can be altered. The ring under design at APS is optimized for this type of device, which has a brightness four to five orders of magnitude greater than that of a bending magnet at the NSLS. With such a device, count rates at high resolution will no longer be a problem. One will, however, have to be much more concerned with problems such as sample heating!

Time structure of ring

For stable operation of a storage ring, the electrons are not distributed about the ring in a uniform current, but are arranged in "bunches" of electrons in the ring. The time required for a given electron to traverse the ring will depend upon the ring diameter and the operating energy. For the NSLS at 2.5 GeV, this time is 570 nanosec. The length of an individual bunch is 10.5 cm, therefore the time for a bunch to pass a beam port will be approximately 0.4 nanosec.

Because photons are present at the sample position only during this time, it is possible, therefore, to design studies that measure transient phenomena with time scales as short as 10^{-9} sec. Most experiments require more photons than can be provided by a single bunch, however, the pulsed nature of the source has an additional ramification. The instantaneous count rate is several orders of magnitude larger than the average. If the detection system is susceptible to saturation, this condition may be obtained, even though the average count rate is well within limits (Glazer et al., 1978).

HIGH-RESOLUTION EXPERIMENTS

Peak shape

Because the peaks in most powder patterns are not fully resolved, a model for the shape of the individual peaks of an envelope must be known before these data can be interpreted. This situation is true both for whole pattern fitting, in the manner of Rietveld (1969), or for the profile-fitting technique of Taupin (1973) or Parrish et al. (1976). Despite the unfortunate choice of name, which suggests fitting to the complete envelope, the latter method depends on extraction of the individual intensities of the component lines.

Experience has shown that for synchrotron powder diffraction data, simple Lorentzian or Gaussian functions are not adequate to describe peak shapes, even though the peaks are symmetric except at very low scattering angles. Various means of convoluting these elementary functions, such as the Pearson type VII distribution (Hall et al., 1977), Voigt (Suortti et al., 1979; Ahtee et al., 1984), or pseudo-Voigt (Wertheim et al., 1974; Young and Wiles, 1982, Cox et al., 1986) have been suggested. Fitch and Murray (1989) have described a Rietveld program using the complex peak shape of Toraya (1986), which has an implicit asymmetry that may be more suitable for conventional sources than for synchrotrons. The main problem with the Toraya formulation is the relatively complicated derivatives of the function, and the difficulty in ascribing a physical meaning to the functional form.

The pseudo-Voigt form (Fig. 8), which seems to be best for synchrotron data, has a calculated intensity $I(\Delta 2\theta)$ at a point displaced by $\Delta 2\theta$ from the Bragg angle that is given by

$$I(\Delta 2\theta) = I_P \left[\frac{2\eta}{\pi \Gamma [1 + 4(\Delta 2\theta/\Gamma)^2]} + \frac{2(1-\eta)(\ln 2/\pi)^{1/2} \exp[-4\ln 2(\Delta 2\theta/\Gamma)^2]}{\Gamma} \right], \quad (1)$$

where I_P is the integrated intensity for the peak, Γ is the full-width at half maximum, and η is a mixing constant. As η ranges from zero to one, the shape of the resulting curve changes from Gaussian to Lorentzian.

The widths of the Gaussian and Lorentzian components of the pseudo-Voigt formulation, which are denoted by Γ_G and Γ_L, respectively, have distinct variation with scattering angle and can be separated in a high-resolution experiment such as conducted on a synchrotron. The angular dependence of the Gaussian component

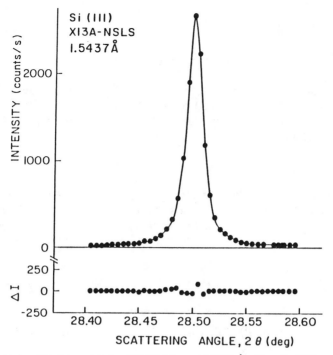

Figure 8. Experimental data points for Si(111) taken at 1.5437 Å with a Ge(111) monochromator and Ge(222) analyzer. Storage ring current was about 30 mA. The solid line represents a least-squares fit to a pseudo-Voigt function ($\Gamma = 0.0172(3)°$, $\eta = 0.77(2)$, $\chi^2 = 1.39$). A difference plot is shown in the lower part of the figure. The true background on either side of the peak is about 3 counts/s (from Cox et al., 1986).

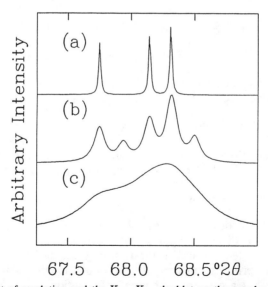

Figure 9. The effect of resolution and the $K\alpha_1$, $K\alpha_2$ doublet on the powder pattern of α-quartz in the vicinity of the (212), (203), and (301) Bragg lines. (a) Pattern calculated for a Γ of 0.02°, a representative value for the NSLS X7A (Fig. 5) instrument in this range of 2θ. (b) Same spacings calculated for the $K\alpha$ doublet and a peak width of 0.1°. (c) Peak width of 0.5°.

was given by Rietveld (1969) as

$$\Gamma_G = (U\tan^2\theta + V\tan\theta + W)^{1/2}, \qquad (2)$$

where U, V, and W are variable coefficients. The effects of Gaussian strain or disorder broadening are expressed by U and the minimum width of peaks for the machine and sample are accomodated by W. The Lorentzian halfwidths are given by Cox et al. (1988) as

$$\Gamma_L = X\tan\theta + Y/\cos\theta, \qquad (3)$$

where the first term describes the spectral and Lorentzian-like strain or disorder broadening and the second term treats particle size broadening. The individual half-widths for the Gaussian and Lorentzian components can be related to the psuedo-Voigt parameters Γ and η by the following polynomial approximations (Cox, 1984):

$$\Gamma = (\Gamma_G^5 + 2.69269\Gamma_G^4\Gamma_L + 2.42843\Gamma_G^3\Gamma_L^2 + 4.47163\Gamma_G^2\Gamma_L^3 + 0.07842\Gamma_G\Gamma_L^4 + \Gamma_L^5)^{1/5}, \qquad (4)$$

and

$$\eta = 1.36603(\Gamma_L/\Gamma) - 0.47719(\Gamma_L/\Gamma)^2 + 0.11116(\Gamma_L/\Gamma)^3. \qquad (5)$$

The peak widths in the pseudo-Voigt formulation can be described by least-squares adjustment of η and a general U, V, and W using equations (1) and (2). The halfwidths Γ_G and Γ_L are obtained from the refined values of Γ and η (Cox et al., 1988) through the use of

$$\Gamma_L/\Gamma = 0.72928\eta + 0.19289\eta^2 + 0.07783\eta^3, \qquad (6)$$

$$\Gamma_G/\Gamma = (1 - 0.74417\eta - 0.24781\eta^2 - 0.00810\eta^3)^{1/2}. \qquad (7)$$

Equations (6) and (7) have been corrected for a typographical error in the original publication (Cox, personal communication).

A better method of determining peak widths is to refine U, V, W, X, and Y in equations (2) and (3), and calculate values for Γ and η using equations (4) and (5). Although there are two extra parameters in the refinement, there is a more direct relationship between the parameters and a physical model.

Determination of lattice geometry

Although a small deviation from a high symmetry condition has major implications for phase diagram topology and may have a major influence on the thermochemistry, detection and verification of these slight distortions is one of the most difficult problems in crystallography. Although a single-crystal diffraction experiment has access to directional information in three dimensions and the intensities of resolved diffraction maxima, departure from ideal symmetry is commonly only a few standard deviations. In a powder experiment, the diffraction from the Miller planes in question overlap. A necessary condition to prove departure from the higher symmetry is to observe peak splitting. With a conventional source, the presence of the $K\alpha$ doublet in the incident radiation is a problem, unless an incident-beam focusing monochromator geometry is employed. When the additional problem of a divergent source is included, peaks become broadened and are not resolved. Figure 9 shows the effect of these two factors on the pattern of α-quartz at the angles at which the (212), (203), and (301) Bragg lines occur. The patterns were calculated for Cu $K\alpha$ radiation. Panel (a) displays the pattern for a Γ of 0.02°, a representative value for the NSLS X7A instrument in this range of 2θ. Panel (b) has the same spacings calculated for the $K\alpha$ doublet and a peak width of 0.1°. If the peaks broaden to 0.5° as in panel (c), the individual components are lost and the envelope assumes the appearance of a doublet of very broad peaks. On the high-resolution synchrotron powder diffractometer at the NSLS (Fig. 5), Sleight and Cox (1986) showed that the pseudocubic superconducting phase $BaPb_{.725}Bi_{.275}O_3$ is actually tetragonal. The resolution of split peaks was nearly complete even though the two distinct axes of the pseudocubic cell differed by only 8×10^{-3}. They rejected a proposed model that involved an orthorhombic modification, because some required splittings did not occur.

The advantages associated with a high-resolution experiment for a mineral sample are shown in Figure 10, which is from a study of benstonite, $Ca_7(Ba,Sr)_6(CO_3)_{13}$, by Post, Finger and Cox (in preparation). At first glance, the diffractogram appears normal; however, if the peaks are expanded as shown in Figure 11, it is obvious that there is a second component on the low-angle side, corresponding to a larger lattice constant. The positions of each component of these doublets were fitted using equation (1), and the peak positions were used as input to program TREOR (Werner et al. , 1985). The two sets of lines could be indexed separately; however, if all lines were included together, the algorithm failed. This pattern corresponds to a rhombohedral lattice that has a = 18.2583(6), c = 8.6848(5) Å in the R-centered form for the major constituent, and a = 18.282(2), c = 8.719(1) Å for the minor component. This difference in lattice constants is believed to arise from a segregation into regions of differing Ba content.

The large peak at 28° in Figure 10 arises from the submicron diamond powder that was mixed with the benstonite to reduce the effects of absorption. This technique is discussed later in this chapter.

Extraction of intensities for structure solution

Solving crystal structures from powder diffraction data is a very exciting recent development. In principal, it has always been possible to extract integrated intensities from a pattern and to use these data with any of the standard techniques for structure solution, such as Patterson syntheses for heavy-atom structures, or direct methods such as the tangent formula. In practice, however, the uncertainties caused by low resolution and the doublet in the incident beam have reduced the effectiveness of the technique. Note the difference between this approach and that of the Rietveld method in which the parameters of a structural model are adjusted to reduce the disagreement between the calculated and observed

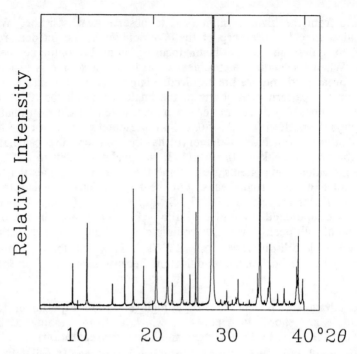

Figure 10. Diffractogram measured for benstonite, $Ca_7(Ba,Sr)_6(CO_3)_{13}$, at the high-resolution instrument at the NSLS beam line X7A. A capillary sample, diluted with diamond powder to reduce absorption, was utilized at a wavelength of 1.4883 Å. The peak at 28° arises from diamond.

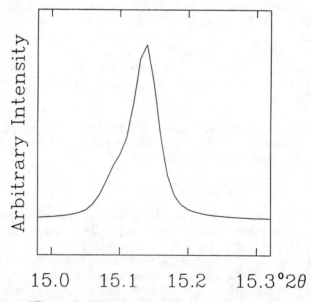

Figure 11. Segment of Figure 10 demonstrating that the peaks are comprised of doublets, which arise from the sample, not from the incident radiation. The two peaks can be fitted to a pseudo-Voigt shape without difficulty, even though the low-angle peak deviates from the stronger by only 0.2% in d.

pattern.

In the case of relatively simple crystal structures such as quartz, corundum, calcite, and ZrO_2 (Toraya, 1986), and Si, CeO_2, Co_3O_4, and Yb_2O_3 (Will et al., 1987), it is possible to extract integrated intensities and refine the structural parameters. For more complicated lattice geometries, peak overlap is a severe problem.

The solution of the structure for complicated materials requires very high resolution and was first accomplished by Attfield et al. (1986) for α-$CrPO_4$, a structure with a shared edge between Cr octahedra and P tetrahedra. This configuration is highly strained; therefore, the material does not grow large crystals. Integrated intensities for 68 well-resolved Bragg reflections were obtained from the synchrotron powder diffraction data. Interpretation of a Patterson map yielded the position of one Cr and one P atom. Successive Fourier cycles resulted in the determination of the positions of all eight atoms in the asymmetric unit. The atomic coordinates were refined to an R of 11% and the resulting structural model was used as the starting point for Rietveld refinement. Additional details of the structure have been elucidated by a joint X-ray and neutron study (Attfield et al., 1988).

A second *ab-initio* crystal structure solution from high-resolution powder data was described by Marezio et al. (1988) on UPd_2Sn. A total of 45 Bragg intensities was extracted from the pattern and input to a Patterson synthesis. Because this structure is composed of relatively heavy atoms, the positions of all four atoms in the asymmetric unit were located. The structure was refined through application of the Rietveld method.

The most complicated structure solved from high-resolution powder diffraction data was determined by McCusker (1988) on Sigma-2, a new clathrasil phase. This structure, which was obtained by direct methods from 268 resolved intensities, has 17 atoms in the asymmetric unit, including six carbon atoms in the cage. Final refinement was accomplished with the Rietveld technique. The power of the technique for important crystal structures has been demonstrated.

Non-isotropic broadening

The instrumental peak widths on conventional sources tend to mask subtle effects due to strain broadening or other factors in the crystallites. On high-resolution synchrotron powder diffractometers, these conditions become much more obvious. In a neutron powder diffraction experiment, Thompson et al. (1987) described the modification of a Rietveld program, because sample treatment had created anisotropic broadening of the powder diffraction lines. The strain broadening contribution to the peak widths is described by a tensor that has the symmetry of the point group. Detection of this line broadening through increased resolution is not only intellectually satisfying, but the angular dependence of the broadening indicates whether it arises from particle size effects or from strain.

LOW- TO MEDIUM-RESOLUTION EXPERIMENTS

Rapid measurement

As discussed above, an experiment can trade resolution for intensity and thereby reduce the integration time needed to reach a given level of counting statistics. Alternatively, it is possible to reduce the counting time by collecting the data in parallel. One means is through the use of linear, position-sensitive detectors to measure powder intensities. Lehmann et al. (1988) described use of a device with a pixel length of slightly more than 0.01°, which collects approximately 2.7° per frame. Finger et al. (1989) used a detector that collects a useful range of 10° with a pixel length of 0.017° (Fig. 12). For such detectors, resolution is reduced relative to the crystal analyzer because the entrance to the detector does not have the ideal geometry.

Time-dependent studies

The study of the kinetics and mechanisms of phase transitions requires time-dependent diffraction analysis. For weakly scattering materials, this type of study is limited by the flux available, whereas detector response is the limiting factor for strongly scattering materials. At present a complete diffraction pattern requires counting times on the order of seconds for energy-dispersive patterns in diamond-anvil cells at relatively low pressure (Skelton et al., 1985). Stephenson (1988) reported collection of single diffraction maxima with counting times of a few milliseconds. In either case, studies of reaction progress are limited to low temperatures or very slow reactions. Most situations will require data collection techniques that are speeded up by several orders of magnitude. At the design specifications of APS, an undulator beam line should provide enough photons from a single bunch of positrons for a complete diffraction pattern with nanosecond time resolution.

UTILIZATION OF SPECTRAL DISTRIBUTION

Experiments near absorption edges

The relatively uniform distribution of intensity with incident energy or wavelength for synchrotron sources has been used to advantage in many diffraction experiments. Single-crystal intensities (Templeton et al., 1980) were measured for experimental determination of the anomalous scattering coefficients as a function of wavelength. Such an experiment is probably not realistic for powdered samples; however, Moroney et al. (1988) described a rather ingenious study utilizing anomalous-dispersion powder diffraction (ADPD).

If zirconia, ZrO_2, is doped with Y_2O_3, the cubic phase is stabilized at room temperature. The two types of cations are disordered on the same position; however, charge balance requires half an anion vacancy for each yttrium ion. As local charge neutrality probably is maintained, it is likely that these defects cluster about the Y^{3+} and that the local environment may be distinguishingly different for Zr and Y, provided that a suitable probe is available. Utilizing the extended X-ray absorption fine structure (EXAFS) technique, Catlow et al.

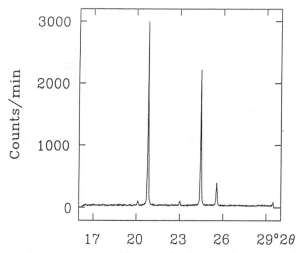

Figure 12. Sample frame from the position-sensitive detector described by Finger et al. (1989). The sample is a very thin dusting of CeO_2 on a Be foil. Counting time for this frame was 60 seconds with a ring current of approximately 100 mA at NSLS beam line X7A. Small peaks at 20 and 23° arise from crystallinity of the Be foil. Wavelength for the experiment equal to 0.685 Å.

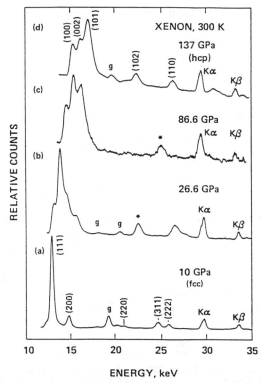

Figure 13. Phase transitions in solid xenon observed by energy-dispersive X-ray diffraction. The face-centered cubic form at low pressures transforms to a hexagonal-close-packed structure at high pressure. An intermediate phase cannot be identified uniquely due to the low resolution of the detector system. Peaks labelled $K\alpha$ and $K\beta$ are fluorescence lines for xenon. All intensities have been corrected for absorption in the diamond-anvil cell and for ring brightness (from Jephcoat et al., 1987).

(1986) concluded that the Zr is displaced from the average, centrosymmetric position, whereas Y was not. In an X-ray experiment, such displacements should be apparent as increased thermal factors due to the static positional disorder. The experiment (Moroney et al., 1988) consisted of measuring the powder intensities at three different energies, (1) well below both K-absorption edges, (2) just below (10 eV) the edge for Y, and (3) above the edge for Y, but just below that of Zr. At the K-absorption edge for these heavy elements, the magnitude of the real part of the anomalous scattering factor changes by approximately 5 electrons. From the three sets of intensities, separate temperature factors for Zr and Y were refined. As would be expected, these parameters correlate rather highly (97%). It was possible, nevertheless, to determine that the mean-square displacement for Zr^{4+} is nearly twice that of Y^{3+}, in agreement with the result obtained in the EXAFS experiment.

Although the previous example involves a case in which the magnitude of the effect is very large, it should be possible to use the approach for other systems. The study described above was conducted on a bending magnet source and was limited in intensity. To increase the counting rate, they accepted a fairly broad bandwidth (12 eV) from their monochromator. Because the peak in the anomalous scattering is fairly sharp, this energy range produces an average effect considerably smaller than the maximum possible. With either a wiggler or undulator source and a bandwidth of 2 eV, this technique should be applicable for any absorption edge with energy greater than 5 keV, which accounts for Ti and all heavier elements. In minerals, it would be of interest to determine distribution coefficients for atoms such as Mn and Fe in a complicated structure such as hollandite (Post et al., 1982).

"White" beam experiments

If the reciprocal relationship between energy and wavelength is noted, Bragg's law can be recast in the form

$$Ed = 6.199/\sin\theta ,\qquad(8)$$

where E is the energy in keV. This equation implies that a complete range of d's could be measured at once provided that one had a powder sample, a detector capable of energy discrimination, and a source containing a broad spectrum of energies. As shown in Figure 7, the source conditions are met for a bending magnet source at a synchrotron. Beam lines with wiggler sources are even better suited. The analysis system for energy-dispersive X-ray diffraction, EDXD, consists of a solid-state detector combined with a multi-channel analyzer. Some of the very earliest synchrotron experiments (Buras et al., 1977) were of this type; however, there is a severe difficulty in this technique that limits its usefulness for most studies. Resolution of a solid-state detector is limited. The best units have a full-width at half maximum on the order of 140 eV at a photon energy of 7 keV. The width increases as $E^{1/2}$, and reaches 300 eV at 30 keV. For a scattering angle of 12°, the resolution at $d = 2$Å, expressed as $\delta d/d = 10^{-2}$, whereas the value for a monochromatic experiment on the same beam line is 3×10^{-4}, nearly two orders of magnitude better. The effect of low resolution is shown in Figure 13, where the structure of a phase of solid xenon, intermediate between face-centered cubic and hexagonal-close-packed forms, could not be determined at low resolution (Jephcoat et al., 1987).

A second problem in doing EDXD arises for bending magnet beam lines at storage rings with relatively low (2–3 GeV) energy. Although the brightness curves (Fig. 7) appear to have high values to 80 keV, a necessary upper limit for good coverage of reciprocal space in a single experiment, the plots have a logarithmic scale. Because background tends to be linear with counting time, signal-to-noise considerations place stringent bounds on the upper energy limit. The minimum energy is determined by absorption in the Be windows, air or helium beam paths, and in the apparatus. For experiments of this type performed on diamond-anvil pressure devices, radiation with energies less then 16 keV is essentially absorbed completely. Figure 14 shows the product of the brightness times the attenuation function of a diamond-anvil cell for a bending magnet (X7) line and a wiggler (X17) beam line at the NSLS.

One means of increasing the resolution of an energy-dispersive detector is to employ a crystal analyzer, rather than a solid-state detector, to determine the energy distribution of the diffracted radiation. By proper adjustment of the analyzer geometry, resolution could be increased to any desired value, up to the limit of $\delta d/d \sim 5 \times 10^{-4}$. This experiment requires increased time and more complex equipment than the ordinary EDXD variety and has not yet been reported. Bourdillon et al. (1978) and Parrish (1988) have performed energy dispersive analysis by scanning an incident-beam monochromater.

SAMPLE PREPARATION

Sample preparation is the critical step in all diffraction experiments; however, the low vertical divergence and intrinsic collimation of synchrotron radiation introduces a new problem. Particles that are only a few seconds of arc away from the position needed to satisfy Bragg conditions will contribute nothing to the scattered intensity, unlike the divergent beam situation. In effect, the sample is composed of fewer crystallites. It is also possible that a particular grain will be in position for only a portion of the total envelope associated with a Bragg peak, thus giving rise to strange shapes. To minimize these difficulties, it is essential to rock the sample about the θ axis, even for a flat-plate specimen. The total motion need not be more than 0.1°; however, a larger motion is usually chosen for capillary mounts to help minimize preferred orientation effects.

Another problem occurs when the amount of material available is too small to prepare a flate-plate specimen; however, if a capillary sample is used, the material may be too absorbing to achieve the optimal condition of $\mu d \approx 1$, where μ is the linear absorption coefficient and d is the inside diameter of the capillary. The seriousness of this problem can be seen in Figure 15 where the variation of intensity with path length is shown. The figure is calculated for a rectangular solid rather than for a cylinder and is valid only in the limit as the scattering angle goes to zero; however, it demonstrates the principal. A solution to this dilemma is to reduce the effective absorption coefficient by diluting the sample with a non-absorbing powder that has a simple, highly-symmetric structure. Amorphous

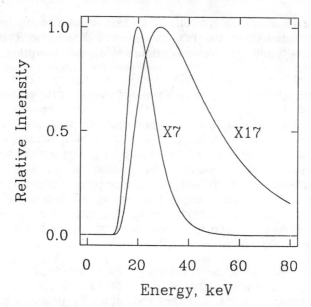

Figure 14. Effective intensity for diamond-anvil cells of the type used by Jephcoat et al. (1987) for energy-dispersive X-ray diffraction at the NSLS. The ring brightness function is convoluted with the transmission function for the diamond-anvil cell and normalized. Useful photon energies are 15–35 keV for a bending magnet (X7), and 18–80 keV for a wiggler (X17) source. The intensity for X17 is several orders of magnitude greater before normalization.

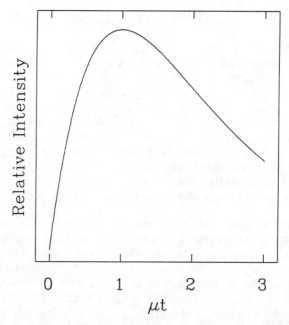

Figure 15. Relative intensity associated with variations in absorption coefficient. The curve plots $te^{-\mu t}$ for a scattering angle of zero as a function of μt in a rectangular solid of thickness t with a linear absorption coefficient of μ.

materials should not be used as the strong diffuse background will degrade signal-to-noise values. Sub-micron diamond powder, which is inexpensive, meets these criteria and is a satisfactory dilutant. Besides reducing the absorption, the equant grains of the diamond powder help to reduce preferred orientation in the mount.

PRESENT AND FUTURE OPPORTUNITIES

Operating beam line locations and characteristics

There is a large number of beam lines at existing synchrotrons that are suitable for high-resolution powder X-ray diffraction. The minimum requirements are for a monochromator that can be tuned in the range from 6–20 keV (2.1–0.6 Å), a diffractometer for mounting the sample, and a diffracted-beam analyzer. Because most such lines are optimized for single-crystal studies, the amount of subsidiary equipment and software for powder diffraction will vary. If lower resolution is satisfactory, a receiving slit can replace the analyzer. Guides to the available beam lines and the procedures needed to gain access to them can be obtained by writing to the Director of the facility.

The number of facilities for energy-dispersive diffraction are somewhat more limited because they require a "white" beam, which implies direct line-of-sight from the experiment to the ring. For this arrangement, a much greater release of radiation would occur in a worst-case accident involving loss of vacuum. Accordingly, beam-line security is more important than for a normal beam line. These systems also require more specialized equipment and software than is generally available for general users. Three such lines are the Dedicated High Pressure wiggler line at CHESS, the general powder diffraction beam line (X7A) at the NSLS that operates in EDXD mode approximately 30% of the time, and the NSLS wiggler line (X17C) that will be dedicated to EDXD.

New techniques

It would be "foolish" to try to predict the experiments that will be done with the next generation of synchrotron sources; however, the imaginative approach to non-traditional samples and scattering methods with beams now available (Dutta et al., 1987; Lim et al., 1987; Moroney et al., 1988) indicates the style of the future. As the brilliance of the sources increases by several orders of magnitude, even more exotic experiments will be performed.

REFERENCES

Ahtee, M., Unonius, L., Nurnela, M. and Suortti, P. (1984) A Voigtian as profile shape function in Rietveld refinement. J. Appl. Crystallogr. 17, 352-357.

Attfield, J. P., Cheetham, A. K., Cox, D. E. and Sleight, A. W. (1988) Synchrotron X-ray and neutron powder diffraction studies of the structure of α-CrPO$_4$. J. Appl. Crystallogr. 21, 452-457.

Attfield, J. P., Sleight, A. W. and Cheetham, A. K. (1986) Structure determination of α-CrPO$_4$ from powder synchrotron X-ray data. Nature 322, 630-632.

Bathow, G., Freytag, E. and Haensel, R. (1966) Measurement of synchrotron radiation in the X-ray region. J. Appl. Phys. 37, 3449-3454.

Buras, B., Olsen, J.S., Gerward, G., Will, G. and Hinze, E. (1977) X-ray energy-dispersive diffractometry using synchrotron radiation. J. Appl. Crystallogr. 10, 431-438.

Bourdillon, A., Glazer, A.M., Hidaka, M. and Buras, J. (1978) High-resolution energy-dispersive diffraction using synchrotron radiation. J. Appl. Crystallogr. 11, 684-687.

Catlow, C. R. A., Chadwick, A. V., Greaves, G. N. and Maroney, L. M. (1986) EXAFS study of yttria-stabilized zirconia. J. Amer. Ceramic Society 69, 272-277.

Cox, D. E., Hastings, J. B., Cardoso, L. P. and Finger, L. W. (1986) Synchrotron X-ray powder diffraction at X13A: a dedicated powder diffractometer at the National Synchrotron Light Source. Mater. Science Forum 9, 1-20.

Cox, D. E., Toby, B. H. and Eddy, M. M. (1988) Acquisition of powder diffraction data with synchrotron radiation. Australian J. Physics 41, 117-131.

Dutta, P., Peng, J. B., Lin, B., Ketterson, J. B., Prakish, M., Georgopoulos, P. and Ehrlich, S. (1987) X-ray diffraction studies of organic monolayers on the surface of water. Phys. Rev. Lett. 58, 2228-2231.

Finger, L.W., Hemley, R.J., Cox, D.E. and Smith, G.C. (1989) High-resolution diffraction at high pressure using a linear, position-sensitive detector (abstract). EOS 70, 472.

Fitch, A. N. and Murray, A. D. (1989) Rietveld refinement of X-ray diffractometer data. J. Appl. Crystallogr., in press.

Glazer, A.M., Hidaka, M. and Buras, J. (1978) Energy-dispersive powder profile refinement using synchrotron radiation. J. Appl. Crystallogr. 11, 165-172.

Gordon, B.M. and Jones, K.W. (1985) Design criteria and sensitivity calculations for multielemental trace analysis at the NSLS X-ray microprobe. Nucl. Instrum. Methods B10/11, 293-298.

Hall, M.M.Jr., Veeraraghavan, Rubin, H. and Winchell, P.G. (1977) The approximation of symmetric X-ray peaks by Pearson type VII distributions. J. Appl. Crystallogr. 10, 66-68.

Jephcoat, A.P., Mao, H.K., Finger, L.W., Cox, D.E., Hemley, R.J., and Zha, C.S. (1987) Pressure-induced structural phase transitions in solid xenon. Phys. Rev. Lett. 59, 2670-2673.

Lehmann, M. S., Christensen, A. Norlund, Nielsen, M., Friedenhans'l, R. and Cox, D. E. (1988) High-resolution synchrotron X-ray powder diffraction with a linear position-sensitive detector. J. Appl. Crystallogr. 21, 905-910.

Lim, G., Parrish, W., Ortiz, C., Bellotto, M. and Hart, M. (1987) Grazing incidence synchrotron X-ray diffraction method for analyzing thin films. J. Mater. Res. 2, 471-477.

Marezio, M., Cox, D. E., Rossel, C. and Maple, M. B. (1988) The *ab-initio* crystal structure determination of UPd_2Sn by synchrotron X-ray powder diffraction. Solid State Commun. 67, 831-835.

McCusker, L. (1988) The *ab-initio* structure determination of sigma-2 (a new clathrasil phase) from synchrotron powder diffraction data. J. Appl. Crystallogr. 21, 305-310.

Moroney, L.M., Thompson, P. and Cox, D.E. (1988) ADPD: a new approach to shared-site problems in crystallography. J. Appl. Crystallogr. 21, 206-208.

Parrish, W. (1988) Advances in synchrotron X-ray polycrystalline diffraction. Australian J. Physics 41, 101-112.

Parrish, W. and Hart, M. (1985) Synchrotron experimental methods for powder structure refinement. Trans. Amer. Crystallogr. Assoc. 21, 51-55.

Parrish, W., Huang, T. C. and Ayers, G. L. (1976) Profile fitting: a powerful method of computer X-ray instrumentation and analysis. Trans. Amer. Crystallogr. Assoc. 12, 55-73.

Parrish, W., Hart, M. and Huang, T.C. (1986) Synchrotron X-ray polycrystalline diffractometry. J. Appl. Crystallogr. 19, 92-100.

Post, J.E., Von Dreele, R.B. and Buseck, P.R. (1982) Symmetry and cation displacements in hollandites: structure refinements of hollandite, cryptomelane and priderite. Acta Crystallogr. B38, 1056-1065.

Rietveld, H. M. (1969) A profile refinement method for nuclear and magnetic structures. J. Appl. Crystallogr. 2, 65-71.

Rivers, M.L. (1988) Characteristics of the Advanced Photon Source and comparison with existing synchrotron facilities. In: Synchrotron X-Ray Sources and New Opportunities in the Earth Sciences: Workshop Report, 5-22, Argonne National Laboratory Technical Report ANL/APS-TM-3.

Sabine, T.M. (1987) A powder diffractometer for a synchrotron source. J. Appl. Crystallogr. 20, 173-178.

Skelton, E.F., Qadri, S.B., Elam, W.T. and Webb, A.W. (1985) A review of high pressure research with synchrotron radiation and recent kinetic studies. In: Solid State Physics under Pressure,

329-334, S. Minomura, Ed., Terra Scientific Publishing Co., Tokyo.
Sleight, A.W. and Cox, D.E. (1986) Symmetry of superconducting compositions in the $BaPb_{1-x}Bi_xO_3$ system. Solid State Commun. 58, 347-350.
Spencer, J.E. and Winick H. (1980) Wiggler systems as sources of electromagnetic radiation. In: Synchrotron Radiation Research, 663-716, H. Winick and S. Doniach, eds., Plenum Press, New York.
Stephenson, G. B. (1988) Time-resolved X-ray scattering studies of phase transition kinetics. In: X-ray Synchrotrons and the Development of New Materials: Workshop Report, 16-17, Argonne National Laboratory Technical Report ANL/APS-TM-1.
Suortti, P., Ahtee, M. and Unonius, L. (1979) Voigt function fit of X-ray and neutron powder diffractometer profiles. J. Appl. Crystallogr. 12, 365-369.
Taupin, D. (1973) Automatic peak determination in X-ray powder patterns. J. Appl. Crystallogr. 6, 266-273.
Templeton, D.H., Templeton, L.K., Phillips, J.C., and Hodgeson, K.O. (1980) Anomalous scatteringcy cesium and cobalt measured with synchrotron radiation. Acta Crystallogr. A36, 436-442.
Thompson, P., Reilly, J.J., Hastings, J.M. (1987) The accomodation of strain and particle broadening in Rietveld refinement; its application to de-deuterized lanthanum nickel ($LaNi_5$) alloy. J. Less-Common Metals 129, 105-114.
Toraya, H. (1986) Whole-powder-pattern fitting without reference to a structural model: Application to X-ray powder diffractometer data. J. Appl. Crystallogr. 19, 440-447.
Werner, P.-E., Eriksson, L. and Westdahl, M. (1985) TREOR, a semi-exhaustive trial-and-error powder indexing program for all symmetries. J. Appl. Crystallogr. 18, 367-370.
Wertheim, G. K., Butler, M. A., West, K. W. and Buchanan, D. N. E. (1974) Determination of the Gaussian and Lorentzian content of experimental line shapes. Rev. Scientific Instruments 45, 1369-1371.
Will, G., Masiocchi, N., Parrish, W. and Hart, M. (1987) Refinement of simple structures from synchrotron radiation powder diffraction data. J. Appl. Crystallogr. 20, 394-401.
Winick H. (1980a) Properties of synchrotron radiation. In: Synchrotron Radiation Research, 11-25, H. Winick and S. Doniach, eds., Plenum Press, New York.
Winick H. (1980b) Synchrotron radiation sources, research facilities, and instrumentation. In: Synchrotron Radiation Research, 27-60, H. Winick and S. Doniach, eds., Plenum Press, New York.
Young, R.A. and Wiles, D.B. (1982) Profile shape functions in Rietveld refinements. J. Appl. Crystallogr. 15, 430-438.

11. NEUTRON POWDER DIFFRACTION
R.B. Von Dreele

INTRODUCTION

It was first recognized by Elsasser in 1936 that neutron motion is determined by wave mechanics and that neutrons can be diffracted by crystalline materials. This was most clearly demonstrated by Mitchell and Powers (1936) using the apparatus diagrammed in Figure 1. The radium-beryllium source was surrounded by paraffin wax which moderated the neutrons to give a thermal distribution of velocities. The corresponding wavelength for the maximum in this distribution was 1.6 Å. Sixteen large single crystals of MgO were placed with their (100) planes inclined at 22° to the incident- and diffracted-beam channels to satisfy the diffraction condition for these planes and 1.6 Å neutrons. Diffraction was then demonstrated by rotating the crystals away from this diffracting condition, resulting in a large drop in the neutron intensity at the detector. Although these experiments demonstrated the diffraction of neutrons, they could not provide any quantitative information. Simultaneously, the theory of thermal neutron scattering was developed with major contributions by Fermi (1936), Schwinger (1937), and Halpern and Johnson (1939). The major advances in neutron scattering occurred after nuclear fission reactors became available in 1945 with sufficient flux to permit selection of a narrow band of neutron energies in a collimated beam. With the possibility of quantitative neutron scattering experiments, the theory was expanded to include most of the features of both elastic and inelastic neutron scattering. The first neutron diffractometer was built at Argonne National Laboratory by Zinn (1947). Major theoretical contributions in this era were made by Placzek (1952) and Van Hove (1954); modern expositions of the theory of thermal neutron scattering have been made by Marshall and Lovesey (1971) and Lovesey (1984). A more basic description of the principles of neutron diffraction and its applications has been given by Bacon (1975). Between 1945 and the present day, several reactors have been constructed with core neutron fluxes up to 10^{15} n/sec-cm^2 which are equipped with as many as 50 neutron scattering instruments (Table 1). More recently, an alternative scheme for producing neutrons in large numbers has been developed. The interaction of a high-energy proton beam with a heavy metal target produces bursts of neutrons by "spallation" reactions (Carpenter et al., 1975). These neutrons can be moderated to thermal velocities and used in scattering experiments, and a few facilities have been constructed that utilize this idea (Table 2). An excellent review of the neutron scattering experiments possible with spallation sources along with special instrument requirements has been published by Windsor (1981).

The applications of neutron powder diffraction to problems of mineralogical interest have developed only since Rietveld (1967, 1969) introduced a least-squares refinement method for analyzing entire neutron powder diffraction patterns. Previously, only very simple structures could be examined by powder diffraction because only a small number of integrated intensities could be extracted from the then generally available low resolution diffraction patterns. At first the method was exclusively applied to neutron diffraction and it was not until much later that it was sucessfully applied to X-ray diffraction data (Wiles and Young, 1981). The areas of principal interest in neutron diffraction (H-atom location and cation ordering analysis) derive from the great difference between neutron and X-ray scattering factors. This chapter will discuss the theory of neutron scattering and compare it with X-ray scattering followed by a discussion of the experimental methods of neutron powder diffraction. Since the major applications of neutron scattering to mineralogy involve the Rietveld method, the theory of this technique will be extensively discussed followed by some examples of its application.

THE NEUTRON -- BASIC PROPERTIES

The neutron is a major constituent of the nucleus and has the properties at rest shown in Table 3.

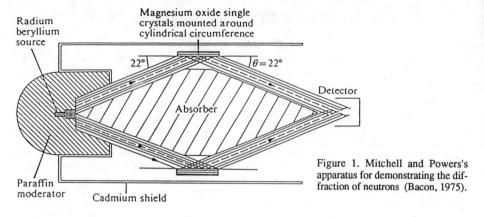

Figure 1. Mitchell and Powers's apparatus for demonstrating the diffraction of neutrons (Bacon, 1975).

Table 1. Neutron Reactor Sources

Name	Location	Flux	No. ports
HFBR	Brookhaven National Laboratory, USA	2.0×10^{14}	9
HFR	Institut Laue Langevin, Grenoble, France	1.0×10^{15}	17
MURR	University of Missouri, USA	4.0×10^{13}	6
OWR	Los Alamos National Laboratory, USA	1.0×10^{14}	2
MITR II	Massachusetts Institute of Technology, USA	2.0×10^{13}	13
NBS	National Bureau of Standards, USA	1.0×10^{14}	13
HFIR	Oak Ridge National Laboratory, USA	1.2×10^{15}	4
ORR	Oak Ridge National Laboratory, USA	1.5×10^{14}	8

Table 2. Spallation Neutron Sources

Name	Location	No. instruments	Proton flux
ISIS	Rutherford-Appleton Laboratory, England	13	100 µA @ 50 Hz
LANSCE	Los Alamos National Laboratory, USA	7	60 µA @ 20 Hz
IPNS	Argonne National Laboratory, USA	9	13.5 µA @ 33 Hz

Table 3. Neutron Properties

mass, m	1.67495×10^{-24} gm
moment, μ_n	6.0311×10^{-12} eV/gauss
charge, q	0
half life, $t_{1/2}$	11.2 min

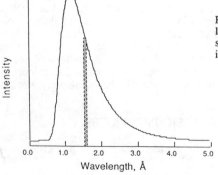

Figure 2. Maxwellian distribution of neutron wavelengths in thermal equilibrium at 273K. A band showing the typical cut by a monochromator is also illustrated.

The relationship between mass, velocity and associated wavelength of any moving particle is given by the de Broglie equation

$$\lambda = \frac{h}{mv} , \qquad (1)$$

where h is Planck's constant. If one considers that the neutrons act as a gas, then the Maxwell-Boltzmann distribution can be used to describe a neutron temperature from the root-mean-square (rms) velocity:

$$E = \frac{1}{2}mv^2 = \frac{3}{2}kT . \qquad (2)$$

Then it follows that

$$\lambda^2 = \frac{h^2}{3mkT} , \qquad (3)$$

from which it can be determined that the neutrons with rms velocity (2200 m/s) in thermal equilibrium at 273K have a wavelength of 1.55 Å and an average kinetic energy of 50 cm^{-1}. This is a fortunate coincidence with the wavelengths required to examine atomic arrangements in materials and the characteristic X-ray wavelengths used for diffraction. It is also a coincidence with the energies of typical atomic vibrations. The entire wavelength distribution can be given as

$$\frac{dN}{N} = \frac{2}{\lambda}\left(\frac{E}{kT}\right)^2 e^{-E/kT} d\lambda = \frac{2}{\lambda}\left(\frac{h^2}{2m\lambda^2 kT}\right)^2 e^{-h^2/2m\lambda^2 kT} d\lambda , \qquad (4)$$

where E is the energy of a neutron of wavelength λ. This distribution for 273K is shown in Figure 2; a typical cut using a monochromator covering a small band of wavelengths is also shown.

NEUTRON POWDER DIFFRACTION THEORY

In a neutron powder diffraction experiment, a polycrystalline sample (typically 10-20 g in a 5-10 mm diameter cylinder 30 mm long) is placed in a beam of neutrons and the resulting scattering is examined by one or more detectors at a range of angles centered about the sample. In order to understand more fully the results of such an experiment we have to first explore the nature of crystals and the interaction between them and neutrons.

The simplest description of a crystal is an infinitely repeating motif or unit cell of atoms in a three-dimensional lattice. Because the lattice is considered to be infinite, i.e. a typical unit cell size is 10 Å (10^{-7} cm) on a side and the crystal is 1,000-100,000 times larger, the distribution of scattering density, $\rho(\mathbf{r})$, within a unit cell can be represented by a Fourier series:

$$\rho(\mathbf{r}) = \frac{1}{V_c}\sum_{\mathbf{h}} F_\mathbf{h} e^{-2\pi i(\mathbf{h}\cdot\mathbf{r})} . \qquad (5)$$

The coefficients of this series, $F_\mathbf{h}$, are known as structure factors and generally are complex numbers. They can be considered to occupy an inverse of real space, reciprocal space, in which the positions are represented by the 3-dimensional vectors, $\mathbf{h}$, whose lengths are in Å^{-1}. The infinite extent of the crystal lattice ensures that the structure factors are δ-functions in reciprocal space, although they may be broadened if the crystal is small enough. The δ-functions are arrayed in a reciprocal lattice occupying only the points represented by 3-dimensional unit translations of a reciprocal unit cell. Each reciprocal vector, **h**, is also the normal to a stack of planes in real space, and the length of the vector is the reciprocal of the interplanar spacing or d-spacing. These planes intercept the three real lattice vectors in integral fractions (0, ±1/1, ±1/2, ±1/3, ±1/4, etc.) whose denominators are the Miller indices (hkl) of the planes. These indices are identical to the vector elements of **h**. This relationship between real space and reciprocal space means that when the crystal is rotated so does the reciprocal lattice. The connection between these two spaces is represented in Figure 3.

Diffraction is a coherent scattering phenomenon that is best described in reciprocal space. The observed scattering intensity at some angle, 2Θ, from the incident beam depends on the orientation of the crystal and hence the reciprocal lattice:

$$\left(\frac{d\sigma}{d\Omega}\right)_{coh} = \frac{N_o}{V_c}\sum_h \delta(\mathbf{S}-\mathbf{h})F_h^2 = I_h \quad . \tag{6}$$

The δ-function means that there must be a coincidence of the scattering vector, **S**, and a reciprocal lattice point, **h**, for scattering to occur. The scattering vector is the difference between the incident and scattered beam vectors, $\mathbf{s_o}$ and **s** (Fig. 4). These vectors are defined in reciprocal space with magnitudes proportional to the reciprocal of the wavelength and thus only part of reciprocal space is accessible with radiation of a particular wavelength. In addition, the intensity depends on both the size of the sample and the magnitude of the structure factor. Also notice that the complex nature of the structure factor is lost, since its square appears in this equation.

In a single-crystal diffraction experiment the scattering angle and orientation of the crystal are adjusted so that each reciprocal lattice point, **h**, is placed in turn to coincide with the scattering vector, and the scattered intensity is measured. As a practical matter, the crystal is either rotated through the (**S**-**h**) coincidence condition or the length of **S** is varied by changing the wavelength so that a scan over each peak is obtained. A single-crystal diffraction data set then consists of a series of these I_h measurements.

An ideal powder diffraction sample consists of a myriad of small crystallites randomly oriented with respect to each other. Thus, there are always some crystallites that satisfy the orientation necessary for scattering and all that is needed to see any intensity is for the length of **S** to match the length of **h**. This simple relationship is then Bragg's Law,

$$\lambda = 2d\sin\Theta \quad , \tag{7}$$

and a powder diffraction pattern consists of a series of peaks or Bragg reflections of varying intensity as a function of 2Θ or λ. The positions of the peaks correspond to the d-spacings for the possible reciprocal lattice points for the crystallites and the intensities are proportional to the square of the structure factors. A powder pattern then contains all the intensity information inherent in the reciprocal lattice but all the directions for the vectors are lost along with the complex part of the structure factors.

The inverse Fourier transform can also be performed to give the structure factors from the scattering density:

$$F(\mathbf{h}) = \int_V \rho(\mathbf{r})e^{2\pi i(\mathbf{h}\cdot\mathbf{r})}d\mathbf{r} \quad . \tag{8}$$

As an alternative way of performing this Fourier transform, we can postulate a set of atom positions and assign a scattering density for each. Then the transform becomes

337

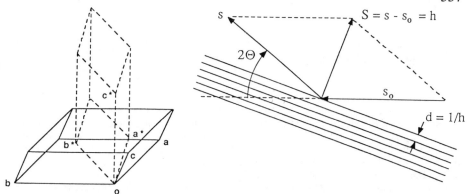

Figure 3 (left). Real and reciprocal lattice unit cells are shown in their respective orientations. The real cell axes are **a**, **b** and **c**; the reciprocal cell axes are **a***, **b*** and **c***.

Figure 4 (right). A set of planes in the diffraction condition where the diffraction vector **S** matches the reciprocal lattice vector **h**.

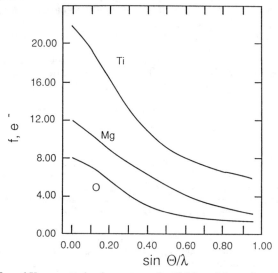

Figure 5. Normal X-ray scattering factor curves for Ti, Mg and O as a function of $\sin\Theta/\lambda$.

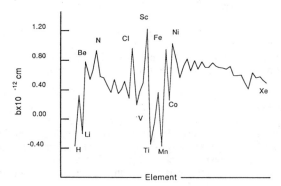

Figure 6. Neutron scattering cross sections for the natural abundance elements H-Xe.

$$F(\mathbf{h}) = \sum_{i=1}^{N} f_i e^{2\pi i (\mathbf{h} \cdot \mathbf{x}_i)} \ . \tag{9}$$

The sum is over the atoms in the unit cell and f_i is the scattering factor for the i-th atom located at position $\mathbf{x}_i$. This scattering factor is the Fourier transform of the scattering density around an atom. However, X-rays and neutrons are scattered by different mechanisms. X-ray scattering occurs primarily by interaction with the electrons that surround an atom. There are two consequences of this interaction. First, the strength of the scattering depends on the number of electrons that surround an atom so that elements at the top of the periodic table are weaker X-ray scatterers than those near the bottom. Second, the spatial extent of the electron cloud around an atom is roughly the same as the X-ray wavelength and the resulting scattering power falls off with increasing angle. Thus, the scattering factors for X-rays look like those shown in Figure 5. On the other hand, neutrons are primarily scattered by atom nuclei. Since the nuclear dimensions are roughly a thousand times smaller than the neutron wavelength, there is only point scattering so neutron scattering factors (lengths or b's) are independent of angle. Also, nuclear scattering is a combination of potential scattering and resonance scattering. Potential scattering depends on the number of nuclear particles and resonance scattering results from neutron absorption by the nucleus. These two factors add and sometimes subtract to give neutron scattering lengths that vary very erratically across the periodic table and from isotope to isotope (Fig. 6). Thus, X-ray and neutron powder diffraction patterns will be very different. Figure 7 shows idealized X-ray and neutron patterns calculated for $MgTiO_3$ "geikielite." These were generated for essentially identical diffractometer experiments (impossible in real life) but are startlingly different. In fact, the strongest peak in the X-ray pattern (at ~32°) is completely absent in the neutron pattern!

The reason for the extreme difference between these two patterns lies in the atomic scattering factors for titanium, magnesium and oxygen for X-rays and neutrons. The X-ray scattering factors are simply proportional to the atomic number, thus $f_{Ti} > f_{Mg} > f_O$. However, the titanium scattering length for neutrons is negative and the value for oxygen is only slightly larger than magnesium, thus $b_O > b_{Mg} > 0 > b_{Ti}$. The consequence is that for each reflection, the neutron structure factor is very different from the X-ray structure factor and gives very different peak heights in a powder diffraction pattern. One result is that phase identification from neutron data alone is generally impossible with the present JCPDS files and search/match routines because they contain only X-ray powder diffraction patterns.

REACTOR NEUTRON SOURCES

Since 1945, nuclear reactors fueled by ^{235}U or ^{239}Pu have been the principal neutron sources for scattering experiments. A typical reactor (Fig. 8) consists of a fuel package, control rods, moderator and shielding. The shielding is penetrated in several places to allow the neutrons formed in the reaction to be directed toward various scattering instruments.

The spectrum seen at each of these beam ports depends largely on the temperature of the moderator which it views. Moderators have been made of graphite held at 2000K to produce a spectrum rich in short-wavelength neutrons as well as liquid hydrogen (20K) which gives many very long-wavelength neutrons. The normal moderator used for a powder diffraction source consists of either light or heavy water near room temperature. The spectrum from such a moderator is shown in Figure 2, and it has maximum flux at ~1.6 Å.

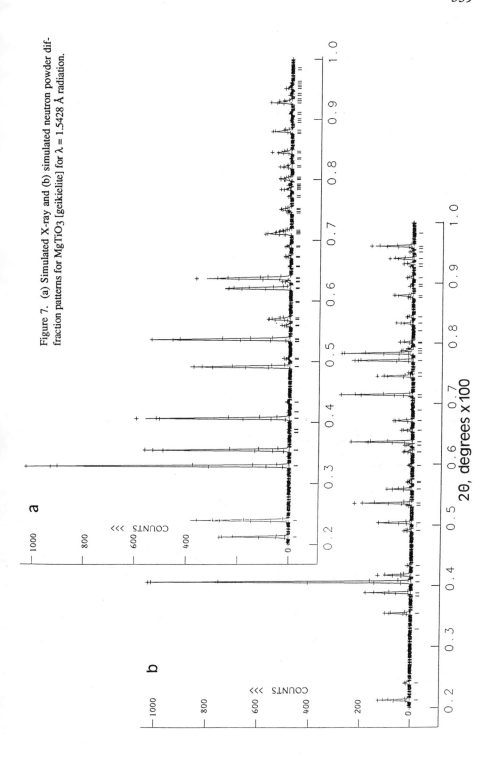

Figure 7. (a) Simulated X-ray and (b) simulated neutron powder diffraction patterns for MgTiO3 [geikielite] for λ = 1.5428 Å radiation.

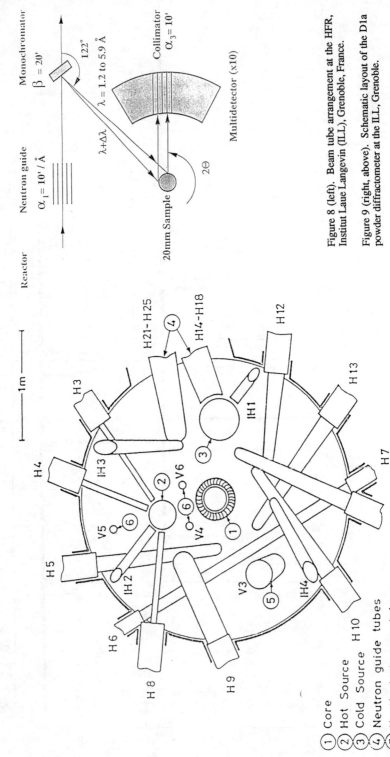

Figure 8 (left). Beam tube arrangement at the HFR, Institut Laue Langevin (ILL), Grenoble, France.

Figure 9 (right, above). Schematic layout of the D1a powder diffractometer at the ILL, Grenoble.

CONSTANT-WAVELENGTH POWDER DIFFRACTOMETER DESIGN

The design of a constant-wavelength (CW) neutron powder diffractometer is a compromise between maximum resolution and maximum flux. Moreover, a detailed consideration of the angle dependence is made to match as best as possible the resolution requirements for high quality structure analysis. Thus, the design criteria include both the monochromator assembly and the diffractometer. The theory for diffractometer design has been covered in great detail by Hewat (1975) and is based on the resolution theory of Cagliotti et al. (1968). The variance or full width at half maximum (FWHM) of the Gaussian line profile for neutron diffractometers is

$$\sigma^2 = \frac{(FWHM)^2}{8\ln 2} = U\tan^2\Theta + V\tan\Theta + W \quad . \tag{10}$$

The coefficients U, V and W are functions of the collimation (α_1 and α_3) and the mosaic spread of the monochromator (β) as well as the monochromator scattering angle (Θ_m). Figure 9 shows a schematic of a high-resolution neutron powder diffractometer and the location of these components. The expressions for the "Cagliotti" coefficients are

$$U = \frac{1}{8\ln 2}(2.5\alpha_1^2 + 2\beta^2)\cot^2\Theta_m \tag{11}$$

$$V = \frac{-1}{8\ln 2}(2\alpha_1^2 + 4\beta^2)\cot\Theta_m \tag{12}$$

$$W = \frac{1}{8\ln 2}(0.5\alpha_1^2 + 2\beta^2 + \alpha_3^2) \tag{13}$$

for the case in which there is no collimation between the monochromator and the sample. When these expressions are differentiated we find that the minimum peak width occurs at

$$\Theta \sim \Theta_m \quad , \tag{14}$$

and the variance of the peak is

$$\sigma^2 \sim \frac{\alpha_1^2 + \alpha_3^2}{8\ln 2} \tag{15}$$

at this point. Clearly the best choice is to use a relatively high monochromator scattering angle (120-150° $2\Theta_m$) and fine collimation in the incident beam before the monochromator and between the sample and detectors ($\alpha_1 = \alpha_3 = 5'-10'$). The intensity of such an instrument is determined largely by the width of the cut, $\Delta\lambda$, from the Maxwellian neutron distribution (Fig. 2) by the monochromator. This is related to the monochromator scattering angle, Θ_m, by differentiating Bragg's law:

$$I \sim \Delta\lambda/\lambda \sim \Delta\Theta_m \cot\Theta_m \quad . \tag{16}$$

Then it is obvious that the intensity is largely dependent on the mosaic spread of the monochromator, $\beta = \Delta\Theta_m$. Although the mosaic spread does not affect the peak width at the focusing angle, it does make a contribution at both lower and higher angles. Thus, a useful choice is to have $\beta = 2\alpha$ so that the resolution curve roughly follows that required for good peak separation. By differentiating Bragg's law with respect to reflection order

$$\lambda\sqrt{p} = 2d\sin\Theta \quad , \tag{17}$$

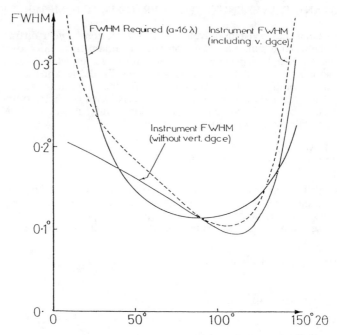

Figure 10. Resolution requirement for cubic lattice ($a = 16\lambda$) for CW diffractometer and the resolution achieved by the D1a diffractometer (Hewat, 1975).

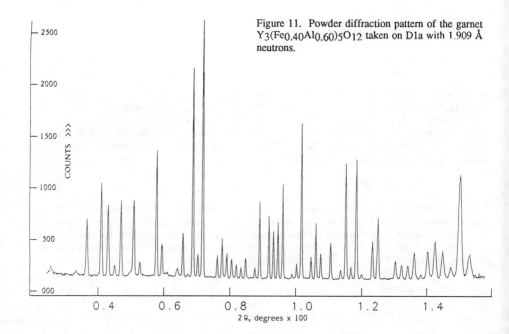

Figure 11. Powder diffraction pattern of the garnet $Y_3(Fe_{0.40}Al_{0.60})_5O_{12}$ taken on D1a with 1.909 Å neutrons.

where $p = h^2 + k^2 + l^2$ for a cubic lattice gives

$$\Delta(2\Theta) = 2\Delta\Theta = \left(\frac{\lambda}{2d}\right)^2 \frac{2}{\sin 2\Theta} \quad . \tag{18}$$

Figure 10 shows this curve for d = 24 Å as compared with the resolution curve for a diffractometer with $\alpha_1 = \alpha_2 = 10'$, $\beta = 20'$ and $2\Theta_m = 120°$.

This design was followed for the D1a powder diffractometer built by Hewat and Bailey (1976) at the Institut Laue Langevin (ILL), Grenoble, France. The monochromator is a specially cut squashed Ge (110) single crystal mounted so that small rotations can give the (551), (331) and (771) planes yielding a range of wavelengths (1.384 Å, 2.268 Å, and 0.994 Å, respectively) at a monochromator angle of $122°2\Theta_m$. Other choices of monochromator planes and scattering angles are also possible. A typical sample used on this instrument is a cylinder 25-50 mm long with a diameter of 5-10 mm. The pattern shown in Figure 11 is of a yttrium-iron garnet taken on this instrument with 1.909 Å neutrons. The slight asymmetry in the diffraction line profiles at both low and high scattering angles result from the vertical divergence in the collimation and makes the real resolution somewhat poorer at those angles than implied by the U, V, and W coefficients for this instrument (Fig. 10). Other instruments at many reactor facilities have also been built using much the same principles giving focusing resolutions from 0.5° to 0.1°2Θ. Each of these instruments maintains a reasonable counting rate by employing several detectors so that a scan moves each one over a relatively small portion of the whole scan range. For example, the D1a diffractometer has 10 detectors placed in 6° intervals.

SPALLATION NEUTRON SOURCES

The second method for producing neutrons begins with a beam of high-energy protons (>500 MeV) produced by an accelerator and allowed to strike a heavy-metal target (e.g. W or ^{238}U). The accelerator is operated in a pulsed mode to give proton pulses <1 μs wide at a repetition frequency of 10-120 Hz. The interaction between the proton and the heavy nuclei causes neutrons to "boil off" while shattering the nuclei. These neutrons are generally of very high energy, with most above 20 Mev. Additional neutrons are produced by "evaporation" in reactions between these high energy neutrons and other atom nuclei. This "spallation" process yields 20-30 neutrons per proton as it interacts with a string of nuclei. The narrow proton pulses result in equally narrow neutron pulses from the target which enable time-of-flight (TOF) energy analysis to be employed on the neutron scattering experiments. To produce the more useful thermal component of the neutron spectrum, small moderators are placed near the target in either "wing" or "slab" geometry. In wing geometry the target is out of the line of sight of the neutron beam lines viewing that moderator, whereas in slab geometry the target can be seen directly through the moderator. The moderators are kept small to avoid spreading the neutron pulse width which would reduce the time resolution of the neutron spectrometers. The entire target-moderator assembly is then surrounded with material that strongly scatters neutrons (Ni or Be) to act as a reflector. The layout of the spallation source at Los Alamos National Laboratory (LANSCE) is shown in Figure 12 and the target/moderator assembly at LANSCE is shown in Figure 13. The moderators in this assembly are arranged in wing geometry about a "flux trap" where the target is split in two and positioned above and below the plane of the moderators. A typical moderator for powder diffractometers is about the size of a book (10x10x6 cm), filled with water or liquid methane and usually "poisoned" by a layer of Gd or Cd foil 1-3 cm below the front surface. This last item is used to reduce the effective moderator thickness for long-wavelength neutrons and hence improve the time resolution.

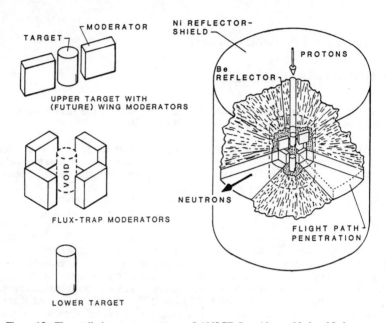

Figure 12. The spallation neutron source at LANSCE, Los Alamos National Laboratory.

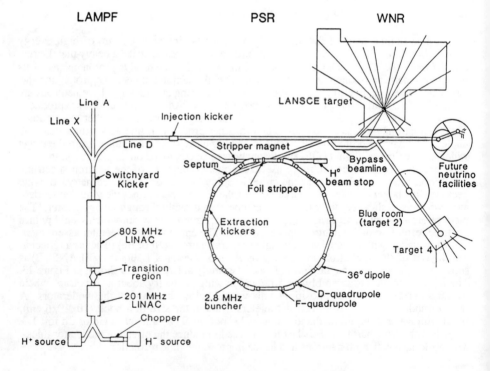

Figure 13. The LANSCE target moderator assembly.

As mentioned above, the neutrons emitted from these moderators have a pulsed time structure where the entire mix of wavelengths is generated within the moderator over a very short time. The neutrons then sort themselves according to their velocity and hence their wavelength so that the time of flight from the moderator to an instrument detector is proportional to the wavelength

$$T = \frac{\lambda L m}{h}, \qquad (19)$$

where L is the total flight path. The time spread for neutrons of a given wavelength is determined by the rate of leakage out of the moderator and is roughly proportional to both the moderator thickness and the wavelength. So that for a given moderator

$$\left(\frac{\Delta T}{T}\right)_\lambda \sim \text{constant} \qquad (20)$$

for all wavelengths viewed at some distance L. The time spectrum for neutrons of a given wavelength is however quite complex and is markedly dependent on wavelength. One complication arises because, unlike a reactor, the neutron spectrum from a spallation source has a large epithermal component (Fig. 14) as well as the usual thermal Boltzmann distribution. The leakage spectra for these two components are very different and can have a considerable effect on the resulting line shapes in a powder diffraction pattern as a function of wavelength. These effects will be discussed in more detail below.

TIME-OF-FLIGHT POWDER DIFFRACTOMETER DESIGN

The design of a TOF powder diffractometer follows along similar lines as for a CW instrument but with some important differences. The two major contributors to the resolution are the time spread for neutrons of a given wavelength and the angle-dependent geometric broadening terms. As noted above the first of these is affected by the moderator design. However, there are limitations such that too small a moderator will not produce any neutrons and, because each moderator is shared among a number of instruments, some accommodation must be made. In addition, Equations 19 and 20 show that the time spread is also dependent on the flight path so that long flight paths are required for high resolution whereas short flight paths give only low resolution. The flight path is not infinitely variable because the collimation and shielding requirements mean that it is necessarily fixed for each instrument. Moreover, the repetition frequency (f) of the source will limit the range of wavelengths accessible at a given flight path by

$$\lambda_{max} = \frac{h}{Lmf}. \qquad (21)$$

Once the choice of flight path has been made, the geometric factors are then set so that each contribution to the resolution is roughly matched thus maximizing the intensity. This optimization can be achieved only at one scattering angle; normally this is a high angle ($2\Theta > 90°$). The overall resolution, expressed as a spread in the reflection, observed for a given d-spacing, is given by

$$\frac{\Delta d}{d} = \left[\left(\frac{\Delta T}{T}\right)^2 + (\Delta\Theta \cot\Theta)^2 + \left(\frac{\Delta L}{L}\right)^2\right] \qquad (22)$$

(Buras and Holas, 1968; Worlton et al., 1976). The first term is the moderator source broadening given in Equation 20, the third term arises from uncertainty in the flight path

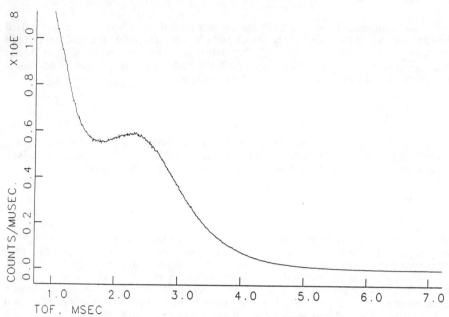

Figure 14. Neutron spectrum from LANSCE water moderator at 283K.

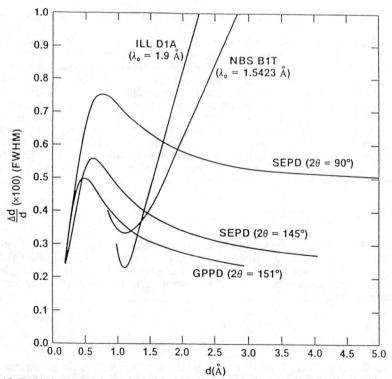

Figure 15. Resolution comparison of the General Purpose Powder Diffractometer (GPPD) and Special Environment Powder Diffractometer (SEPD) at IPNS, Argonne National Laboratory with the D1a and the B1T diffractometer at NBS.

due to sample and detector sizes (~10 mm diameter), and the middle term is the angular dispersion term. It comes from the wavelength spread (Eq. 16) due to angular widths of moderator, sample, and detector. This term dominates at low 2Θ and can nearly vanish at very high 2Θ. There are additional minor terms from the vertical divergences in the instrument, but these are easier to assess by Monte Carlo methods. This resolution function is vastly different from that of a CW powder diffractometer; all components are constants for a given 2Θ so that the fractional resolution is essentially constant for all d-spacings. For a CW instrument, the fractional resolution is dramatically poorer at large d (small 2Θ) than at the focusing angle defined in Equation 14 (Fig. 15).

Most TOF powder diffractometers have scores of detector tubes positioned at a wide variety of 2Θ angles. They are collected into "banks" each with 4-32 tubes positioned near some fixed 2Θ value. There are many different ways these tubes can be positioned relative to each other, some of which represent various types of focusing. For example one of the earliest versions of a TOF powder diffractometer (Jorgenson and Rotella, 1982) positioned each detector in a bank so that d-spacing was proportional to TOF identically for all tubes in a bank. This "geometric time focusing" was achieved by changing the secondary flight path, L_2, to compensate for the change in 2Θ. At a nominal 2Θ of $160°$ the 15 detectors were arrayed in a nearly straight line tilted ~30° to the scattered beam, however the bank of 12 detectors at the nominal 2Θ of $90°$ had to be nearly parallel (~8° tilt) to the scattered beam to achieve focusing at a considerable cost in solid angle. A simpler arrangement that gives much greater solid-angle coverage was employed in the two diffractometers built at the Intense Pulsed Neutron Source (IPNS) at Argonne National Laboratory. The detectors were arrayed on the circumference of a circle bunched at $150°$, $90°$ and $60°2\Theta$. The focusing was achieved electronically by computing a time offset for each neutron count, which depended on the 2Θ for the detector tube, where the event occurred, and the time-of-flight. The combination of these events for each bank of detectors produced a single time-focused powder pattern. A weakness of both of these detector arrangements is that the resolution observed for a given d-spacing is different for each detector in a bank. When the patterns are combined to give a single pattern for the bank, the resulting diffraction line shape is the summation of slightly different line shapes. However, it is possible to arrange the detectors in a bank so that all give identical diffraction line widths and this "resolution focusing" condition yields less restrictive arrangements of detectors than geometric time focusing. For example, the tilt of a straight line of tubes at 2Θ of $150°$ is only ~35° from the normal and at $90°2\Theta$ the tilt is ~60°. At low 2Θ angles the bank of resolution-focused detectors tends to be parallel to the incident beam (Fig. 16). This arrangement has been employed on several powder diffractometers built at LANSCE and one of them was used to produce the TOF neutron powder pattern of $MgTiO_3$ geikielite shown in Figure 17.

NEUTRON DETECTION

Neutrons are detected by inducing a nuclear reaction with a highly absorbing isotope and sensing the resulting gamma or heavy-particle emission. For example, a common neutron detector is a proportional counter filled with ^3He or $^{10}BF_3$ gas (Fig. 18) operated at a potential of 1-3 kV. The reactions are

$$^1n + {}^3He \longrightarrow {}^1H + {}^3H + 0.77 \text{ MeV} , \tag{23}$$

$$^1n + {}^{10}B \longrightarrow {}^7Li + {}^4He + 2.3 \text{ MeV} . \tag{24}$$

The detection is by sensing the charge deposited on the thin wire anode resulting from gas ionization by the heavy particles (^1H, ^3H or ^4He) in the reaction; the resulting pulse height is ~5 V. Gamma rays are also detected by these counters but the charge is much less and can be discriminated against because of the lower pulse height. Because the neutron is detected via an absorption process, these detectors are quite efficient (>90%) at

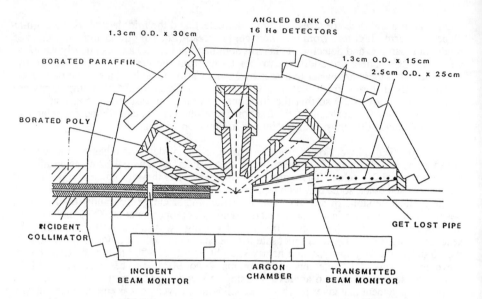

Figure 16. Schematic layout of a resolution-focused powder diffractometer.

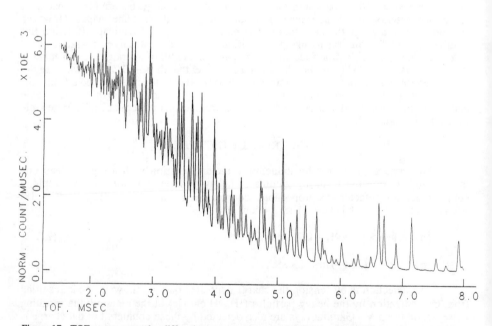

Figure 17. TOF neutron powder diffraction pattern taken at 151°2Θ on a 10m diffractometer at LANSCE (Wechsler and Von Dreele, 1989). The intensities have been normalized by the incident spectrum.

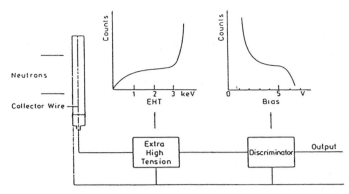

Figure 18. The counting system for a gas counter. The high voltage serves to drift the secondary electrons to the collector wire. The pulse voltage is compared with a bias voltage to reject small noise pulses.

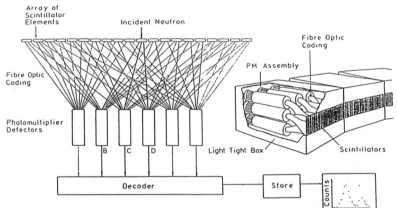

Figure 19. The principle of a coded scintillator array using triple coincidence to uniquely identify the location of a neutron detection event as used in HRPD at ISIS.

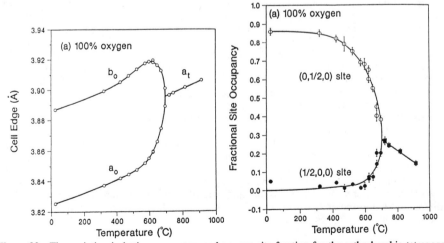

Figure 20. The variation in lattice parameters and oxygen site fraction for the orthorhombic-tetragonal phase transition for $YBa_2Cu_3O_{7-x}$.

long neutron wavelengths but much less so (<30%) at short wavelengths. Generally, the solid angle covered by a detector is maximized by positioning it so that the neutron beam is perpendicular to the tube axis.

An alternative neutron detection scheme uses a ^{6}Li-enriched scintillator which emits a flash of light arising from the interaction between the reaction products and a phosphor such as Ag-doped ZnS or a rare earth-doped glass. The reaction is

$$^1n + {}^6Li \longrightarrow {}^3He + {}^4He + 4.79 \text{ MeV} \ . \tag{25}$$

The scintillator is coupled by a light guide to a photomultiplier which senses the light flash. These detectors are generally more efficient than gas tubes because higher atom densities of ^{6}Li can be achieved in the solid scintillator, but they are more sensitive to gamma rays. Gamma ray discrimination is possible by either pulse-shape analysis or coincidence methods using detectors in pairs. One other advantage of these detectors is that they can be used to cover a large area operating as a position-sensitive detector. The high resolution powder diffractometer (HRPD) at the Rutherford-Appleton Laboratory, England, (ISIS) has such a detector (Fig. 19) consisting of rings of paired glass scintillator chips each connected by optical fiber light guides to three photomultipliers out of the full set of 64. The detection of light in all three uniquely identifies the location of the event. The paired scintillators allows discrimination of gamma rays because they are detected in both scintillators while a neutron is seen by only one.

SPECIAL SAMPLE ENVIRONMENTS IN NEUTRON DIFFRACTION

Apart from a few special cases, most elements and isotopes display very small absorption of thermal neutrons. The absorption coefficients are generally 3-4 orders of magnitude smaller than those for X-rays. This circumstance is fortunate in two ways. The low absorption allows the use of much larger samples for neutron scattering than for X-ray diffraction. A typical neutron diffraction sample is a powder filled cylinder 5-10 mm in diameter and 25-50 mm long compared to the 20-100 µm active part of an X-ray sample. This compensates for the fact that neutron sources have very much lower intensity than even a conventional X-ray tube so that counting times are comparable for both types of radiation. This large sample size also alleviates for a neutron diffraction experiment much of the preferred orientation problems commonly encountered with X-ray diffraction samples. A disadvantage is that a large mass of material (10-20 g) is usually required for a neutron powder diffraction experiment. Secondly, the construction of special apparatus for both high and low temperature diffraction studied is facilitated because a large number of materials can be interposed in the incident and scattered neutron beams with little loss of intensity. Thus, it is a relatively simple matter to build liquid-He cryostats for neutron scattering as well as 2500K furnaces. The construction is even more simplified for TOF diffractometers because the fixed detector positions require smaller fields of view in the apparatus. The construction of moderate high pressure cells for neutron diffraction is also possible and cells for pressures up to 30 kbar have been constructed. The sample size requirements for neutron scattering at present make very high pressure (>50 kbar) experiments difficult so that regime is still best studied by X-ray diffraction. The details of the design of special environment apparatus, especially for high pressure, have been reviewed by Carlile and Salter (1978). Some examples of work under non-ambient conditions include the work by Lager et al. (1982) on the thermal expansion of α-quartz and the study of α-quartz compression in SiO_2 and GeO_2 by Jorgensen (1978). Recently, the tetragonal-orthorhombic phase transition for the high T_c superconductor, $YBa_2Cu_3O_{7-x}$, was examined by a series of variable temperature and variable oxygen pressure TOF neutron powder diffraction experiments (Jorgensen et al., 1987) which detailed the structural changes near the transition at 700°C (Fig. 20).

RIETVELD REFINEMENT

About 20 years ago, H.M. Rietveld (1967,1969) recognized that a mathematical expression could be written to represent the observed intensity at every step in a neutron powder diffraction pattern:

$$I_c = I_b + \Sigma Y_h \quad . \tag{26}$$

This expression has both a contribution from the background (I_b) and each of the Bragg reflections (Y_h) which are in the vicinity of the powder pattern step (Fig. 21). Each of these components to the intensity is represented by a mathematical model which embodies both the crystalline and noncrystalline features of a powder diffraction experiment. The adjustable parameters for this model are refined by a least-squares minimization of the weighted differences between the observed and calculated intensities. This approach to the analysis of powder patterns has been so successful that it has lead to a renaissance in powder diffraction and this technique of treating powder diffraction data is now known as "Rietveld refinement."

A more complete description of the powder pattern model includes the possibility of multiple phases in the sample and the possibility that the incident intensity is not the same for all data points. Thus, the observed intensities are normalized for neutron time-of-flight data by

$$I_o = \frac{I'_o}{W I_i} \quad , \tag{27}$$

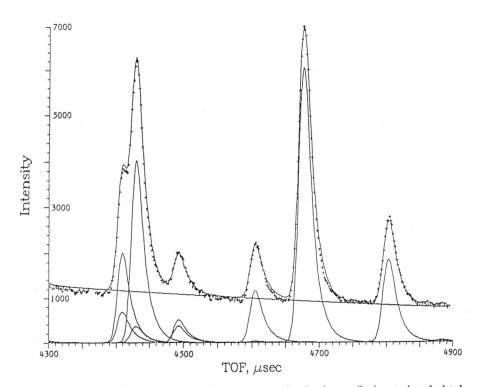

Figure 21. Portion of a TOF neutron powder diffraction pattern showing the contributions to the calculated pattern from background and nine reflections.

or for constant wavelength data by

$$I_o = \frac{I'_o}{I_i} \quad , \tag{28}$$

where I'_o is the number of counts observed in a channel of width W; I_i is the incident intensity for that channel. The sum in Equation 26 is over those reflections from all phases in the sample that are sufficiently close to the profile point to make a significant contribution. Then the function

$$M = \Sigma[w(I_o-I_c)^2] \tag{29}$$

is minimized by least-squares, where ideally the weight, w, is computed from the variances in both I'_o and I_i. The quality of the least-squares refinement is indicated by some residual functions,

$$R_p = \Sigma|I_o-I_c|/\Sigma I_o \quad , \tag{30}$$

and $\quad R_{wp} = \sqrt{M/\Sigma[wI_o^2]} \quad . \tag{31}$

The reduced χ^2 or "goodness of fit" is defined from the minimization function as

$$\chi^2 = M/(N_{obs}-N_{var}) \quad , \tag{32}$$

and the "expected R_{wp}" from

$$\text{expected } R_{wp} = R_{wp}/\sqrt{\chi^2} \quad . \tag{33}$$

Background intensity contribution

In the original program written by Rietveld, the background in a powder pattern was visually estimated from relatively clear portions and subtracted by linear interpolation from the observed intensity. These modified data were then fitted by the program to a refined structural model. Recent modifications to the method now include a description of the background with a set of refinable coefficients. There are several different functions in use, one of the most successful is the entirely empirical function, 'Chebyshev polynomial of the first kind,'

$$I_b = \sum_{j=1}^{n} B_j T'_{j-1} \quad , \tag{34}$$

where T'_{j-1} are the coefficients of the Chebyshev polynomial and B_j are refinable. Each coefficient is a polynomial expansion of the form

$$T'_n = \sum_{m=0}^{i-1} C_m X^m \quad . \tag{35}$$

The coefficients, C_m, of these expansions define the form of the Chebyschev polynomial. The TOF or 2Θ is converted to X to make the Chebyshev polynomial orthogonal by

$$X = (2/T) - 1 \tag{36}$$

(T in milliseconds TOF or degrees 2Θ). Given that the usual range of TOF is 1 to 100 millisec and 2Θ is from 0 to 180 then X ranges from about -1 to +1 which is the orthogonal range for this function.

A second empirical background function is a cosine Fourier series including a leading constant term.

$$I_b = B_0 + \sum_{j=1}^{n} B_j \cos(P(j-1)) \quad . \tag{37}$$

In the case of CW data, P is in degrees 2Θ and is the detector position for the step. For TOF data the times are scaled by $180/T_{MAX}$, where T_{MAX} is the maximum TOF in the pattern. This function appears to be more robust than the Chebyshev function for both CW and TOF data.

A third function uses a straight line for the first part and the major features of a real space correlation function (RSCF) for the remainder. The RSCF for an amorphous phase has peaks which correspond to the interatomic distances and is closely related to the radial distribution function but is easier to compute. Thus this background function can model diffuse background from an amorphous phase in the sample or the background from a silica sample holder. The function is

$$I_b = B_0 + B_1 T + \sum_{i=1}^{n} B_{2i} \sin(QB_{2i+1})/(QB_{2i+1}) \quad , \tag{38}$$

where $Q = 2\pi/d$. \hfill (39)

The values of B_{2i+1} correspond to interatomic distances in the amorphous phase.

A fourth method for fitting the background is a modification of the original method used by Rietveld(1969). The user selects a set of specific points across the powder pattern and the background is then fitted in the least-squares refinement by adjusting the background intensity at these points calculating the background contribution for all other points by linear interpolation. This method is somewhat less satisfying that the previous three because the selection of the interpolation points can be subjective.

The first two functions are normally quite flat. There is usually only a slight curvature arising from the thermal diffuse scattering from the sample. Larger background contributions will be fitted by these functions given sufficient terms, but short period fluctuations not associated with an amorphous phase cannot be fit very well. Examples of background problems include frame overlap peaks in a TOF pattern, miscellaneous peaks from sources outside the experiment and unaccounted extra phases.

Bragg intensity contribution

The contributed intensity, $Y_h(T)$, from a Bragg peak to a particular profile intensity will depend on several factors. Obviously the value of the structure factor and the amount of that particular phase will determine the contribution. In addition the peak shape and width in relation to its position will have an effect. The intensity is also affected by extinction and absorption as well as some geometric factors. Thus,

$$Y_h(T) = SF_h^2 H(T - T_h) K_h \quad , \tag{40}$$

where S is the scale factor for the particular phase, F_h is the structure factor for a particular reflection, $H(T - T_h)$ is the value of the profile peak-shape function for that reflection at the position, T, which is displaced from its expected position, T_h, and K_h is the product of the various geometric and other intensity correction factors for that reflection. Each of these contributions will be discussed in turn.

Profile functions for CW and TOF

The contribution a given reflection makes to the total profile intensity depends on the shape function for that reflection profile, its width coefficients and the displacement of the peak from the profile position. The locations of the peak are usually given in microseconds of TOF or in centidegrees 2Θ. Discussion of these values is given first followed by details of some of the peak shape functions presently in use.

Reflection positions in powder patterns

For a neutron time-of-flight powder diffractometer the relationship between the d-spacing for a particular powder line and its TOF is given by the simple quadratic

$$T = C d + A d^2 + Z . \tag{41}$$

The three parameters C, A, and Z are characteristic of a given counter bank on a TOF powder diffractometer. C may be calculated with good precision from the flight paths, diffraction angle, and counter tube length by use of the de Broglie equation:

$$C = 252.816*2\sin \Theta (L_1 + \sqrt{L_2^2+L_3^2/16}) , \tag{42}$$

where Θ is the Bragg angle, L_1 is the primary flight path, L_2 is sample to detector center distance and L_3 is the height of the detector; all distances are in meters. The units of C are then μsec/Å. A is a small second-order correction and Z depends on the various electronic delays in the counting system. Precise values for constants C, A and Z must be obtained by fitting to a powder diffraction pattern of a standard material.

The reflection position in a constant-wavelength experiment is obtained from Bragg's Law.

$$T = 100 \arcsin (\lambda/2d) + Z , \tag{43}$$

where Z is the zero-point error on the counter arm position and the result is in centidegrees. Both λ and Z for a constant wavelength neutron diffractometer are obtained from refinement of a standard material.

TOF profile functions

The best known profile function for TOF neutron data is the empirical convolution function of Jorgensen et al. (1978) and Von Dreele et al. (1982):

$$H(\Delta T) = N[e^u \text{erfc}(y) + e^v \text{erfc}(z)] , \tag{44}$$

where ΔT is the difference in TOF between the reflection position, T_h, and the profile point, T; the terms N,u,v,y and z are dependent on the profile coefficients. The function erfc(x) is the complementary error function. This profile function is the result of convoluting two back-to-back exponentials with a Gaussian:

$$H(\Delta T) = \int G(\Delta T-\tau)P(\tau)d\tau \quad , \tag{45}$$

where $P(\tau) = 2Ne^{\alpha\tau}$ for $\tau < 0$ \hfill (46)

and $P(\tau) = 2Ne^{-\beta\tau}$ for $\tau > 0$ \hfill (47)

for the two exponentials; α and β are the rise and decay coefficients for the exponentials and are largely characteristic of the specific moderator viewed by the instrument. The Gaussian function is

$$G(\Delta T-\tau) = \frac{1}{\sqrt{2\pi\sigma^2}} e^{-(\Delta T-\tau)^2/2\sigma^2} \quad . \tag{48}$$

The Gaussian variance is the coefficient σ^2 and is largely characteristic of the instrument design (scattering angle and flight path) and the sample. These functions when convoluted give the profile function shown above. The normalization factor, N, is

$$N = \frac{\alpha\beta}{2(\alpha+\beta)} \quad . \tag{49}$$

The coefficients u,v,y and z are

$$u = \frac{\alpha}{2}(\alpha\sigma^2 + 2\Delta T) \tag{50}$$

$$v = \frac{\beta}{2}(\beta\sigma^2 - 2\Delta T) \tag{51}$$

$$y = \frac{(\alpha\sigma^2 + \Delta T)}{\sqrt{2\sigma^2}} \tag{52}$$

$$z = \frac{(\beta\sigma^2 - \Delta T)}{\sqrt{2\sigma^2}} \quad . \tag{53}$$

Each of the three coefficients α, β and σ^2 all show a specific empirical dependence on the reflection d-spacing.

$$\alpha = \alpha_0 + \alpha_1/d \tag{54}$$

$$\beta = \beta_0 + \beta_1/d^4 \tag{55}$$

$$\sigma^2 = \sigma_0^2 + \sigma_1^2 d^2 + \sigma_2^2 d^4 \tag{56}$$

Interpretation of TOF profile coefficients

The profile coefficients from a TOF neutron powder pattern Rietveld refinement can give information about the microtexture of the sample. This discussion will describe how this information can be extracted from the coefficients.

The strain in a lattice can be visualized as a defect-induced distribution of unit cell dimensions about the average lattice parameters. In the reciprocal space (Fig. 22) associated with a sample with isotropic strain, there is a broadening of each point which is proportional to the distance of the point from the origin, i.e.,

$$\frac{\Delta d^*}{d^*} = \text{constant} \ . \tag{57}$$

In real space (the regime of a TOF experiment) the peak broadening from strain is

$$\frac{\Delta d}{d} = \text{constant} \ . \tag{58}$$

Thus, examination of the function for the Gaussian component of the peak shape from a TOF pattern (Eq. 56) implies that the second term contains an isotropic contribution from strain broadening. The other major contribution to σ_1^2 is from the instrument (Eq. 22); because it is expressed as a variance, σ_{1i}^2, it can simply be subtracted. The remaining sample-dependent contribution is then converted to strain (S), a dimensionless value which is frequently expressed as percent strain or fractional strain as full width at half maximum (FWHM):

$$S = \frac{1}{C}\sqrt{8\ln 2(\sigma_1^2 - \sigma_{1i}^2)} \cdot 100\% \ , \tag{59}$$

where C is the diffractometer constant from Equation 41. The term $\sqrt{8\ln 2}$ is the ratio between FWHM and the standard deviation, σ, of a Gaussian distribution.

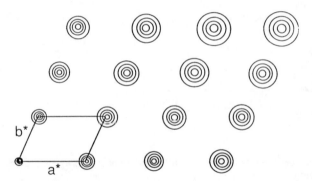

Figure 22. Illustration of the strain broadening of the reflection maxima in reciprocal space.

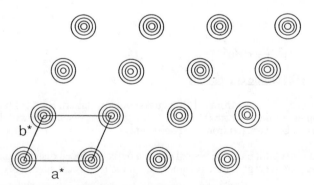

Figure 23. Illustration of the particle-size broadening of the reflection maxima in reciprocal space.

For small particles the assumption that the lattice is infinite no longer holds so that the reciprocal lattice points, **h**, are not δ-functions but are all smeared out uniformly depending on the average particle size. Thus, all the points are the same size independent of the distance from the origin (Fig. 23), and

$$\Delta d^* = \text{constant} . \tag{60}$$

The reciprocal of this quantity is the average particle size. In real space (for TOF) the broadening is

$$\frac{\Delta d}{d^2} = \text{constant} . \tag{61}$$

From the functional form for the Gaussian broadening of a TOF peak (Eq. 56), the particle size affects the third term (σ_2^2) in the expression. As implied by Equation 22, this term generally has no instrument contribution and is used directly to calculate particle size, p, by

$$p = \frac{CK}{\sqrt{8\ln 2 \sigma_2^2}} , \tag{62}$$

where C is the diffractometer constant, K is the Scherrer constant and the units for p are Å.

CW profile functions

The original function described by Rietveld (1967) for the shape of neutron powder diffraction peak shapes was Gaussian (Eq. 48) using the Cagliotti et al. (1968) function (Eq. 10) for the variation of FWHM with scattering angle. Several modifications of this function have been made including asymmetry (Rietveld, 1969) and the introduction of other broadening functions (Young et al., 1977; Young and Wiles, 1982) especially for X-ray diffraction.

The most successful function for CW data employs a multi-term Simpson's-rule integration described by Howard (1982) of the pseudo-Voigt, F(ΔT), described by Thompson et al. (1987):

$$H(\Delta T) = \sum_{i=1}^{n} g_i F(\Delta T') , \tag{63}$$

where the pseudo-Voigt is

$$F(\Delta T) = \eta L(\Delta T', \Gamma) + (1-\eta) G(\Delta T', \Gamma) , \tag{64}$$

and the Lorentzian function is

$$L(\tau) = \frac{\gamma}{2\pi} \left[\frac{1}{(\gamma/2)^2 + \tau^2} \right] . \tag{65}$$

The Gaussian expression is the same as used for the TOF function above (Eq. 48). The mixing factor, η, is given by

$$\eta = 1.36603(\gamma/\Gamma) - 0.47719(\gamma/\Gamma)^2 + 0.11116(\gamma/\Gamma)^3 , \tag{66}$$

and the FWHM parameter is

$$\Gamma = \sqrt[5]{\Gamma_g^5 + 2.69269\,\Gamma_g^4\gamma + 2.42843\,\Gamma_g^3\gamma^2 + 4.47163\,\Gamma_g^2\gamma^3 + 0.07842\,\Gamma_g\gamma^4 + \gamma^5} \quad (67)$$

where the Gaussian FWHM is

$$\Gamma_g = \sqrt{8\ln 2\,\sigma^2} \ . \quad (68)$$

The 2Θ difference modified for asymmetry, A_S, and sample shift, S_S, is

$$\Delta T' = \Delta T + \frac{f_i A_S}{\tan 2\Theta} + S_S \cos\Theta \ , \quad (69)$$

where the sums have an odd number of terms depending on the size of A_S. The corresponding Simpson's-rule coefficients, g_i and f_i, depend on the number of terms in the summation. The sample shift can be interpreted as a physical shift of the sample, s, from the diffractometer axis by

$$s = \frac{-\pi R S_S}{36000} \ , \quad (70)$$

where R is the diffractometer radius (Klug and Alexander, 1974). The variance of the peak, σ^2, varies with 2Θ as

$$\sigma^2 = U\tan^2\Theta + V\tan\Theta + W + \frac{P}{\cos^2\Theta} \ , \quad (71)$$

where U, V, and W are the coefficients described by Cagliotti et al. (1958) and P is the Scherrer coefficient for Gaussian broadening. The Lorentzian coefficient, γ, varies as

$$\gamma = \frac{X}{\cos\Theta} + Y\tan\Theta + Z \ . \quad (72)$$

<u>Interpretation of CW profile coefficients</u>

In the case of a CW experiment the strain broadening in real space is related to 2Θ broadening from

$$\frac{\Delta d}{d} = \Delta 2\Theta \cot\Theta = \text{constant} \ , \quad (73)$$

or $\quad \Delta 2\Theta = \frac{\Delta d}{d}\tan\Theta \ . \quad (74)$

In this expression $\Delta 2\Theta$ is in radians. Examination of the expression for the Gaussian broadening (Eq. 71) indicates that the first term contains a strain broadening component. As for the TOF expression, this is a variance and the instrument contribution can be subtracted. This variance must be converted to radians to yield strain, thus

$$S = \frac{\pi}{18000}\sqrt{8\ln 2(U - U_i)} \cdot 100\% \ . \quad (75)$$

Alternatively, the strain term is the one that varies with tanΘ in the Lorentzian component of a CW peak shape (Eq. 72). Again any instrumental or spectral contribution can be subtracted to yield the strain component. This is in centidegrees and is already a full width at half maximum so the strain is

$$S = \frac{\pi}{18000}(Y - Y_i) \cdot 100\% \quad . \tag{76}$$

For the case of a CW experiment the particle-size broadening can be obtained from

$$\frac{\Delta d}{d^2} = \frac{\Delta\Theta \cot\Theta}{d} = \text{constant} \quad . \tag{77}$$

From Bragg's law and $\Delta 2\Theta = 2\Delta\Theta$, then

$$\frac{\Delta d}{d^2} = \frac{\Delta 2\Theta \cot\Theta \sin\Theta}{\lambda} \quad . \tag{78}$$

The broadening is

$$\Delta 2\Theta = \frac{\lambda \Delta d}{d^2 \cos\Theta} \quad . \tag{79}$$

The first term in the expression for the Lorentzian broadening is of this form, where

$$X = \frac{\Delta d}{d^2} \tag{80}$$

The particle size can be obtained by rearrangement of this expression and converting from centidegrees to radians by

$$p = \frac{18000 K \lambda}{\pi X} \quad , \tag{81}$$

which includes the Scherrer constant, K. The units are Å.

The corresponding term in the Gaussian expression is the fourth one. Converting from centidegrees to radians gives the expression

$$p = \frac{18000 \lambda}{\pi \sqrt{8 \ln 2P}} \quad , \tag{82}$$

and again the units are Å.

Reflection peak widths

In most Rietveld refinement programs, the reflection peaks are considered to have a definite cutoff at the wings where the magnitude of the function falls below some value. The cutoff positions depend on the nature of the profile function and its coefficients. Generally, for line shapes that are mostly Gaussian a cutoff of 1% of the peak height is adequate but for Lorentzians a cutoff of 0.5% or less should be used.

Systematic effects on the intensity

The intensity correction factors, K_h, consist of those factors which are dependent on the sample, the instrument geometry, and the type of radiation used.

$$K_h = \frac{E_h A_h O_h M_h L}{V_h} \quad , \tag{83}$$

where E_h is an extinction correction, A_h is an absorption correction, O_h is the preferred orientation correction, M_h is the reflection multiplicity, L is the angle-dependent Lorentz correction, and V_h is the unit-cell volume for the phase. Some of these will be discussed in turn.

Extinction in powders

The extinction in powders is a primary extinction effect within the crystal grains and can be calculated according to a formalism developed by Sabine (1985) and Sabine et al. (1988). It is only of importance for TOF neutron data because extinction is strongly dependent on wavelength and not on scattering angle, and if it is not allowed for, the refined values of the atomic temperature factors can be seriously in error. From the Darwin energy transfer equations Sabine (1988), by following the formalisms of Zachariasen (1945, 1967) and Hamilton (1957), developed intensity expressions for both the symmetric Laue and Bragg cases of diffraction by an infinite plane parallel plate. The extinction correction E_h for a small crystal is a combination of Bragg and Laue components

$$E_h = E_b \sin^2\Theta + E_l \cos^2\Theta \quad , \tag{84}$$

where $E_b = 1/\sqrt{1+x}$ \hfill (85)

and $\quad E_l = 1 - \dfrac{x}{2} + \dfrac{x^2}{4} - \dfrac{5x^3}{48} \ ... \ \text{for } x < 1$ \hfill (86)

or $\quad E_l = \sqrt{\dfrac{2}{\pi x}}\left[1 - \dfrac{1}{8x} - \dfrac{3}{128x^2} \ ...\right] \ \text{for } x > 1$ \hfill (87)

where $x = E_x(\lambda F_h/V)^2$, \hfill (88)

F_h is the calculated structure factor, and V is the unit cell volume. The units for these expressions are such that E_x is in μm^2 and is a direct measure of the block size in the powder sample. Sabine et al. (1988) demonstrated this by examining the extinction effects for neutron powder diffraction by hot pressed MgO samples characterized by electron microscopy. The extinction effects in some of these samples decreased the observed intensity in low-order reflections by nearly 70%. The Sabine model allowed correction of these intensities and gave temperature factors which were independent of grain size and matched both theoretical models and other experimental results. Moreover, the refined extinction coefficients correlated very well with the measured particle size distribution in these samples.

Powder absorption factor

The absorption, A_h, for a cylindrical sample is calculated for powder data according to an empirical formula (Hewat, 1979; Rouse et al., 1970). It is assumed that the linear absorption of all components in the sample vary with λ and is indistinguishable from multiple scattering effects within the sample.

$$A_h = e^{(-T_1 A_B \lambda - T_2 A_B^2 \lambda^2)} \quad , \tag{89}$$

where $T_1 = 1.7133 - 0.0368 \sin^2\Theta$, \hfill (90)

and $T_2 = -0.0927 - 0.3750\sin 2\Theta$.   (91)

For a fixed wavelength, this expression is indistinguishable from thermal motion effects and hence can not be refined independently of atomic thermal parameters. Therefore, it is only of importance for the analysis of TOF neutron data.

<u>Preferred orientation of powders</u>

After a review of the available preferred orientation models, Dollase (1986) selected a special case from the more general description by March (1932) and developed what he regarded as the best preferred orientation correction, O_h. For this model, the crystallites are assumed to be effectively either rod or disk shaped. When the powder is packed into a cylindrical neutron diffraction sample holder, the crystallite axes may acquire a preferred orientation that is approximated by a cylindrically symmetric ellipsoid. In the usual diffraction geometry for neutron powder diffraction, the unique axis of this distribution is normal to the diffraction plane. Integration about this distribution at the scattering angle for each reflection gives a very simple form for the correction,

$$O_h = \sum_{j=1}^{n}(R_o^2\cos^2 A_j + \sin^2 A_j/R_o)^{-3/2}/M) ,$$ (92)

where A_j is the angle between the preferred orientation direction and the reflection vector **h**. The sum is over the reflections equivalent to **h**. The one refinable coefficient, R_o, gives the effective sample compression or extension due to preferred orientation and is the axis ratio for the ellipsoid. If there is no preferred orientation then the distribution is spherical and $R_o=1.0$ and thus $O_h = 1.0$.

The other preferred orientation models as reviewed by Dollase (1986) are one or two-coefficient variations on a Gaussian distribution model originally employed by Rietveld (1969). For weak preferred orientation these functions perform just about as well as the March model; this is the usual case for neutron diffraction. However, the problem with these other models is that they do not satisfactorily describe the intensity changes associated with strongly developed preferred orientation and their coefficients are difficult to interpret in a physically meaningful way.

<u>Other angle-dependent corrections</u>

The only other angle-dependent correction for neutron powder diffraction data is the Lorentz factor. For TOF neutron data there is an additional factor for the variation of scattered intensity with wavelength or

$$L = d^4\sin\Theta ;$$ (93)

for constant wavelength neutrons:

$$L = \frac{1}{2\sin 2\Theta \cos\Theta} .$$ (94)

APPLICATIONS OF NEUTRON POWDER DIFFRACTION

There are two main areas of interest for the mineralogist in neutron powder diffraction aside from basic structural information. These are the placement of hydrogen atoms and the distribution of cations on various sites. Two recent reviews of the Rietveld method

(Cheetham and Taylor, 1977; Hewat, 1985) showed that the bulk of the materials studied by this method are oxides virtually all of which are synthetic. For example, much work summarized by the latter review has been on the study of various aspects of synthetic zeolite structures including Si/Al ordering, location of hydrocarbon absorbates, and cation location. All of these are particularly well studied by neutron scattering because of the relative magnitudes of the scattering lengths. For example, the study of a thallium zeolite-A with a Si/Al ratio ~1.0 by Cheetham et al. (1982a) takes advantage of the ~20% difference in the Si and Al neutron scattering lengths (0.415 and 0.345 x 10^{-12} cm, respectively) to determine the ordering. They found that the Si and Al atoms strictly alternate in the framework giving the space group Fm3c rather than having each Si linked to 3 Si and 1 Al atoms in the space group Pn3n. This pattern of Si/Al ordering also is consistent with intrepretation of ^{29}Si magic angle spinning NMR spectra (Cheetham et al., 1982b). However, another zeolite, Na-ZK-4 with a Si/Al ratio of 1.667 showed no evidence of ordering (Cheetham et al., 1983; Eddy et al., 1985). The more usual case is exemplified by the two studies by Fischer et al. (1986a,b) of the zeolites D-ZK-5 and D-RHO where the Si/Al distribution could only be inferred by ^{29}Si NMR and not from the neutron diffraction data. In a more recent zeolite study, Wright et al. (1985) located the guest molecule of pyridine in gallozeolite-L from a partial structure refinement of the framework atoms. The location of the pyridine at one side of the zeolite cavity at the temperature of the diffraction measurement (4K) coincided with the minimum enthalpy position found by mapping the pyridine-zeolite interaction energy. Included in the earlier review is a comparison between a single-crystal X-ray study and a neutron powder study of the structure of $GeO_2 \cdot 9Nb_2O_5$ (Anderson et al., 1975). The refined values of the 17 atomic coordinates for this block structure agreed within 1-3 esd's for the two methods demonstrating that the Rietveld refinement technique can produce results comparable with single-crystal studies. A more recent study by Lager et al. (1981a) compared the refinement of both coordinates and anisotropic thermal parameters for synthetic forsterite, Mg_2SiO_4, from both single-crystal X-ray and neutron refinements with a neutron TOF powder study. The 11 positional parameters for all three refinements are in excellent agreement with each other and exhibit similar values of esd's. The same is also generally true for the thermal parameters although the early form of the TOF Rietveld refinement program used in this study inadequately handled some systematic effects. Thus, considerable confidence can be placed on the results of a Rietveld refinement of a crystal structure from neutron powder diffraction data and it is clear that a carefully done refinement in which the all the systematic effects are modeled will give a structural description with meaningful esds that is as precise as that obtained from a single-crystal study.

Hydrogen atom location

The hydrogen atom neutron scattering is dominated by the very large incoherent scattering cross section arising from the combination of scattering by the two possible nuclear spin states for ^{1}H. This effect makes a high-quality powder diffraction study of hydrogenous materials nearly impossible because of the very high incoherent background contribution. Thus, hydrogen atom studies are nearly always on deuterated materials where the incoherent scattering is minimal. For example, the recent study of a hydrogarnet by Lager et al. (1987) examined the deuterium atom positions in $Ca_3Al_2(O_4D_4)_3$ at 100K, 200K and 300K. This material is the Si-free end-member of the hydrogrossular series and has the space group Ia3d where all the SiO_4^{4-} groups in a normal garnet are replaced by $O_4D_4^{4-}$. They also examined at 300K a ~125 µm single crystal of the compound by X-ray diffraction to provide a comparison with the neutron work. Apart from the D atom position, all atom coordinates and thermal motion parameters agreed for the two types of analyses. The D atom was easily located from difference Fourier maps calculated from the 100K neutron powder data fitted with just the heavy atoms. The heavy-atom-only refinement had a residual of R_{wp} = 17%; after the D atom was introduced and the structure refined the residual was R_{wp} = 3.1%. This large change in residual is not surprising considering that D has the largest neutron scattering length of any of the constituent atoms in this hydrogarnet. The fit in the final refinement can be seen in Figure 24. The substitution for Si^{4+} by $4D^+$ increases the size of the O_4 tetrahedra (Fig. 25) considerably; the D-O distance is 1.950 Å as compared to a normal Si-O bond of 1.645 Å. The D atoms are located slightly outside the faces of the O_4 tetrahedron 0.91 Å from their respective O atom. The D atom location in this work is quite different from that reported for katoite, $Ca_3Al_2(SiO_4)_{0.64}(O_4H_4)_{2.36}$, as determined from single-crystal X-ray data (Sacerdoti and Passaglia, 1985). The H atom in that structure was found from difference-Fourier maps to be 0.68 Å from the O atoms directed toward the center of the O_4 tetrahedron. However, Basso et al. (1983) found the H atom in plazolite, $Ca_3Al_2(SiO_4)_{1.53}(O_4H_4)_{1.47}$, by similar techniques to be close to where it was found in the neutron study by Lager et al. (1987). It might have been expected that all three hydrogrossulars would have very similar H atom positions. However, the various reported positions for H illustrate the difficulty in locating this atom from X-ray data and is clearly a result of the very low scattering power by H atoms in comparison with the other atoms in the structure.

A second example of the power of neutron diffraction in locating hydrogen atoms is the study of the dehydration products of gypsum by Lager, et al. (1984). In this study $CaSO_4$, hemihydrate, and γ-$CaSO_4$ were examined by TOF neutron diffraction. The structure of γ-$CaSO_4$ was found to be hexagonal, $P6_322$, with the SO_4^{2-} tetrahedra arranged in columns forming large channels along the c axis. The hemihydrate structure could not be refined in this study because of considerable line broadening evidently caused by a combination of microtwinning and possible nonstoichiometry. However, a model of the hemihydrate structure was derived as a orthorhombic distortion of the hexagonal γ-$CaSO_4$ structure with the channels partially occupied by water molecules and the calculated powder pattern from this model was roughly the same as the observed one. This is an example of one of the difficulties in using the Rietveld method for structure analysis. Adequate models for the diffraction peak profiles are essential for the refinement to proceed and in some cases sample broadening effects can severely hamper the refinement.

Cation distribution

A great strength of neutron diffraction is the extremely variable scattering lengths for the elements (Fig. 6). This allows one to distinguish easily between certain adjacent pairs of elements and determine site compositions with much higher precision than could be accomplished by X-ray diffraction. For example, Nord and Ericsson (1982) examined by ^{57}Fe Mössbauer spectroscopy the cation ordering in the entire range of solid solution for $(Fe,Mn)_3(PO_4)_2$ graftonites and performed a neutron powder diffraction study on $(Fe_{0.5}Mn_{0.5})_3(PO_4)_2$ to determine the cation preference for the three sites in this structure. This is a particularly easy task because the scattering lengths for Fe and Mn are quite different (0.954 and -0.373 x 10^{-12} cm, respectively). The results of analyzing the Mössbauer spectra gave a strong preference of Mn for one of the three sites tentatively assigned as the octahedral site. The six possible arrangements of Fe/Mn site compositions were tested by Rietveld refinement of the neutron diffraction data. The model with the Mn-rich octahedral site gave the lowest residuals and confirmed the Mössbauer assignment.

Another case where there is strong discrimination between the neutron scattering lengths was the determination of the Ti/Nb ordering in two niobium oxide block structures, $TiNb_2O_7$ and ortho-$Ti_2Nb_{10}O_{29}$, by one of the earliest applications of Rietveld refinement (Cheetham and Von Dreele, 1973; Von Dreele and Cheetham, 1974). These structures are made up of columns of ReO_3 type structure of fixed n x m dimensions which abut each other via edge sharing of octahedra (Fig. 26). For $TiNb_2O_7$ the blocks are 3 x 3 octahedra joined to form a monoclinic face-centered (A2/m) structure and in ortho-$Ti_2Nb_{10}O_{29}$ the blocks are 3x4 arranged to give the orthorhombic face centered (Amma) structure, as determined previously by Wadsley (1961a,b) by X-ray single-crystal diffraction. Each structure has several metal sites some of which only share corners and others which share various numbers of edges and corners and Wadsley had determined from limited X-ray data that all

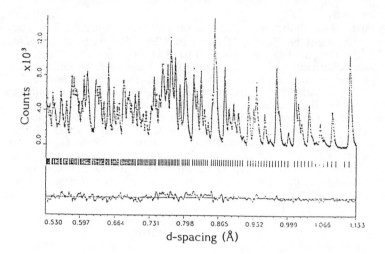

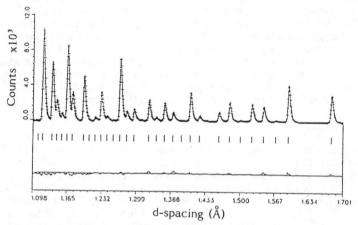

Figure 24. Rietveld refinement profile for $Ca_3Al_2(O_4D_4)_3$ at 100K for TOF neutron data taken at 160°2Θ on GPPD. Plus marks are the data. Solid line in the best-fit profile. Tick marks below the profile indicate the positions of allowed reflections. A difference curve appears at the bottom. The background was fit as part of the calculation but has been subtracted for clarity (Lager et al., 1987).

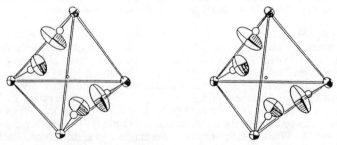

Figure 25. Stereodiagram of the atomic environment about the $\bar{4}$ site (D position) showing the oxygen tetrahedron and associated hydrogen atoms in $Ca_3Al_2(O_4D_4)_3$ from the neutron Rietveld refienment (Lager et al., 1987).

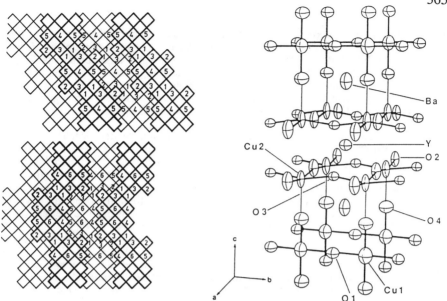

Figure 26 (left). Idealized diagrams of the structures of $TiNb_2O_7$ and ortho-$Ti_2Nb_{10}O_{29}$ showing the block structure generated by crystallographic shear (c.s.) planes. The numbers indicate metal atom positions. The heavy and light outlines for the octahedra distinguish M layers at y = 0 and y = 1/2, respectively (Von Dreele and Cheetham, 1974).

Figure 27 (right). Structure of orthorhombic $YBa_2Cu_3O_{7-x}$ obtained from combined Rietveld refinement of TOF neutron and CuKα powder diffraction data. The ellipsoids correspond to 90% probability surfaces for the atomic thermal motion.

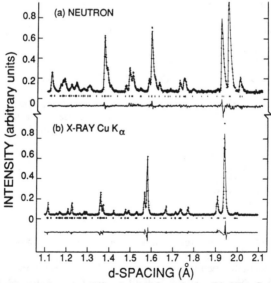

Figure 28. Part of the X-ray and neutron diffraction data for orthorhombic $YBa_2Cu_3O_{7-x}$. Data are shown by plus marks in (a) represent data collected on the +153°2Θ bank of the HIPD diffractometer at LANSCE and (b) represent data taken on a Siemens D500 X-ray diffractometer with CuKα radiation. The solid lines are the best fit to the data from the combined refinement. Tick marks show positions for allowed reflections; those for CuKα$_2$ are shown shifted from the d-spacings calculated for CuKα$_1$. The lower curve in each panel represents the difference between the observed and calculated profiles (Williams et al., 1988).

sites have the same composition of Ti and Nb even though a nonrandom distribution might have been expected. The neutron scattering lengths for the two metals are very different (-0.36 and 0.69×10^{-12} cm for Ti and Nb, respectively) and give much better site-fraction determination than possible from X-ray data. The resulting pattern of site fractions (4.5-64.5% Ti) from the Rietveld analysis was in reasonable agreement with that expected from a site-potential calculation. In general, the atom sites with exclusively corner sharing were rich in Nb whereas those with the most edge sharing were rich in Ti.

In a more recent study of some titanium-magnesium oxides, Wechsler and Von Dreele (1989) examined by neutron TOF powder diffraction the cation distributions in Mg_2TiO_4 spinel, $MgTiO_3$ (geikielite) ilmenite and $MgTi_2O_5$ (karrooite) pseudobrookite as a function of annealing temperature. The great disparity in neutron scattering lengths (-0.36 and 0.54×10^{-12} cm for Ti and Mg, respectively), as in the other examples, gave good discrimination for the site distribution analyses. The distributions found in the various materials were in accord with what had been expected on the basis of some thermodynamic measurements (Wechsler and Navrotsky, 1984) and also conformed to Vegard's law for the average M-O bond lengths in each octahedral site. Specifically, the high-temperature annealed Mg_2TiO_4 cubic spinel was completely inverse. Upon annealing at 773K, Mg_2TiO_4 gave an ordered tetragonal structure ($P4_122$) with half the volume of the cubic spinel. $MgTiO_3$ has the ilmenite structure and was completely ordered. The pseudobrookite showed increasing cation disordering as the annealing temperature increased.

However, in some cases neither neutron or X-ray diffraction can distinguish between certain pairs of elements. For example, in their study of the (Ni,Fe)-sarcopsides, $(Ni_{1-x}Fe_x)_3(PO_4)_2$, Ericsson and Nord (1984) used Mössbauer spectroscopy to determine the Ni/Fe ratio in the M1 and M2 sites. The X-ray scattering factors are virtually identical because Fe and Ni are nearly adjacent elements in the periodic table, and their neutron scattering lengths, 0.96 and 1.03×10^{-12} cm respectively, are also nearly identical. An intermediate case was the determination of cation ordering in the ternary orthophosphate, γ-$(Zn_{0.70}Fe_{0.20}Ni_{0.10})_3(PO_4)_2$ farringtonite (Nord and Ericsson, 1985). They used Mössbauer spectroscopy to determine the Fe site fraction for the two sites and Rietveld refinement of a neutron powder pattern for the Zn and Ni site fractions.

Combined X-ray/neutron Rietveld refinement

Because the complex nature of the structure factors is lost in a diffraction measurement and the directional character of reciprocal space is also lost in a powder diffraction experiment, there is a question of uniqueness for the result of a Rietveld refinement of a single powder pattern. Clearly, the crystal structure model of atom positions, etc. that produces calculated patterns that would match both a neutron powder pattern and an X-ray powder pattern is more likely to be unique (and correct). To accommodate this idea Larson and Von Dreele (1986) developed a computer program that performs the necessary calculations to allow combined X-ray and neutron Rietveld refinement of a crystal structure. A similar program has also been devloped by Maichle et al. (1988). One of the first applications of this method was the work by Williams et al. (1988) to the problem of cation disorder in the high T_c superconductor $YBa_2Cu_3O_{7-x}$. This material had been investigated at great length by many groups throughout the world and the orthorhombic structure (Fig. 27) had been established with little ambiguity within a few months of its discovery (Wu et al., 1987). Almost all of these structural studies were the result of Rietveld refinements of neutron powder diffraction data obtained at either reactor or pulsed spallation sources. By unfortunate coincidence the neutron scattering lengths of yttrium and copper are virtually identical so that it was possible that the assignment of these two atom locations were in fact reversed. However, the X-ray scattering factors for these two atoms are very different and a refinement combining both X-ray and neutron diffraction data would remove the ambiguity.

The Rietveld refinement of this structure consisted of fitting four neutron time-of-flight patterns taken on the High Intensity Powder Diffractometer (HIPD) powder diffractometer at LANSCE and two X-ray powder patterns collected on a commercial instrument with Cu Kα radiation. The entire set of six powder patterns contained ~25,000 data points and was fitted with ~120 adjustable parameters. The parameters included those needed to describe the crystal structure of YBa$_2$Cu$_3$O$_{7-x}$, namely, atomic positions, fractional occupancies, thermal parameters and lattice parameters; these totaled 33 parameters. The rest characterized details of the powder diffraction patterns. These included coefficients for the background functions needed for each powder pattern, coefficients for the powder diffraction peak shapes and intensity correction coefficients for absorption, preferred orientation and extinction as well as the six scaling factors. The resulting structure as illustrated in Figure 27 was dramatically more precise than any of the previous single measurement results. Some idea of the fit for part of the data is seen in Figure 28 which shows the same d-spacing range from one X-ray and one neutron pattern. Most importantly, there was no evidence of any interchange between the metal atoms on their respective sites. This result had been expected because of crystal chemical considerations based on comparison of interatomic distances and ionic radii. In addition, this refinement gave a complete description of the thermal motion for all the atoms which was in some ways different from the previous studies. Because of the requirement to fit both X-ray and neutron diffraction data, the description obtained in this study is probably more accurate as well.

CONCLUSION

Although the practice of neutron powder diffraction is limited to only the few locations equipped with high intensity neutron sources, the applications are of considerable mineralogical interest. The possibility of discerning between near neighboring elements and the high quality of the structural parameters from Rietveld refinements particularly in combination with X-ray powder diffraction data makes neutron diffraction an important structural tool. Moreover, the relative ease of variable sample conditions makes neutron powder diffraction the method of choice for determining structural thermodynamic properties of materials.

REFERENCES

Anderson, J.S., Bevan, D.J.M., Cheetham, A.K., Von Dreele, R.B., Hutchison, J.L. and Strähle, J. (1975) The structure of germanium niobium oxide, an inherently non-stoichiometric 'block' structure. Proc. Roy. Soc. London A346, 139-156.

Azaroff, L.V. (1955) Polarization correction for crystal-monochromated X-radiation. Acta Cryst. 8, 701-704.

Bacon G.E. (1975) Neutron Diffraction, 3rd Ed. Clarendon Press, Oxford.

Basso, R., Della Giusta, A., and Zefiro, L. (1983) Crystal structure refinement of plazolite, a highly hydrated natural hydrogrossular. N. Jahrb. Mineral. Monat., 251-258.

Buras, B. and Holas, A. (1968) Intensity and resolution in neutron time-of-flight powder diffractometry. Nukleonika 13, 591-619.

Cagliotti, G., Paoletti, A. and Ricci, F.P. (1958) Choice of collimators for a crystal spectrometer for neutron diffraction. Nucl. Inst. 3, 223-228.

Carlile, C.J. and Salter, D.C. (1978) Thermal neutron scattering studies of condensed matter under high pressure (A review). High Press. - High Temp. 10, 1-27.

Carpenter, J.M., Mueller, M.H., Beyerlein, R.A., Worlton, T.G., Jorgenson, J.D., Brun, T.O., Sköld, K., Pelizzari, C.A., Peterson, S.W., Watanabe, N., Kimura, M. and Gunning, J.E. (1975) Neutron diffraction measurements on powder samples using the ZING-P pulsed neutron source at Argonne. Proc. Neut. Diff. Conf., Petten, Netherlands, 5-6 Aug., RCN Report, no. 234, 192-208.

Cheetham, A.K., Eddy, M.M., Jefferson, D.A. and Thomas, J.M. (1982a) A study of silicon, aluminium ordering in thallium zeolite-A by powder neutron diffraction. Nature 299, 24-27.

____, Fyfe, C.A., Smith, J.V. and Thomas, J.M. (1982b) On the space group of zeolite-A. J. Chem. Soc., Chem. Commun. 1982, 823-825.

____, Eddy, M.M., Klinowski, J. and Thomas, T.M. (1983) A study of silicon and aluminium ordering and cation positions in the zeolites sodium ZK-4 and sodium Y. J. Chem. Soc. Chem. Commun. 1983, 23-25.
____ and Taylor, J.C. (1977) Profile analysis of powder neutron diffraction data: its scope, limitations, and applications in solid state chemistry. J. Solid State Chem. 21, 253-275.
____ and Von Dreele, R.B. (1973) Cation distributions in niobium oxide block structures. Nature Phys. Sci. 244, 139-140.
Dollase, W.A. (1986) Correction of intensities for preferred orientation in powder diffractometry: Application of the March model. J. Appl. Cryst. 19, 267-272.
Eddy, M.M., Cheetham, A.K. and David, W.I.F. (1986) Powder neutron diffraction study of zeolite Na-ZK-4; an application of new functions for peak shape and asymmetry. Zeolites 6, 449-454.
Elsasser, W.M. (1936) Diffraction of slow neutrons by crystalline substances. Comptes rendus 202, 1029-1030.
Ericsson, T. and Nord A.G. (1984) Strong cation ordering in olivine-related (Ni,Fe)-sarcopsides: a combined Mössbauer, X-ray and neutron diffraction study. Amer. Mineral. 69, 889-895.
Fermi, E. (1936) Motion of neutrons in hydrogenous substances. Ricerca sci. 7-2, 13-52.
Fischer, R.X., Baur, W.H., Shannon, R.D., Staley, R.H., Vega, A.J., Abrams, L. and Prince, E. (1986a) Neutron powder diffraction study and physical characterization of zeolite D-ZK-5 deep-bed calcined at 500°C and 650°C. Zeolites 6, 378-387.
____, Baur, W.H., Shannon, R.D., Staley, R.H., Vega, A.J., Abrams, L. and Prince, E. (1986b) Neutron powder diffraction study and physical characterization of zeolite D-RHO deep-bed calcined at 773 and 923K. J. Phys. Chem. 90, 4414-4423.
Halpern, O. and Johnson, M.H. (1939) On the magnetic scattering of neutrons. Phys. Rev. 55, 898-923.
Hamilton, W.C. (1957) The effect of crystal shape and setting on secondary extinction. Acta Cryst. 10, 620-34.
Hewat, A.W. (1975) Design for a conventional high-resolution neutron powder diffractometer. Nuclear Inst. and Meth. 127, 361-370.
____ (1979) Absorption corrections for neutron diffraction. Acta Cryst. A35, 248-250.
____ (1985) High-resolution neutron and synchrotron powder diffraction. Chemica Scripta 26A, 119-130.
____ and Bailey, I. (1976) D1a, a high resolution neutron powder diffractometer with a bank of mylar collimators. Nucl. Inst. and Meth. 137, 463-471.
Howard, C.J. (1982) The approximation of asymmetric neutron powder diffraction peaks by sums of Gaussians. J. Appl. Cryst.15, 615-620.
Jorgensen, J.D. (1978) Compression mechanisms in α-quartz structures - SiO_2 and GeO_2. J. Appl. Phys. 49, 5473-5478.
____, Beno, M.A., Hinks, D.G., Soderholm, L., Volin, K.J., Hitterman, R.L., Grace, J.D., Schuller, I.K., Segre, C.U., Zhang, K. and Kleefisch, M.S. (1988) Oxygen ordering and the orthorhombic-to-tetragonal phase transition in $YBa_2Cu_3O_{7-x}$. Phys. Rev. B 36, 3608-3616.
____, Johnson, D.H., Mueller, M.H., Worlton, J.G. and Von Dreele, R.B. (1978) Profile analysis of pulsed-source neutron powder diffraction data. Proc. Conf. on Diffraction Profile Analysis, Cracow, 14-15 Aug., 20-22.
____ and Rotella, F.J. (1982) High-resolution time-of-flight powder diffractometer at the ZING-P' pulsed neutron source. J. Appl. Cryst. 15, 27-34.
Klug, H.P. and Alexander, L.E. (1974) X-ray Diffraction Procedures for Polycrystalline and Amorphous Materials. Wiley-Interscience, New York, p. 302.
Lager, G.L., Armbruster, T. and Faber, J. (1987) Neutron and X-ray diffraction study of hydrogarnet $Ca_3Al_2(O_4H_4)_3$. Amer. Mineral. 72, 756-765.
____, Armbruster, T., Rotella, F.J., Jorgensen, J.D. and Hinks, D.G. (1984) A crystallographic study of the low-temperature dehydration products of gypsum, $CaSO_4 \cdot 2H_2O$: hemihydrate $CaSO_4 \cdot 0.50H_2O$, and γ-$CaSO_4$. Amer. Mineral. 69, 910-918.
____, Jorgensen, J.D. and Rotella, F.J. (1982) Crystal structure and thermal expansion of α-quartz SiO_2 at low temperatures. J. Appl. Physics 53, 6751-6756.
____, Ross, F.K., Rotella, F.J. and Jorgenson, J.D. (1981) Neutron powder diffraction of forsterite, Mg_2SiO_4: A comparison with single-crystal investigations. J. Appl. Cryst. 14, 137-139.
Larson, A.C. and Von Dreele, R.B. (1987) GSAS - Generalized Crystal Structure Analysis System, Los Alamos National Laboratory Report No. LA-UR-86-748.
Lovesey, S.W. (1984) Theory of Neutron Scattering from Condensed Matter, Vols. 1 & 2. Oxford University Press, Oxford.
Maichle, J.K., Ihringer, J. and Prandl, W. (1988) Simultaneous structure refinement of neutron, synchrotron and X-ray powder diffraction patterns. J. Appl. Cryst. 21, 22-27.

March, A. (1932) Mathematische Theorie der Regelung nach der Korngestalt bei affiner Deformation. Z. Kristallogr. 81, 285-297.
Marshall, W. and Lovesey, S.W. (1971) Theory of Thermal Neutron Scattering. Oxford University Press, Oxford.
Mitchell, D.P. & Powers, P.N. (1936) Bragg reflection of slow neutrons. Phys. Rev. 50, 486-487.
Nord, A.G. and Ericsson, T. (1982) The cation distribution in synthetic $(Fe,Mn)_3(PO_4)_2$ graftonite-type solid solutions. Amer. Mineral. 67, 826-832.
____ and Ericsson, T. (1985) Cation distribution studies of some ternary orthophosphates having the farringtonite structure. Amer. Mineral. 70, 624-629.
Placzek, G. (1952) The scattering of neutrons by systems of heavy nuclei. Phys. Rev. 86, 377-388.
Rietveld, H.M. (1967) Line profiles of neutron powder-diffraction peaks for structure refinement. Acta Cryst. 22, 151-2.
____ (1969) A profile refinement method for nuclear and magnetic structures. J. Appl. Cryst. 2, 65-71.
Rouse, K.D., Cooper, M.J., York, E.J. and Chakera, A. (1970) Absorption corrections for neutron diffraction. Acta Cryst. A26, 682-691.
Sabine, T.M. (1985) Extinction in polycrystalline materials. Aust. J. Phys. 38, 507-18.
____ (1988) A reconciliation of extinction theories. Acta Cryst. A44, 368-373.
____, Von Dreele, R.B. and Jorgensen, J.-E. (1988) Extinction in time-of-flight neutron powder diffractometry. Acta Cryst. A44, 374-379.
Sacerdoti, M. and Passaglia, E. (1985) The crystal structure of katoite and implications within the hydrogrossular group of minerals. Bull. Minéral. 108, 1-8.
Schwinger, J. (1937) On the magnetic scattering of neutrons. Phys. Rev. 51, 544-552.
Thompson, P., Cox, D.E. and Hastings, J.B. (1987) Rietveld refinement of Debye-Scherrer synchrotron X-ray data from Al_2O_3. J. Appl. Cryst. 20,79-83.
Van Hove, L. (1954a) Time-dependent correlations between spins and neutron scattering in ferromagnetic crystals. Phys. Rev. 95, 1374-1384.
____ (1954a) Temperature variation of the magnetic inelastic scattering of slow neutrons. Phys. Rev. 93, 268-269.
____ (1954a) Correlations in space and time and Born approximation scattering in systens of interacting particles. Phys. Rev. 95, 249-262.
Von Dreele, R.B. and Cheetham, A.K. (1974) The structures of some titanium-niobium oxides by powder neutron diffraction. Proc. Roy. Soc. London A338, 311-326.
____, Jorgensen, J.D. and Windsor, C.G. (1982) Rietveld refinement with spallation neutron powder diffraction data. J. Appl. Cryst. 15, 581-589.
Wechsler, B.A. and Navrotsky, A. (1984) Thermodynamics and structural chemistry of compounds in the system $MgO-TiO_2$. J. Solid State Chem. 55, 165-180.
____ and Von Dreele, R.B. (1989) Structure refinements of Mg_2TiO_4, $MgTiO_3$ and $MgTi_2O_5$ by time-of-flight neutron powder diffraction. Acta Cryst. (in press).
Wiles, D.B and Young, R.A. (1981) A new computer program for Rietveld analysis of X-ray powder diffraction patterns. J. Appl. Cryst. 14, 149-151.
Williams, A., Kwei, G.H., Von Dreele, R.B., Larson, A.C., Raistrick, I.D. and Bish, D.L. (1988). Joint X-ray and neutron refinement of the structure of superconducting $YBa_2Cu_3O_{7-x}$: Precision structure, anisotropic thermal parameters, strain, and cation disorder. Phys. Rev. B37, 7960-7962.
Windsor, C.G. (1981) Pulsed Neutron Scattering. Halsted Press:New York.
Worlton, T.G., Jorgenson, J.D., Beyerlein, R.A. and Decker, D.L. (1976) Multicomponent profile refinement of time-of-flight neutron powder diffraction data. Nucl. Inst. Meth. 137, 331-337.
Wright, P.A., Thomas, J.M., Cheetham, A.K. and Nowak, A.K. (1985) Localizing active sites in zeolitic catalysts: neutron profile analysis and computer simulation of deuteropyridine bound to gallozeolite-L. Nature 318, 611-614.
Wu, M.K., Ashburn, J.R., Torng, C.J., Hor, P.H., Meng, R.L., Gao, L., Huang, Z.J., Wang, Y.Q. and Chu, C.W. (1987) Superconductivity at 93K in a new mixed-phase Y-Ba-Cu-O compound system at ambient pressure. Phys. Rev. Lett. 58, 908-910.
Young, R.A., Mackie, P.E. and Von Dreele, R.B. (1977) Application of the pattern-fitting structure-refinement method to X-ray powder diffractometer patterns. J. Appl. Cryst. 10, 262-269.
____ and Wiles, D.B. (1982) Profile shape functions in Rietveld refinements. J. Appl. Cryst. 15, 430-438.
Zacharaisen, W.H. (1945). "Theory of X-ray Diffraction in Crystals" Wiley:New York.
____ (1967) A general theory of X-ray diffraction in crystals. Acta Cryst. 23, 558-64.
Zinn, W.H. (1947) Diffraction of neutrons by a single crystal. Phys. Rev. 71, 752-757.